Reclamative SOIL SCIENCE

by
I I Plyusnin

2004
BIOTECH BOOKS
DELHI - 110035

First Indian Impression 2000
Second Indian Impression 2004
Third Indian Impression 2008

ISBN 81-7622-047-7

Published by : **BIOTECH BOOKS**
1123/74, Tri Nagar,
Delhi - 110 035
Phone: 27382765
e-mail: biotechbooks@yahoo.co.in

Showroom : 4762-63/23, Ansari Road, Darya Ganj,
New Delhi - 110 002
Phone: 23245578, 23244987

Printed at : **Chawla Offset Printers**
New Delhi - 110 052

PRINTED IN INDIA

CONTENTS

PART TWO

ELEMENTS OF SOIL GEOGRAPHY

PART THREE

IMPROVEMENT OF SOILS

INTRODUCTION

The soil is a huge reservoir of material wealth. Our task is to make good use of it. This requires a thorough knowledge of the processes governing the formation of soil and of the ways in which soil can be ameliorated. Any step we take towards ameliorating it brings about in soil deep changes, which it is necessary to know in order that we may channel them towards a progressive imporovement of the soil.

Soil, according to the Russian scientist V. V. Dokuchayev, is a body subjected to a natural and historical development, which came into being on the surface of the earth as a result of a complex combination of the interactions of rocks, the organic world (macro- and microorganisms of vegetable and animal origin), the climate, the local relief and the production activities of man.

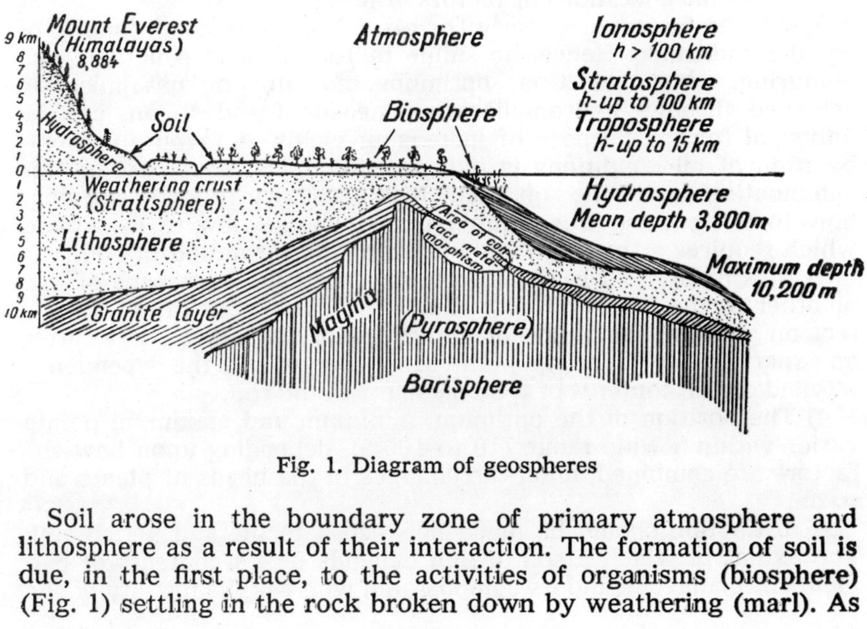

Fig. 1. Diagram of geospheres

Soil arose in the boundary zone of primary atmosphere and lithosphere as a result of their interaction. The formation of soil is due, in the first place, to the activities of organisms (biosphere) (Fig. 1) settling in the rock broken down by weathering (marl). As a result of the dying of plants and microorganisms and of the decomposition of organic residues, the surface layer of the earth becomes enriched with new compounds: organic substances containing the ash elements of plant food (P, K, S, Ca, Mg, Fe, Mn and others), and nitrogen, which is not found in rocks.

Soil is the thick (up to several metres) top layer of the earth's crust (lithosphere), the habitat of roots, endowed with fertility, which is the seat of complex biological and mineral soil-forming processes.

Fertility is the significant and integral property of soil, which differentiates it from rock.

Fertility develops in the process of soil formation. It can be regarded as a result of the combination of the other properties of soil. On the whole, the fertility of a soil is expressed by its aptitude to provide plants simultaneously with water and nutrients, which form the soil factors governing the life of plants. But the needs of plants are not limited to that. They must at the same time dispose jointly of at least the following basic factors: elements of ash food (nutrients), water, air, warmth, space for the roots (friability) and a favourable biochemical regime (absence of injurious substances).

The needs of plants with regard to the basic factors of life can be expressed by the following rules, well-known in agriculture:

1) No single factor of plant life can be replaced by another factor, or in other words: plants require the simultaneous and joint presence of all the conditions or factors of life.

2) All the factors necessary for plant life are unconditionally of equal significance. Hence, no single factor taken in isolation, viz., manuring, plant selection, optimum moisture or desalinisation achieved through hydromeliorative measures and so on, can be universal for the purpose of increasing yields. A favourable combination of all conditions is indispensable. The main part of this combination lies in the soil itself. It is necessary to know not only how to bring about this combination, but also how to control it, which requires a thorough knowledge of the formation of soil.

3) Maximum yield is obtained upon "optimum" presence of one or other of the factors. Fig. 2 shows the curve illustrating the effect on yield of one single plant life factor (moisture curve, after an experiment by Hellriegel). The same law governs the dependence of yield on the contents of food elements in the soil.

4) The position of the optimum, minimum and maximum points varies within a wide range (10 to 100%), depending upon how the factors are combined, upon the changes in the needs of plants and so on.

A distinction should be made between a fertile and a rich soil. The extent to which a soil is rich depends on its contents of elements of plant food and its composition, whereas the fertility of the soil is governed, in addition, by all the conditions of life of plants and microorganisms. Relatively rich soils, like swamped and boggy soils, or saline and dry soils of arid regions, possess little or no fertility due to excess or lack of water. Swampy soils are devoid of fertility due to the fact that the food elements they contain are in the shape of undecomposed organic substances and lower valent compounds unavailable to plants, and saline soils (solonchaks) are infertile due to the high concentration of their solutions, which have a harmful (toxic) effect on plants. Relatively poor podzolised soils or other types of sandy loams or loams may possess sufficient fertility due to optimum water and physical conditions.

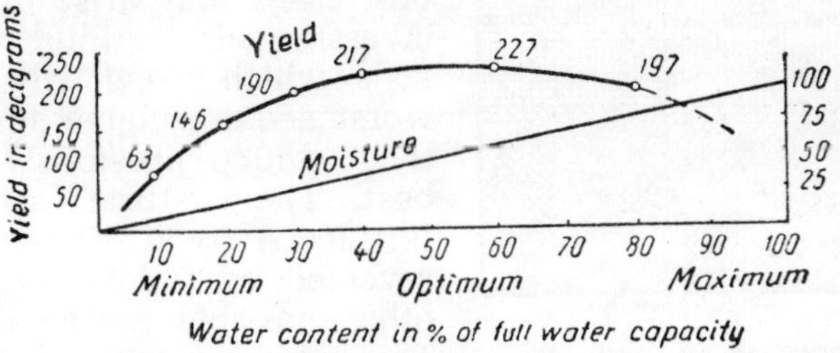

Fig. 2. Efficiency of successive, equal increments of water on barley yields

Any soil can become fertile if we remove the causes of infertility or low fertility. Simple and easily applicable measures, such as the drainage of swamped or boggy soils, or the irrigation of dry soils and the leaching of saline ones, the correct cultivation and manuring of poor soils renders them fertile.

Fertility can be: a) natural, i.e., potential (latent) fertility, which may materialise upon changes in the conditions under which the soil develops; b) artificial, or effective fertility, materialising in the process of utilisation of the soil for production purposes, whereupon man's labour alters and improves it.

Potential fertility develops together with the soil and reflects its condition at a given stage of development. It always precedes effective fertility, in other words the soil, no matter what stage of development it has reached, always holds within itself inexhaustible resources with regard to increase of fertility through taming of the soil.

Soils in different areas may possess the same chemical composition, yet differ in their effective fertility at a given period of time due to differences in the water and physical properties, biological and production particularities.

Differences in natural fertility are governed by factors and conditions of soil formation, as well as the composition (chemical composition, organic matter, colloids, gases), properties (physico-chemical and biological) and the soil structure in the given concrete condition of one soil or another. These differences appear in general lines in a certain way on any soil map, which gives a representation of the disposition of the various groups, subgroups, classes and varieties of soils.

Once the soils have been put to use, the differences in natural fertility smooth down on the whole, but some persist, due to differences in the geological, hydrological, geomorphological and climatic conditions.

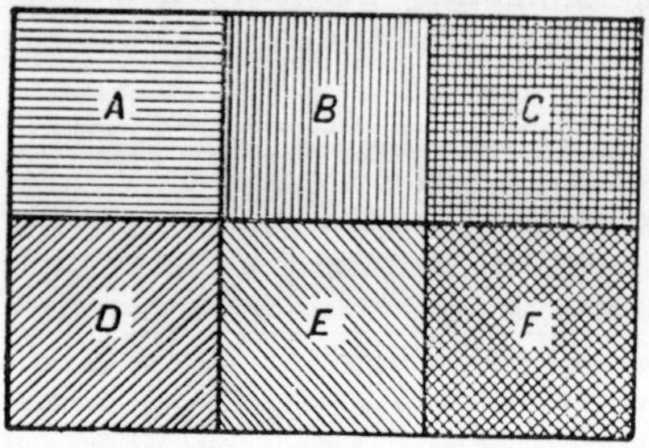

Fig. 3. Disposition of plots with best and poorest soils

As a result of the progressive development of agricultural methods, cases may arise of a kind of inversion of fertility whereupon soils which previously were the worst acquire higher fertility than those which previously were the best. The natural sequence in fertility A<B<C<D<E<F may be reversed or follow one of many other possible patterns (Fig. 3). This may become possible in the case of a radical improvement and transformation of soils. As a result of man's labour, the soil acquires an exceptional capacity for displaying effective fertility. The productivity of a soil at any given time is of finite but progressively increasing magnitude. Over a long period, the soil becomes richer and its fertility increases. To raise the fertility up to the required level is possible only through the adoption of a combination of measures, worked out by science and confirmed by practice.

The natural fertility of soils of the native soil zones changes, increasing from the soils of the podzol-taiga zone to the chernozems and decreasing down to the soils of semideserts. In the chernozem zone we find the most favourable combination of factors and conditions for soil formation, which deserve to be re-created in their best form under conditions of production.

The raising of soil fertility constitutes the most important problem facing soil science.

Soil fertility is amenable to radical regulation through extensive ameliorations, comprising agricultural melioration, forest improvement and hydromelioration.

Meliorative measures serve as a mighty means for radically improving soils and raising their fertility. For instance, irrigation coupled with manuring makes it possible to cultivate highly valuable agricultural crops. Drainage, manuring and a high level of management play an enormous role with regard to the mobilisation of the unused natural resources of swamped lands and marshes, when drained land is first brought under cultivation. No lesser results are obtained through the melioration of saline soils (desalinisation, leaching of solonchaks, gypsuming of solonetzes).

Meliorative measures are aimed at transforming low fertility and infertile soils into highly fertile ones.

Meliorative soil science, relying on agricultural meliorations and general soil science, gives a special meliorative characteristic of soils and indicates ways and means for their progressive improvement and transformation.

Meliorative soil science is confronted with the task of considerable importance: to find and realise new resources and reserves for increasing yields, which, in turn, raises the problem of how to control fertility by meliorative methods, making the best possible use of land brought under melioration, using old neglected meliorative systems and raising their efficiency, putting irrigation systems to the best possible use.

The irrigation, watering and drainage of vast areas with the object of converting them into fields, orchards, plantations, vegetable gardens and rich pastures through the erection of dams, hydroelectric power stations and drainage systems, put before meliorative soil science tasks of unprecedented magnitude and importance. But the main general task is the raising of soil fertility in every possible way with the object of ensuring high and stable yields of all crops. Modern soil science has reached such a level as to make it possible not only to bring about a radical improvement of soils but also their transformation and the creation of new types of soils endowed with the highest possible productive capacity.

Soil science came into being and developed to meet the requirements of agricultural production. We find evidence of practical interest in soil already with the most ancient peoples of the East: the Egyptians, Babylonians, Assyrians, Chinese, Indians, and later, the Greeks and Romans. The first attempts to draw some general conclusions from the information accumulated by agriculturists can be found in ancient Greece and Rome, in the works of Aristotle, Theophrastus, Titus Lucretius Carus, Mago, Cato, Varro, Columella, Pliny the Elder, Virgil and many others. But with the fall of ancient culture, the development of agricultural knowledge and information on soil came to a standstill lasting many centuries.

Up to the beginning of the 19th century, soil had been but very scantily studied and there was no soil science. It was not even clearly realised that plants depend on soil for their development. Up to as recently as 1800 it was maintained that plants drew only water from the soil and that the substances making up the ash were produced spontaneously as a result of growth, which accounts for the total lack of interest in soil even as the substratum for plants. Interest in soil arose at the beginning of the 19th century when the theory was put forward that plants derived their food from humus. That theory was disproved in 1840 by Liebig who proved that plants, in addition to water, obtain from soil the elements of their ash food. The theory according to which plants obtain their ash food from soil stimulated much work on the chemical analysis and study of the chemical properties of soil. The attention of the investigators was focused on the composition of the elements of ash food in plants and in soil, so much so that the importance of the mineral nutrition of plants was overestimated. The growing demand for agricultural produce led to the mechanisation of soil cultivations and stimulated the study of the chemical and physical properties of the soil. Towards the middle of the 19th century, the new ideas concerning the evolution in geology (K. Hoff and C. Lyell) and the theory of evolution of living organisms (A. Wallace and C. Darwin) promoted the development of soil science. In the second half of the 19th century, interest in the genesis of soil was further stimulated by geographical and geological investigations (Lepyokhin, Ruprecht, Pallas, Walter, Sven-Hedin, V. A. Obruchev).

At the end of the 19th century, the teaching on soil science developed into a precise, well defined branch possessing its own methods of investigation, its theory, tasks and prospects. Soil science is thus a new field of knowledge if we compare it with the other sciences. Soil science is a purely Russian science, which was first worked out in Russia. This priority of Russian soil science is universally acknowledged and is due to the following reasons:

a) the immense spaces of the Soviet land with its diverse natural conditions of soil formation all the way from polar regions down to the subtropics, allowing the study of the most diverse natural soils, from the tundras to the krasnozems;

b) the dialectical-materialistic foundation on which Russian soil science developed from its very inception.

At the very end of the last century, the great Russian soil scientist, the founder of soil science, V. V. Dokuchayev set up the science of soil on a really scientific basis. Dokuchayev worked out the main principles of this new branch of the natural sciences and developed them into a coherent original theory. He proved that soil is a special independent body subject to natural development.

The history of soil science is given in general courses on this subject. We shall therefore confine ourselves to a reference to several outstanding scientists and their work in the field of soil science.

Vasily Vasilyevich Dokuchayev (1846-1903) was the first in the world to develop genetic soil science and to establish the main laws of soil formation and of the geography of soils. He discovered the law of the zonality of soils, which is exposed in his famous book

The Russian Chernozem published in 1883. *The Russian Chernozem* is rightly considered as the starting point of scientific pedology. V. V. Dokuchayev developed the teaching of soil as of a special natural body, the product of the interaction of the mineral and biological worlds. He considered soil as the result of the combined activities of such factors as the parent rock, vegetable and animal organisms, the climate, the relief and the age of the land. Dokuchayev discovered the law of the constant change of soils in time

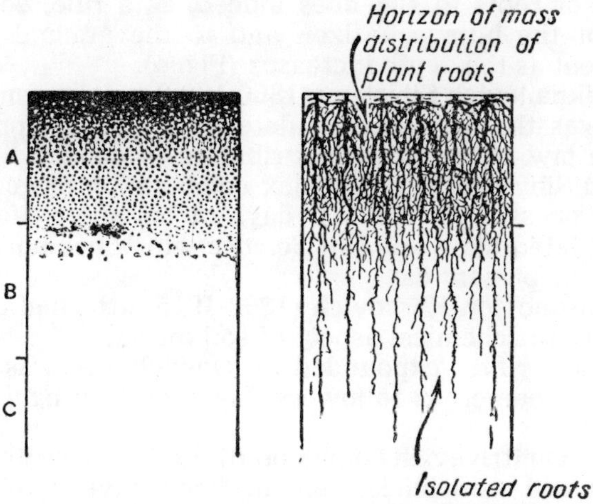

Fig. 4. The depth of the humus horizon corresponds to that of the mass distribution of roots

and space. V. V. Dokuchayev also wrote a book entitled *Our Steppes Now and Before* published in 1892. In this work, the author gives a detailed plan for fighting drought, as well as a thorough analysis of how drought and the drying up of steppes are brought about. The author appeals for a planned fight not against the consequences of drought but against the causes which engender it. Ways are outlined for regulating the water regime in open steppes. The author points to the urgent necessity of regulating rivers, snow retention, afforestation, advocates the creation of artificial water reservoirs, the fight against the formation of gullies, the selection and creation of varieties of agricultural plants to suit different conditions, the adoption of suitable methods of working soils with a view to a better utilisation of the soil's moisture, etc. Dokuchayev suggested setting up trial plots and took part in the carrying out of this idea.

Pavel Andreyevich Kostychev (1845-95) exerted a major influence on the subsequent trend of scientific research in the field of soil science. A notable feature of P. A. Kostychev's research work was his approach to soil science from the agronomical standpoint. He was the first to bring to light the true significance of the structure of soil; he attached great importance to grass cultivation and worked out an appropriate system for cultivating soils and fighting erosion. Kostychev pointed to the indissoluble connection of soil and plant, their unity. He was of the opinion that we must call soil "the top layer of earth down to the depth to which reaches the main mass of plant roots". "The accumulation in soil of organic substances may depend on the plant roots." And the depth of the mass distribution of roots in soil does indeed, as a rule, correspond to the depth of the humus horizon and to the gradual decrease of humus content as the depth increases (Fig. 4).

Nikolai Mikhailovich Sibirtsev (1860-1900), a follower of V. V. Dokuchayev, was the first to formulate, in the most comprehensive manner, the law of the zonal distribution of soils. Soil formation, according to Sibirtsev, is a complex combination of geological and biological processes. In the last days of his short life, N. M. Sibirtsev completed his work *Soil Science* which has not lost its significance to the present day.

Pyotr Samsonovich Kossovich (1862-1915) attached considerable importance to the chemical aspect of soil formation.

Soil science as first expounded by Dokuchayev was further developed by his numerous followers. The most distinguished among them were:

Konstantin Dmitriyevich Glinka (1867-1927), a prominent Russian soil scientist and academician, one of Dokuchayev's followers, laid the foundation of soil mineralogy, of paleopedology. Wrote *Soil Science*, a textbook reprinted many times.

Sergei Semyonovich Neustruyev (1874-1928), a soil scientist, geographer and geologist, investigated the soils of the desertic steppes and mountains of Central Asia, is the author of a remarkable *Elements of Soil Geography*.

Georgy Nikolayevich Vysotsky (1865-1940), a soil scientist and sylviculturist, investigated the biology and water regime of soils. He determined the main types of water regimes of soils and the so-called "dead horizon" of soil with minimum moisture, persisting through the yearly cycle.

Vladimir Ivanovich Vernadsky (1863-1945), a soil scientist and mineralogist, discovered new soil minerals. He showed that the peculiarity of the mineralogical composition of soil is the result of the biogeochemical processes of soil formation. He wrote that "the living substance not only regulates the chemical processes in soil, but is itself one of the components of soil". V. I. Vernadsky pointed out the great importance of soil gases for plants and soils.

Lev Parmyonovich Rozov (1886-1942) worked out the principles of meliorative soil science.

Konstantin Kaetanovich Gedroitz (1872-1932) studied the laws governing the behaviour of colloids in soil and worked out the teaching concerning the absorbing power of soils. He indicated a series of well-founded ways for the treatment of soil and practical measures for raising the fertility of acid podzolic soils through liming, and worked out the problem of the melioration of alkaline solonetzic soils through the application of gypsum. K. K. Gedroitz brought to light the special role played by the exchangeable calcium in the life of soil and the creation for it of favourable physical properties. Gedroitz's teaching concerning the absorbing power of soils was a great contribution towards the understanding of the chemical properties of soil. It helped a great deal in the study of soil formation processes and the raising of the fertility of soils. This teaching has gained universal recognition. Gedroitz showed that by altering the composition of the exchangeable cations, it was possible to bring about the most favourable conditions for the life of plants, raising or lowering the effect of fertilisers on the physical properties of soil. He suggested that the soil's absorbing colloidal complex and the composition of exchangeable cations be utilised for the classification of soils.

The works of K. K. Gedroitz, A. N. Sokolovsky on the colloidal complex of soils and the works of S. P. Kravkov, V. R. Williams and I. V. Tyurin on the organic substances of soils made it possible to determine the interdependence between the chemism of soils and their physical properties. Thus was broadened the field of possible active influence on soil in order to change its physical and chemical properties with a view to raising the productivity of soils (which provided the scientific basis of liming, gypsuming, acidification, silicatisation, enrichment of soils with humus, etc.).

The ideas of Dokuchayev, Kostychev, Kossovich, Izmailsky and others were taken up and further studied by Vasily Robertovich Williams (1863-1939). V. R. Williams put forward new scientific principles regarding soil formation, the study of soil fertility as the significant character of soils, the evolution of the soil covering, the duality of the biological processes in soil. He worked out the teaching regarding the indivisibility of the soil formation process, the leading importance of vegetation in the development of soils and soil fertility. Quantitative changes in soils, maintained V. R. Williams, lead to qualitative changes: some soils are transformed into different ones. He regarded the geographical distribution of soils as a series of stages following one another in regular fashion and passing one into the other in time and space.

Russian and Soviet soil science exerted a great influence on the development of world soil science. On the basis of Russian ideas, after the First World War, soil science made great strides in all countries, wrote the well-known American soil scientist C. Kellogg.

In this connection we must note the great part played by international congresses of soil science. At these congresses, world science acquainted itself with the successes of Soviet scientists. The ideas and methods of Soviet soil science spread rapidly throughout the whole world, and in all countries of the world soil investigations were begun, using the methods worked out by Soviet soil scientists, and even adopting the Russian nomenclature. The designation of soils by the names of "chernozem", "podzol", "solonetz", "solonchak", "solod", "sierozem" were adopted by all nations and these terms became international.

A great number of soil scientists trained in the U.S.S.R. are now engaged in research under the direction of the leading scientists of the country.

PART ONE

SOIL FORMATION. COMPOSITION AND PROPERTIES OF SOIL

Chapter I

WEATHERING

Weathering is the inevitable process of breakdown and change (transformation) undergone by rocks on the surface of the earth, in its bowels and under water, brought about by various agencies (solar radiation, daily and seasonal changes of temperature, the action of natural atmospheric precipitations, ground water, water vapour, carbon dioxide, oxygen, living organisms and the products of their activity and decay, etc.).

Prior to the formation of soil, solid rock passes through prolonged stages (phases) of weathering. Any rock which weathers and breaks up, first turns to marl, keeping its petrographic and mineralogical composition. Marl becomes permeable to air and water. Breaking up still further, marl turns to eluvium, an earthy, porous mass, which gradually acquires capillary properties and water capacity, i.e., the capacity to withhold a certain reserve of water. In this fashion we get the formation of a marl-eluvium of the rock. As a result of further weathering, the eluvium becomes a suitable substratum for the establishment of living organisms. The marl of the soil-forming rock possesses rudiments of the significant property of soil, viz., fertility.

Weathering is accompanied by soil formation; the way in which this formation proceeds depends, to begin with, on the character of the weathering processes, and later on the prevailing agencies of soil formation. But weathering processes go on uninterruptedly, they do not stop even in intensely cultivated soil, playing a part in the complex processes of soil formation.

The weathering of rocks on the earth's surface is attended by a disintegration of the complex alumosilicates and other minerals; these are split off into their constituents, go into solution and are washed out, chiefly by water. An eluvial residue is formed on the seat of weathering. The process continues until the formation of a relatively inert residue, stable under the given conditions.

During the first weathering phase, there occur a breaking up, decomposition and dissolution, a softening and degeneration, as it were, of the rocks, a simplification of their composition. All the substances split off and remaining on the seat of weathering, together constitute the products of weathering; they can be in the colloidal and crystalloidal state. In the second phase of weathering, side by side with the disintegration of the initial products, occurs a mineral synthesis of secondary minerals.

In order better to understand the complex weathering processes, let us consider them in the shape of simplified schemes conformably to certain conditions of soil formation.

As a result of prolonged physical, or mechanical, weathering, occurs a progressive fragmentation of the rock or mechanical breakdown to fragments of various sizes, from large stones to colloids. Periodical (daily, annual, seasonal) fluctuations of the temperature of the air and rocks, reaching several tens of degrees, Centigrade, bring about irregular changes in the bulk of the rocks, with the formation of cracks and a mechanical breakdown.

The mechanical comminution (rupture and splitting) of rocks is also brought about by crystallisation forces which are set up when salts and ice are crystallised in pores and cracks. Mechanical comminution is also caused by the repeated swelling and drying up of rocks or of substances which get into their pores and cracks.

Mechanical breaking up of rocks takes place in various fashions, depending upon the properties of the rock, the conditions under which it was deposited, its position, the insolation, etc., and according to the native zones.

The fragments of rocks of various forms, sizes and composition formed as a result of weathering, constitute what is known as marl: a polydispersive system. Polyhedral fragments coming in contact with one another form pores of irregular shapes: interstices, such as separate uneven cavities or ducts with numerous distensions, narrows and obstructions, only remotely resembling irregular capillaries.

In marl consisting of fine fragments, attraction forces are set up between the grains of rock and water. When wet, a marl of this type is amenable to moulding. In marly rocks in which particles larger than 0.1 mm predominate, water moves in the shape of dropping-liquid water under the force of gravity (gravitational movement) or under the influence of hydrostatic pressure. Water trickles down freely without hindrance. There is no capillary movement of water in a rock of this type. Such a marl possesses minimum water capacity. In a marly rock in which particles smaller than 0.1 mm predominate and especially smaller than 0.01 mm, a capillary movement of water can already be observed, from the more to the less humid zone. As capillarity increases, so does the water capacity. With a further decrease in the diameters of the pores, the rate of capillary movement progressively decreases until, in a heavy clay, it practically stops.

As the colloidal part in marl goes up, so does what is termed the absorbing power of the rock, due to the progressive increase of the specific surface. The ions of salts, retained on the surface of fine colloidal particles and circulating in solutions, become involved in exchange reactions. As a result, there occurs an exchange of cations and anions in equivalent quantities. A considerable part of the salts are removed from the marl. Only such sparingly soluble salts as the calcium carbonates remain. Calcium carbonate together with humus and the iron salts govern the formation of what are known as microaggregates.

Physical (mechanical) weathering is attended by chemical weathering, but the separation between the two types of weathering is purely conventional. The progressive mechanical breaking up to which the rock is subjected prepares it for the subsequent prevailing chemical weathering.

As a result of further breaking up of the rock and decrease of the particles' size (down to 0.1-0.01 mm), physical weathering is slowed down, whereas together with the increase of the specific surface of the marl, there is an increase of the energy of the chemical action until it finally completely predominates.

Chemical weathering involves the interaction of atmospheric masses (the gases of the air: nitrogen, oxygen, carbon dioxide, water vapour, etc.), the biosphere (the living organisms, the products of their vital activities and organic matter in the shape of their remains) and the lithosphere (principally the salts of K, Na, Mg, Ca, Fe, Al, acids: silicic (metasilicic) H_2SiO_3 and orthosilicic: H_4SiO_4, alumino-silicic: $H_2Al_2Si_6O_{16}$ in the shape of quartz, silicates, alumosilicates, as well as carbonates, sulphates, chlorides, etc.).

Water plays quite an important role in chemical weathering, acting as a chemical reagent and solvent as a result of the hydrolytic dissociation of water into H^+ and OH^- ions and the electrolytic dissociation of salts and their hydrolysis. Hydrolysis is attended by an exchange decomposition and a change in the chemical composition of salts. It consists in the replacement in the crystalline structure of the alumosilicates of the ions of alkalis and alkaline earths by hydrogen ions. Hydrolysis is attended by the formation of alumino-silicic acids and the liberation of hydrated oxides of alkalis and alkaline earths. The presence in water of carbon dioxide in solution raises the concentration of hydrogen and intensifies hydrolytic action. Chemical compounds are dissociated into ions, which react with water, forming new compounds, for example: $Na_2CO_3+H_2O=NaHCO_3+NaOH$.

Many minerals absorb water (orthoclase, albite, the micas, hornblende, magnetite, haematite, etc.) undergoing thereby major changes in chemical composition and physical properties. Water,

2*

which contains carbonic acid and other acids, acts on the alumosilicates, displacing the bases and forming soluble salts, which are removed from the sphere of reaction. There remain, the insoluble products, such as SiO_2, which is extremely resistant to chemical change, or the hydrated oxides of Fe, Mn, Al and their combinations with SiO_2, and alumina.

Chemical weathering is an exothermic process in which there is a liberation of heat. It leads on the whole to a simplification in chemical composition, but the formation of more complex combinations is also possible. The hydrolysis of primary silicates is quite possible, with the Vattendant synthetic production of secondary minerals.

As a result of the interaction between rocks, gaseous and other products of weathering, there is a production of minerals of weathering or secondary minerals: a) soluble ones: gypsum, calcite; b) insoluble ones: montmorillonite, nontronite, kaolinite, halloysite, opal, limonite, bauxite, etc. Weathering continues until the primary silicates and alumosilicates are completely dissociated. Of the products of dissociation remain the hydrated oxides of Al and Fe, partly Mn, SiO_2. As a result of intense weathering, i.e., complete dissociation, there is a production of allites (rocks containing aluminium): allitic impoverished eluvium, viz., laterite, as well as clayey richer products: kaolin and allophane, which retain a notable amount of acid silicates of aluminium and silicates and partly a certain amount of primary minerals.

In nature, in the course of chemical weathering of minerals and rocks, the following processes take place: hydration, dehydration, oxidation, reduction, carbonatation, decarbonatation, kaolinisation.

The process of hydration consists in the addition of water to anhydrous minerals, transforming them into aqueous ones. Thus, haematite (red ironclay—Fe_2O_3) turns to limonite (brown haematite —$Fe_2O_3.3H_2O$), quartz (SiO_2) to opal ($SiO_2.nH_2O$), anhydrous aluminium oxide (Al_2O_3) to bauxite ($Al_2O_3.3H_2O$), anhydrite ($CaSO_4$) to gypsum ($CaSO_4.2H_2O$), etc. The hydrates of the sesquioxides may contain various amounts of water ($Fe_2O_3.H_2O$, $Fe_2O_3.2H_2O$, $Fe_2O_3.3H_2O$, $Al_2O_3.H_2O$, $Al_2O_3.2H_2O$, $Al_2O_3.3H_2O$) and are differentiated, in addition, by different colorations, which give to the products of weathering or soil-forming rocks, various colours and tints.

Hydrous minerals can lose all or part of their water, turning to anhydrous or less hydrous minerals. This process, which is the opposite of hydration, is called *dehydration*.

The addition of oxygen, or its removal, characterises the processes of *oxidation* or *reduction*, respectively. Oxidation and hydration may take place simultaneously. Thus, magnetite (Fe_3O_4) may turn to brown haematite ($Fe_2O_3.3H_2O$). The processes of hydration and dehydration, as well as oxidation or reduction, cause changes not only in the chemical composition but also in the physical properties: volume, density, porosity, etc. These changes are attended by considerable mechanical displacement of the reacting masses.

When carbon dioxide dissolved in water penetrates within the marl, it causes the production of carbonates and bicarbonates (carbonatation) which are, in turn, soluble in water and, consequently, removed from the marl. Due to a decrease in the CO_2 content of the solution, as, for instance, upon a rise in temperature, a removal of carbonates from the solution may occur according to the following pattern: $Ca(HCO_3)_2 \leftrightarrows CaCO_3+H_2O+CO_2$. Any increase of the CO_2 content in the solution causes an increase of the solubility of the carbonates and their transformation into other compounds, i.e., *decarbonatation.*

An important part in the natural processes of weathering and soil formation is played by the process of *kaolinisation,* i.e., the disintegration of alumosilicates to clays. Upon weathering, the ions of certain elements are replaced by others. One mineral gives rise to another. For instance, limestone, through the partial exchange of calcium to magnesium, turns to dolomite, and then to $MgCO_3$ and subsequently: $MgCO_3+2H_2O=Mg(OH)_2+CO_2+H_2O$.

The weathering of minerals takes quite a number of different forms. Furthermore one and the same mineral weathers differently and gives different products of weathering, in accordance with changes in the environment conditions and the trend of the process. In accordance with the conditions of the environment, there is a formation of intermediate products of weathering, which sometimes cannot be expressed by the usual chemical formulas, i.e., soil minerals reflecting the unstable conditions of existence of certain minerals on the earth's surface. In the end, the weathering of soil-forming rocks comes to the weathering of the minerals making up the rocks. The weathering of the minerals, as well as that of the rocks, is closely dependent on climate, the relief of the area, hydrology and other conditions.

In a cold climate, physical weathering predominates, where we get hydration, oxidation, desilicatisation, i.e., the first stage of kaolinisation. Under these conditions, a gleisation of rocks and a formation of ortsteins occur, but the process of leaching is weak. In a temperate climate, the process of kaolinisation goes on intensely, there is an accumulation of clays and loams of eluvial origin. In a temperate humid climate, there is marked leaching, i.e., the removal of carbonates and other salts and the accumulation of SiO_2 in podzolic soils.

In the continental climate of the temperate belt, there is an accumulation of certain mobile products of weathering, migrating into the upper layers of weathering rocks, moving downwards in wet and upwards in hot periods. In steppe regions, there is an accumulation in soils and soil-forming rocks of $CaCO_3$ and $CaSO_4.2H_2O$ and, in places, of readily soluble salts.

In a warm and wet subtropical climate (in the Chakva region, where the annual rainfall reaches 2,500 mm) leaching is very pronounced. There is a very marked hydration, oxidation and desilicatisation. The weathering rock loses bases and SiO_2, and acquires much R_2O_3 and particularly Al_2O_3, but even the Fe_2O_3 content in the products of weathering often exceeds by 3.5 times the quantity contained in the parent rock. The residual rock is coloured mainly in red, orange and yellow.

In a humid tropical climate, the weathering crust loses its alkaline and alkaline earths elements: SiO_2 up to 50% and Al_2O_3 up to 20%.

In deserts there is a formation of hydrates of R_2O_3 and a crust: the so-called "desert tan".

Upon weathering of magmatic crystalline rocks, there is usually a formation of kaolin and clay, with very varied colorations. The weathering products of the basic rocks have brown and reddish colours, those of the acid rocks, which contain less iron, have pinkish colours of lighter shades and hues. The most widely distributed typical massively crystalline rock, granite, disintegrates to give quartz, kaolin, biotite, muskovite, orthoclase, albite and others. Upon weathering, calcite, dolomite, iron oxide, halite, sylvinite and others are washed away in solution. On the seat of weathering remain quartz, kaolin, clayey minerals, limonite, partly feldspar and others.

The weathering of sedimentary rocks, such as shaley clays, sandstones, limestones, dolomites, clays, goes on more rapidly and in a somewhat simpler fashion. Thus, as a result of weathering, shaley clays turn to clays.

Weathering of sandstones begins with the breaking up of the cementing material and the formation of a new marl, consisting of sand, clay and silt. As a result of the weathering of limestone, dolomite, marl and other rocks containing carbonates, there is a formation of clayey and marly products of weathering. The character of the weathering products depends on the composition of the initial rocks and the presence in them of impurities. The weathering of clays and loess rocks begins with their bleaching, i.e., the removal of iron salts and soluble material.

The weathering of eluvium, deluvium, alluvium, moraines and other formations is attended by a smoothing out of the primary stratification, a change in the mechanical composition, a breaking up of inclusions, etc.

Microorganisms (bacteria, fungi), which give out acids and salts, also play a part in the disintegration of rocks. In particular, nitrifying bacteria obtain nitrogen from the air and, in the course of their vital activities, produce a strong mineral nitric acid (HNO_3). The latter transforms carbonates into nitrates and disintegrates relatively stable rocks. Blue-green algae, which produce pectic substances, also contribute to the disintegration of rocks. Diatoms destroy silicates, removing SiO_2 from them. The lichens and fungi which grow on the surface of cliff rocks give out oxalic acid which destroys the rocks. Algae, mosses and bacteria, which give out CO_2 and salts, disintegrate rocks. Upon their death, there is a formation of humic acids and CO_2, which carry still further the disintegration of the rocks. As is well-known, microorganisms penetrate deep into soil, to a depth of several dozens of metres and more.

Plant roots exert a splitting action, penetrating within rocks along cracks. Furthermore the acid substances by root hairs also exert a chemical action on the rocks. Natural water, enriched with humic acid, together with the mineral (HNO_3, HCl, H_2SO_4) and organic acids which it contains, exerts a disintegrating effect on the rocks in which it circulates.

Animals exert a direct and indirect action on the disintegration of the rocks which come to the surface of the earth, by ramming them, or burrowing in them holes and subterranean passages, and due to the effect on the rocks of the products of their vital activities, etc.

Finally, 'the productive activity of man (application of fertilisers, agricultural operations) plays an important part in the weathering of rocks, exerting an action on the top layers of the earth's crust.

Major (Geological) and Minor (Biological) Cycles of Changes

Two cycles of changes take place in the upper horizons and on the surface of the earth: a major one, or geological, and a minor one, or biological, which together form a grandiose biogeochemical process. External (exogenous) and internal (endogenous) forces on the earth, which are engaged in the formation of its upper part, are at work in the continuous geological cycle of changes in the earth's crust.

The surface of the earth is the seat of an uninterrupted transference from relatively high points to low ones of broken up material, found as dust in the air, in suspension in water and dissolved in it. Owing to this fact, there is an uninterrupted lowering of the absolute marks of higher places (by an average of 0.01 to 0.5 mm every year) and a raising of the absolute marks of low lying areas, which become filled up by deposited material. Enormous quantities of sediments accumulate on the bottom of seas and oceans, deposited chiefly near their shores.

The pressure exerted by the enormous masses of deposits, as well as the geochemical processes and the work of endogenous forces, including intra-atomic energy, account for the secular land rises and mountain-forming processes. As a result of the latter, the beds of seas and oceans rise, in places, from under the water, whereas dry land is lowered and goes under water. Deposits, which at one time were removed from dry land and deposited on the bottom of seas and oceans, appear once again on dry land, forming new mountains and high ground on the surface of the earth. Rocks raised above sea level are disintegrated anew and are displaced once again in a broken-up state, fragmented down to colloidal size, as well as in the form of solutions and gases; they find their way once more into rivers, seas and oceans, get deposited on the bottom which, in time, may again become dry land and so on.

The products of weathering removed from dry land, having passed through prolonged stages, return once again, as it were, from the bottom of the sea to dry land. This, in the main, constitutes what is known as the major, or geological, cycle of changes, which goes on during quite a prolonged period over vast areas of the earth's surface. Upon repeated redeposition (diagenesis) in the course of the major geological cycle, sediments of the earth's crust pass to a considerable extent into the colloidal state. The latter circumstance, which makes itself felt within the complex of the other conditions, favoured, in the geological past, the formation of organic matter, then of organisms, at first archaeobions, then prototrophic ones, i.e., microorganisms with a mineral type of nourishment (bacteria, algae) and still later, organisms which were more discriminating towards food: mosses, fungi and lichens, the forerunners of higher plants.

The higher plants appeared on earth later, after the formation of primitive soil and the accumulation in it of a certain reserve of elements capable of providing plants with ash food.

Having reached its maximum development, at a given stage, the major, or geological, cycle of changes on earth engendered the minor, or biological, cycle of changes, developing in the opposite direction. The biological cycle of changes comes, in the end, to a biogenic accumulation in soil of nitrogenous and ash food for plants.

The accumulation of elements of ash and nitrogenous food for plants in soil is governed by the selective absorbing power of living organisms, particularly of plants. Green plants are able, through photosynthesis, to capture from the mighty flow of solar radiation an enormous amount of energy, which they draw into the biological cycle of changes, introducing into it C, N, H, O, P, S and many other biogenic elements. Solar radiation is the main source of the energy which sustains life on the earth. Thanks to the energy obtained through sunlight and chlorophyll, plants are capable of synthesising organic matter from CO_2 and H_2O, utilising, at the same time, elements of ash food from the soil. Photosynthesis is the process of primary production of organic living substances from simple mineral substances, CO_2 and H_2O. The absorption of solar energy sets off processes of oxidation and reduction (with the liberation of O) of such oxidised and stable compounds as CO_2 and H_2O, which are inert from the chemical point of view.

Viewed as a biogeochemical process of previous epochs, photosynthesis resulted in the formation of biogenic deposits in the bowels of the earth, such as graphite, coal, petroleum, peat, sapropel, soil humus, etc.

The uninterrupted process of synthesis and decay of organic matter, which goes on in soil, is inevitably accompanied by the concentration in it of the elements of plant food.

Soil formation is on the whole, a biogenic phenomenon. It goes on in a direction opposite to that of weathering and leads not to the loss of matter from dry land, but to its consolidation within the biological cycle going on between soil and the organisms which live in it, particularly plants and bacteria. The biological cycle of changes displays itself on part of the trajectory of the major geological cycle, in the opposite direction.

The organisms developing on the earth wrestle, as it were, from the major geological cycle of changes, for a long time or forever, a series of quantitatively limited elements, drawing them more and more into the minor, progressively widening, biological cycle of changes. An uninterrupted accumulation goes on in soil of biogenic elements (C, N, P, K, S and others) due to their concentration in the remains of organisms.

In comparison with the soil-forming rocks, soils are relatively enriched with radium and thorium. Soluble elements of ash food, potassium and others, are maintained in the upper horizons of the earth's crust on a relatively high level, corresponding to the biological processes. Phosphorus, one of the most important biogenic elements, accumulates in stable compounds of the earth's crust. Phosphorus, according to Vernadsky,* is a good index of the historical development of life (Table 1). The irregular distribution of organisms leads to a corresponding displacement and redistribution of phosphorus on the surface of the earth.

---

* In the present and in other similar cases, the author quotes the name of the Soviet scientist who was the first to establish a certain principle in science.

Table 1

Mean Contents of Certain Chemical Elements in the Earth's Crust, Soils and Organisms (according to A. P. Vinogradov, 1949) (in weight percentages)

	O	Si	Al	Fe	C	Ca	K	Na
Earth's crust	49.4	27.6	8.5	5.0	0.01	3.5	2.5	2.6
Soil	49.0	33.0	7.12	3.8	2.0	1.37	1.36	0.63
Organisms	70.0	0.05	0.02	0.02	18.0	0.5	0.2	0.05

(continued)

	Mg	N	H	P	S	Mn	Sr	Cl
Earth's crust	2.0	0.02	—	0.08	0.05	0.09	0.04	0.048
Soil	0.6	0.1	—	0.08	0.05	0.085	0.03	0.01
Organisms	0.07	0.5	8.0	0.7	0.2	0.0005	0.001	0.1

Naturally, the problem arises of the regulation of the phosphorus balance in the biosphere. The same goes for certain other chemical elements of the biosphere (S and others) utilised by living matter.

The geological and biological cycles of changes on earth govern a fairly complex but regular migration of atoms which it is most important for us to follow up: this migration, viewed within a complex diagenesis, reflects the whole history of the biosphere and of the sedimentary layer of the earth's crust.

The geochemical history of the biogenic elements and of the biosphere, evidently begins before the formation of the Archeozoic and Archean suites of rocks of the earth's crust. Between the chemical composition of living substances and that of the mineral matter of the earth's crust, there is a link which is revealed by a bulk analysis of any of the bodies. The chemical composition of relic vegetable organisms, which differs considerably from that of the present-day ones (for instance the high Al content in ancient horsetail and licopodium) points to the existence on earth of living conditions which did not resemble those of today. This fact points also to the unity and interdependence of the biosphere, soil, and lithosphere. Organisms which settle on rock loosened by weathering (marl) transform it into soil and later evolve together with it. Soil formation is thus one of the clear indications of the evolution of life on earth. The biological processes, which endlessly mobilise newer and newer substances and draw them into the spirally widening biological cycle of changes, create the conditions for a progressively increasing fertility of the soil, leading to a progressive development of life on the earth.

Our task consists, by controlling soil formation, in regulating the cycle of changes on the earth, particularly with regard to the quantitatively limited biogenic elements.

Chapter II

FACTORS AND CONDITIONS OF SOIL FORMATION

Soil-Forming Rocks

Soil-forming rocks may be the products of the weathering of massively crystalline (granite, basalts, porphyries, diorites, diabases, etc.), metamorphic (schists, shaley clays, phyllites, mica slots, hornblende shales, chlorite slates, steatites, quartzites, marbles, calcites, etc.) and sedimentary rocks (clays, loams, sands, sandstones, dense aluminous clays, limestones, chalk, marl, etc.). Parent rocks are more frequently the products of the weathering of sedimentary rocks. The most ancient sedimentary rocks are, however, usually covered by the newest quaternary deposits. The latter, which lie on the surface of the earth, form the main parent rocks.

The most widespread parent rocks are the continental quaternary deposits: ancient and recent glacial formations (moraines), loess and loess-like rocks, alluvium, deluvium, eluvium and others. An ancient moraine consists of boulder clay deposits of a depth of 50 to 60 or more metres.

Moraines are associated with what are known as periglacial (extraglacial) formations of sanders plains, i.e., fluvio-glacial deposits: pebbles, gravel, sands, passing into ancient alluvial deposits of river terraces. To glacial formations also belong the ancient lacustrine stratified deposits.

Adjacent to the fluvio-glacial deposits, to the south of them, lies an area of distribution of loess-like rocks and loesses. Loess is a pale yellow, macroscopically homogeneous, dust-like, porous rock containing carbonates, which can be separated into vertical column-like parts. Loess, as rock, is widely distributed in the valleys of the Danube, Dniester, Bug, Dnieper, Don, Volga. It is also distributed in Western Siberia, in the Tien-Shan and Pamir regions and in Northern China, where it reaches a depth of 400 to 500 m. The loess found in Southern Russia reaches a depth of 20 m and more and is separated into several layers by fossil soils.

Several theories have been put forward to explain the origin of loesses: the aeolian, deluvial, glacial (fluvio-glacial), soil-eluvial and others. The stratification of loess and its progressive transition into fluvio-glacial sands and moraine point to an aqueous origin. Terraced (ancient flood plain) deposits, lying on ancient alluvial riverbed deposits are also often regarded as loesses. The extraordinary variety of the loesses from the point of view of their structure can hardly be ascribed to aeolian processes. However, the possibility of an aeolian precipitation of loess cannot be ruled out. Loess and loess-like rocks may arise as a result of various processes. But the same processes may give rise to different loessy rocks. Loess, like other rocks (sands, clays) is a collective term. Loess designates rocks of various origin.

In mechanical and chemical composition, loess exhibits changes in a horizontal direction from mountains to plain and, on a plain, downwards and downstream.

Weathered, as well as redeposited loess and loess-like loams are distinguished from typical loess.

Loess-like loams contain less carbonate, their mechanical composition is characterised by larger grains, they are more distinctly stratified. They are quite evidently of aqueous origin. They are closely linked by their progressive transition into loess.

Red and yellow watershed clays are seldom encountered in recent valleys due to the fact that in such locations they are destroyed by erosion. Typical red clays are very dense, devoid of stratification, contain gypsum (druses) and numerous calcareous inclusions or growths (concretions) often hollow, black inside, sometimes even marble-like. Red clays are often covered up or replaced by yellow clays of the eluvial formation and deluvial deposits types. Clays of this kind are also encountered outside the distribution areas of red clays and represent rock eluvium.

Widely distributed on the left bank side of the Lower Volga region are the so-called syrt clays of the Trans-Volga region. They occupy a rising undulating plain between the Volga and the Obshchy Syrt hills to the east, the Samarka River to the north and the mouth of the Yeruslan River to the south, where they present the outline of a ledge, reminding one of a dry high bank. The syrt clays are deposited in such a way that the pliocene (high tertiary) depressions are levelled out. In composition these clays are relatively homogeneous, stratified, carbonate formations of heavy mechanical composition (the amount of particles <0.01 mm$=50$ to 75%). The SiO_2 content is lower than in the loesses, not exceeding 50 to 60%. They are of a dense constitution, their filtration capacity is low. These clays contain sulphates and chlorides in varying quantities. The upper horizons of the northern part of the massif are leached to a depth of 2 to 3 m, those of the southern part to a depth of less than 1 m. The western syrt deposits, which contain more sand, are more thoroughly leached. The syrt clays represent deposits of the delta-alluvial type. Under the syrt clays lie sand of river-bed deposits. Adjacent to the syrt clays, in a transgressive fashion, leaning against them, lie ancient Caspian sea deposits. Embedded within the syrt clays are encountered humic and calcareous horizons of fossil soils with concretions, krotowinas and other clear indications of soils.

All the cited ancient quaternary deposits do not appear everywhere directly as parent rocks in view of the fact that they often lie under recent geological deposits of a genetic type, such as eluvium, deluvium and alluvium (Fig. 5).

Eluvium. The term eluvium designates continental geological formations produced as a result of considerable changes and breaking up undergone by rocks on the seat of their primary deposition. It also covers the weathering products of rocks which retain vestiges of structural and petrographic characteristics, a genetic tie and an uninterrupted progressive transition from the initial rocks.

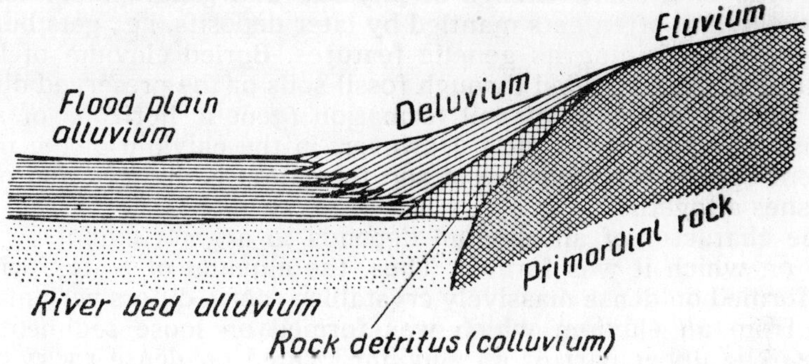

Fig. 5. Diagram showing the correlation between and disposition of eluvium, colluvium, deluvium and alluvium on bedrock

Eluvium can be formed anywhere and on any rock showing no marked erosion and washing out, nor any intense accumulation of deposits, and where there are no significant gravitational displacements to disturb their structure. The formation of eluvium may be hampered by the accumulation of deposits and erosion proceeding at a faster rate than the formation of the eluvium. Light erosion proceeding at a slower rate than the weathering of rocks, does not hamper the formation of eluvium but only rejuvenates it.

There are no rocks on the earth's surface which have not, in one way or another, been altered and affected by an eluvial process, i.e., the sum-total of the numerous denudation, weathering and soil formation phenomena. However, slight changes undergone by a rock do not make an eluvium of it. The eluvial process proceeds and eluvium is formed, in the main, directly in the upper horizons of the earth. The most important and the most widespread is surface eluvium, in the formation of which soil formation takes a part.

Eluvium is differentiated into eluvium of plains and eluvium of slopes. The eluvium of plains is formed on relatively uninclined, levelled off surfaces of the earth. In the upper horizons, such an eluvium usually passes into soil. On slopes and convex watersheds, we get the formation of what is called residual eluvium. Residual eluvium is constantly rejuvenated as a result of sheet and linear erosion. Eluvium is also formed at a certain depth from the earth's surface, usually in the zone of contact of rocks, when one of them is subjected to more pronounced breaking up. Eluvium can occur underwater, directly on the bottom of a lake or river, in places where there is no noticeable accumulation of deposits, or in places of slow washing out and ablation, which do not hamper its development. In contradistinction to land eluvium, this type of eluvium is produced without the formation of soil and is not brought to completion by the appearance of soil in its upper part.

In case of a sharp change of physical and geographical conditions, eluvium often gets mantled by later deposits, i.e., gets buried (fossilised), retaining its genetic features. Buried eluvium of land origin is easily identified through fossil soils or the preserved direct and indirect evidence of soil formation (genetic horizons of soil, inclusions, neogeneses, etc.). However, in the eluvium of the most ancient epochs the evidence of soil formation is dimmed and even vanishes altogether.

The character of an eluvium depends to a great extent on the rock on which it was formed. Thus, the eluvium of plains, which was formed on dense massively crystalline rocks, differs very markedly from an eluvium which was formed on loose sedimentary rocks. The upper part of an eluvium located on dense rocks consists of the loose products of the disintegrated dense rock, frequently so much changed by weathering and soil formation as to become unrecognisable, and appears as an earthy mass without the slightest reminders of the massively crystalline rocks. Further down, at a certain depth from the surface of the earth, there is a very gradual admixing, at first of an insignificant amount of small fragments of these rocks and then, as we go deeper, the amount and size of these fragments increase sharply, and at a depth of some 2 to 3 m and more, there is a gradual passage to the same but thoroughly broken up rock penetrated by cracks. At a depth of several metres, this massively crystalline rock lies in a hardly changed and further down still, in a practically unchanged state (Fig. 6). Quite different is the character of an eluvium which was formed on a loose, sedimentary rock. In view of its more pronounced permeability to air and water, a rock of this kind disintegrates substantially faster and down to a greater depth. Water solutions penetrate into a loose rock right down to the first water-bearing horizon which sometimes lies at a depth of dozens of metres. In connection with all this, we get the formation of an eluvium of maximum possible thickness; furthermore, the looser the initial sedimentary rock, the greater the thickness reached by the eluvium. The average thickness of an eluvium on sedimentary rocks is approximately determined by the depth of the seasonal infiltration of water and the range of the relatively sharp seasonal temperature fluctuations.

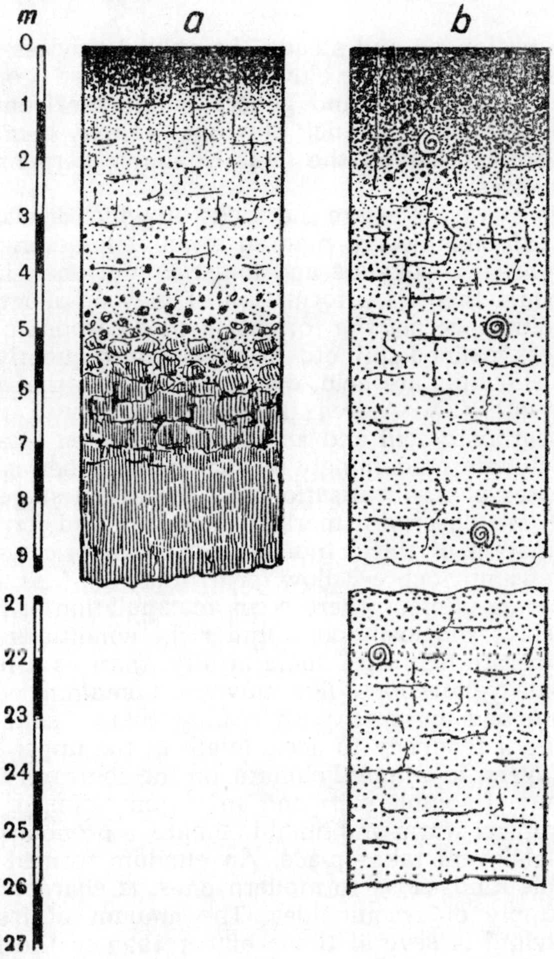

Fig. 6. Diagram illustrating the structure of eluvium:
a—on a dense rock, *b*—on a loose rock

The most intense formation of eluvium occurs in the root zone, which is involved in soil formation.

The occurrence of exogenous and endogenous processes and the constant development of the forms on the earth's surface which leads to a change of situation of the basis of erosion and denudation frequently brings about, in time, an increase of the thickness of the eluvium. But depending upon the correlation between the intensity of disintegration of the rocks on the spot and the removal of the products of weathering from the earth's surface, the thickness of the eluvium may decrease right down to its ablation and the uncovering of only slightly altered rock.

The more prolonged and steady the eluvial process, the greater the thickness reached by the eluvium and the more pronounced the difference in composition and general character from the initial rock. As an example of a thick eluvium, we may mention the ancient weathering crust on the Orsk-Khalilovo serpentines, which reaches 50 metres.

Eluvium is as varied as are the rocks on which it was formed.

Eluvium exhibits clearly marked zonal characteristics. Under arid conditions, it is alkaline and under humid ones, it is acid. In an acid medium there is a formation of gley or of laterite. In an alkaline medium, we get the formation of a carbonate eluvium of the marl, loess, salined soil, etc., type. Not infrequently, the upper horizons of eluviums are acid, owing to the fact that water there is enriched with CO_2 whereas lower down, occurs a neutralisation of the carbonic acid and an alkaline reaction sets in.

In a cold climate predominate mechanical weathering, hydration, occur processes of desilicatisation and the initial stage of kaolinisation. Here, we observe a marked gleisation and ferruginisation, i.e., the formation of thick, bluish grey, viscous, clay-like eluvial masses and swampy ochre-yellow formations.

In a temperate climate, there is an accumulation of red and yellow-brown clays and loams and, under the conditions of the continental temperate belt with some aridity, there is a formation of carbonate pale yellow loess-like eluvium containing occasionally some gypsum and enriched with readily soluble salts. In places, the salts have a tendency to accumulate in the upper horizons of the eluvial layers. In a humid climate, on the contrary, there occurs a leaching out of soluble salts and an accumulation of silicic acid.

In a subtropical warm and humid climate, a pronounced leaching out of eluvial layers takes place. An eluvium formed under conditions of this kind, close to modern ones, is characterised by an abundant supply of sesquioxides. The amount of iron oxide in such an eluvium is several times higher than in the initial rock. In the humid tropics, eluvium loses alkaline and alkaline-earth bases and also, progressively, SiO_2. There is an accumulation of Al_2O_3 with Fe_2O_3 and a formation of red-coloured laterite-like and bauxite-like rocks. Under conditions of this kind, very pronounced hydration, oxidation and desilicatisation occur.

Under desertic conditions, eluvium becomes enriched with hydrates of sesquioxides and with readily soluble salts. In semi-deserts and deserts, in places, there is a marked accumulation of salts in the uppermost horizons of the eluvium. On the surface of compact rocks a desert porous (corroded) or dense crust is sometimes formed. This crust possesses various colours: cinnamon, black and red due to the presence of iron and manganese oxides, yellow due to iron hydroxide and grey due to iron and manganese hydroxide.

In the temperate belt, an eluvium with a thick illuvial horizon is widely distributed. The latter arises in the areas of primary depressions as a result of prolonged infiltration. In such conditions we get the formations of sinks,.minor depressions, hollows of the estuarine type, etc. The appearance of such sinks and patelloid depressions is tied, in the main, with the occurrence of suffosion,* complicated, in places, by loessial karst. The cause of the formation of sinks may be the solothisation of solonetzes and solonetzic soils, their subsequent sinking, leaching, and the growth of shrubbery and trees.

On slopes, in places, under the force of gravity, eluvium with surplus moisture is, to some or other extent, affected by the movement of the weathering masses of rocks without any marked disturbance of their structure and genetic bond with the initial rocks which lie at the base. But strongly pronounced disturbances in the deposition of an eluvium as a result of gravitational displacement shift the eluvium into formations of the boulder-stones, landslip or landslide type (colluvium). The development of eluvium proceeds in interdependency with the formation of the weathering crust of the earth. Until the Archeozoic, the eluvial process proceeded abiotically. Since the appearance of life on earth, eluvium is the joint product of weathering and soil formation.

In the period of the establishment of life on earth, the eluvial process, preceding the formation of soils, gave rise to soil formation. Subsequently, the latter, developing in interdependency with eluvium formation, started to serve as the most important factor of the formation of eluvium. But soil itself forms incomparably faster than eluvium, that is why right beneath the soil, eluvium proper is frequently absent or, to be more accurate, is combined with it.

The eluvial process creates favourable conditions for the production of deluvium, alluvium and aeolian formations. It prepares, so to speak, the material for the formation of various geologic deposits. Through eluvium and soils, on the path followed by diagenesis, proceed all the formations of the earth's crust up to sea deposits, which derive their initial material from eluvium. The eluvium formed through various epochs usually settles in the earth's crust in the form of layers and horizons, represented by the products of disintegration of various rocks.

In spite of the fact that the eluvial process on the earth's surface proceeds without interruption and everywhere, thick layers of distinct eluvium are found only in places particularly favourable for its formation and preservation. This is due to the fact that erosion and accumulation hamper the formation of eluvium and even prevent its formation. As the products of the eluvial process are being formed, they are constantly being removed by water and

* Suffosion means the removal by water of soluble and suspended substances.

wind or are subjected to displacement under the force of gravity and biological factors.

The eluvial process proceeds on all the elements of the relief and affects all geological deposits, not excluding the most recent deluvial, alluvial, aeolian, glacial and sea deposits, on which eluvium has not yet had time to form. In the latter case, eluvium formation is reduced to a minimum, in view of the fact that it is overpowered by other prevailing geological processes and is masked, disappearing in a fashion similar to aeolian dust which is constantly settling everywhere and sinks in the mass of other deposits without giving rise to loess.

On ancient deluvium and alluvium, as well as on moraines and other geological deposits, eluvium is more clearly marked, although the traces of previous deposits have not yet been completely effaced. Eluvium formation on the very ancient deluvium of gentle slopes and the alluvium of high terraces leads to loess-like eluvial formations. The prolonged process of eluvial loessiation of deluvium and alluvium may bring them close to typical loess of aeolian origin.

It also follows from the above that the eluvial process affects, in varying degrees, any sedimentary rock on the surface of the earth and it is difficult to find deposits which have not undergone chemical and mechanical changes. As a result of this process, on the surface of the earth arise the following rocks of continental origin with strongly zonal characters: gley, ferruginous sandy loams, loams and clays, loesses, loess-like and marl-like loams, laterites, laterite-bauxites and others. The eluvial process on earth proceeds without interruption. In the course of time, eluvium constantly acquires new properties and characteristics.

With the progressive development of life on earth as a whole and the development of the productive activity of man in particular, the development of eluvium entered upon a new phase, i.e., the anthropogenic phase. The eluvial process can be controlled and directed so as to raise the usefulness of the earth's surface for agriculture, mainly through its consolidation, by checking excessive erosion and accumulation. The consolidation of the earth's surface, its protection from washout, ablation and deflation, the creation of a cultivated vegetative cover and of soils, the regulation of rivers and surface flow, the control of the ground water regime, all lead the present-day eluvial process in the needed direction.

Deluvium. This term designates mostly loamy deposits, varying in colour and mechanical composition, usually porous, the origin of which is due to the action, variable in force, power and time, of flowing water currents, which have no definite beds but flow down slopes and produce what is known as sheet wash and the precipitation of deposits on gently sloping surfaces. The production of deluvium proceeds until an elementary system of water flow is formed and linear erosion sets in.

Deluvium should be regarded as the resultant of a whole series of factors: a) climatic (zonal) conditions, b) rocks, c) the shape and incline of the slope, d) the catchment area of the slope and e) soil formation and the development of the vegetation.

In a downward direction along the slope, the mechanical composition of the deluvium changes successively from coarse-grained to fine-grained; the same thing is observed in a vertical direction: at the bottom of the deluvial blanket lie coarser-grained deposits, towards the top finer-grained ones. The latter occurrence is connected with the smoothing out of the incline and the slowing down of the flows.

Frequently, in the profile of the entire deluvial layer, fossil soils are observed which, in most cases, correspond to relative interruptions in the intensity of deposition of the deluvium. Fossil soils divide the whole thickness of the deluvium into horizons of various ages. Furthermore, the deluvium of such horizons often varies in mechanical and lithological composition, depending on changes undergone by the factors of deluvium formation.

Deluvium lies symmetrically, sometimes asymmetrically, reaching in this case maximum development only on one slope of the valley. Deluvium of convex slopes, in contradistinction to that of concave and straight ones, is characterised by a somewhat lighter mechanical composition. Deluvium of long gentle slopes is characterised by a heavier mechanical composition than that of short steep slopes. Differences in the orientation of slopes are reflected in the chemical composition of the deluvium: that of southern slopes is relatively less leached, richer in salts, particularly carbonates. An examination of the profile of a deluvial layer reveals the ancient fossil relief and the previous positions of the bases of erosion and denudation.

In connection with the smoothing out of the slope and as its inclined surface gets closer to the horizontal, the deluvial process gradually slows down and the eluvial process prevails.

In mechanical composition, the main mass of the deluvium consists, in most cases, of loams. A thick sandy deluvium cannot arise on broad slopes upon a relatively slight water flow, in view of the fact that the water of the atmospheric precipitations has the time to percolate into sandy rocks without running down the surface of the slope. Where the disintegration of hard rocks is in progress, coarse fragments of it get into the deluvium in the shape of breccia and broken stone, often forming whole horizons at the base of the deluvial layers. The study of the deluvial deposition of slopes has shown that the main mass of deluvium was formed after the formation of the deep valleys of an old hydrographic network. The upper horizons of deluvial layers are crowned, in places, with newest deluvium reaching sometimes a thickness of several metres and much stained with humus.

The appearance of layers of newest deluvium tainted with humus, as well as the formation of alluvial-deluvial dark grey sediments of new river terraces, must necessarily be linked in time with the beginning of intense cultivation.

Alluvium. This term designates any type of deposit of rock debris forming in the regions of flood plains, in young tracts of land.

The area of any flood plain consists of an alluvial complex of deposits arranged in two floors. The first (lower) floor of alluvial deposits usually consists of sandy and sandy-pebbly river bed deposits. The second, top floor, consists of flood plain, usually loamy-clayey, deposits, lying more or less horizontally. The arrangement in two floors of the alluvium is linked with the phases of the alluvial process: the low water phase and the flood phase. River bed and flood plain deposits, taken together, represent a single complex (Fig. 7). They are formed in the same physical-

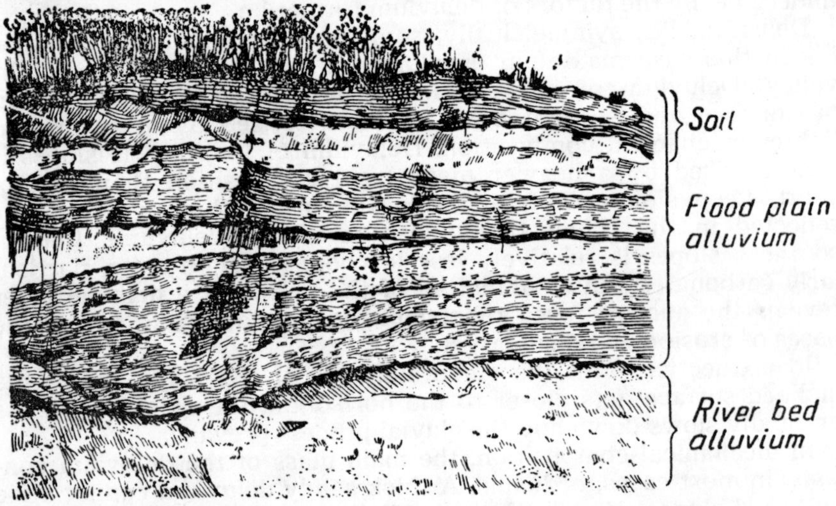

Fig. 7. Flood-plain soil and alluvial complex

geographical conditions. Previously they were erroneously considered as deposits of different ages.

The twofold alluvial complex is clearly marked not only in flood plains but also in the structure of river terraces. The alluvial complex varies greatly in character and in thickness of deposition. The total depth of the alluvial complex is conditioned by the range between the highest level reached by flood water and the lowest depth of the erosive incision made by water flows. The respective thicknesses of river bed and flood plain deposits are conditioned by the average height of normal flow waters, which corresponds to the position of the contact between them. The overall thickness of alluvial deposits sometimes reaches 50 m and more.

Flood plain sediments are deposited on the whole surface of the flood plain which gets submerged by flood waters whereas river bed formations arise as a result of the activity of the water flowing in the bed. One of the constituents of flood plain alluvium is a variety of alluvium referred to as old age alluvium, which forms in the old age area of rivers, i.e., where meanders are formed. Old age alluvium usually represents clayey lacustrine and swampy deposits. River bed deposits are delimited at their base by residual alluvium composed of coarse debris, which predominates in the valleys of mountain rivers. River bed deposits are formed chiefly due to the accumulation of transported material in the bed, and flood plain sediments, due to the precipitation of suspended matter of the flow in the flood plain. It follows that the mechanical composition of the alluvial complex is dependent upon the mechanical composition of the solid portion of the load (Table 2).

Table 2

Mechanical Composition of the Solid Portion of the Load, in Percentages
(averaged for 20 rivers; after Samoilov, 1952)

Solid portion of load	Fractions, mm				
	>0.5 coarse sand	0.5-0.25 medium sand	0.25-0.5 fine sand	0.05-0.01 coarse dust	<0.01 physical clay
Bed load	9	46	38	7	—
Suspended load	—	3	17	25	55

The water flow of rivers also transports a large amount of dissolved substances (Table 3).

Table 3

Amount of Dissolved Substances (in mg/l) in Rivers

Composition	Volga		Syr-Daria		Amu-Daria		Nile	
	winter	summer	winter	summer	winter	summer	winter	summer
Dense residue	306	—	500	290	530	320	205	116
Cl	18	7	42	23	110	50	17	6
SO_4	78	2	145	74	103	68	29	18
K_2O	—	—	11	9	12	8	10	15
Na_2O	20	10	55	24	90	40	13	6
CaO	64	35	100	70	120	78	52	44
MgO	14	7	45	25	28	17	10	10

Suspended sediments contain up to 0.14% P_2O_5 and 0.06% N (gen.) of absolutely dry weight.

The composition and classification of parent rocks. The physical and chemical properties of soils depend to a large extent on the properties of the parent material. But owing to the fact that the prevailing factors of soil formation are the biological processes, the same soils can often be observed on different soil-forming rocks. Chernozem, for instance, arises on moraine, eluvium, loess and other rocks. Conversely, different soils may be found on the same rocks. For instance, on moraine may arise podzolic, soddy-podzolic, chernozemic and other soils. But this does not minimise the great importance played by the rocks in soil formation.

Loose soil-forming rocks constitute the physical medium for the root system of plants, in view of the fact that they possess sufficiently high porosity and water permeability to satisfy the needs of plants in water and air.

An idea of the mechanical and chemical composition of some quaternary deposits constituting soil-forming rocks can be obtained from the data given in tables 4 and 5.

Table 4

Mechanical Composition of Soil-Forming Rocks, in Percentages

Names of rocks	Size of particles, mm			
	>0.25	0.25-0.05	0.05-0.001	<0.001
	sand		dust	silt
Syrt clays	0.69	12.44	75.1	9.6
Moraine clays	2.34	6.10	67.3	22.8
Sandy moraine	30.1	30.0	29.1	10.0
Fluvio-glacial sands	46.6	50.0	2.1	0.74
Old alluvial sands	1.28	94.9	2.9	0.4
Loess like deposits	0.2	16.6	66.7	14.0

Table 5

Chemical Composition of Soil-Forming Rocks, in Percentages

Names of rocks	SiO_2	SO_3	P_2O_5	Al_2O_3	Fe_2O_3	CaO	MgO	K_2O	Na_2O
Boulder loams	72.13	0.14	0.11	13.76	3.82	0.92	1.06	2.97	1.26
Varve clays	62.70	0.02	0.01	18.02	6.95	1.29	2.24	3.35	1.38
Permian clays eluvium	55.28	—	0.09	19.68	6.02	4.42	2.85	1.66	1.72
Boulderless clays of Vologda region	73.0	0.03	0.1	10.5	5.0	1.3	1.8	2.3	1.9
Loess-like clays of Tambov region	71.0	0.2	0.3	16.1	6.1	2.2	3.3	3.0	1.7
Upper terrace sand (Kazan)	95.92			2.31	0.65	0.33	0.15		0.3
Ukrainian loess	58.89	0.09	0.16	12.91	5.27	7.32	2.01	2.39	1.08
Yellow-brown loam of left bank of Volga region	62.70	0.73	0.07	11.12	4.21	6.95	2.24	1.92	2.05

With respect to the supply of plants in mineral food, water and air, all soil-forming rocks are divided into six groups:

1. Mixed, marly, coarse-grained. Unweathered minerals predominate. The rock is poor in available elements of mineral food. Such are the pebbles, gravel, breccia, the coarse-grained sands, not sorted out by water. These rocks are characterised by an insignificant water-retaining capacity (water capacity).

2. Medium-grained rocks, sorted out by water. Usually they are the washed quartz-sand rocks, where the SiO_2 content reaches 90-95%. They represent the residual product of the disintegration of many silicate rocks. Quartz being a relatively stable mineral is more resistant to weathering than the other minerals.

3. Small-grained, frequently mixed medium and fine-grained sandy-clayey (loamy) deposits. These soil-forming rocks are characterised by a relatively slow circulation of the solutions and a medium water-retaining capacity.

4. Fine-grained clayey soil-forming rocks in which clayey products predominate. However, in the rock, $Al_2O_3 < 18\text{-}20\%$ and $SiO_2 > 60\%$. These rocks are the various sediments and eluvium of the temperate latitudes. They possess the capacity to absorb bases. Are characterised by a high water-retaining capacity. Contain a large reserve of moisture unavailable to plants (dead reserve).

5. The products of the disintegration and redeposition of clayey formations (allites) enriched with alumina [$Al(OH)_3$]. Contains more Al_2O_3 than SiO_2, the ratio of Al_2O_3 to SiO_2 being superior to one. Usually these rocks are leached (allito-allophanic) but they may contain carbonate and be saliniferous, containing readily soluble salts in amounts which are injurious to plants.

6. Dust-like (aleuritic) rocks devoid of clayey and sandy particles. They are formed as the result of the redeposition and sorting out of the products of weathering. Apparently sand remains on the seat of disintegration of the rocks or in the vicinity, dust or aleurite accumulates somewhat further, where these rocks are deposited, and the clay particles travel still further. Examples of such rocks are the loesses, the flood plain deposits lying close to the river bed, dust-like clays and small-grained sand or quicksands.

Climate and Soil Formation

Soil can be regarded as the mirror of climate, reflecting the course followed by soil formation in the past. All climates are grouped as follows: nival climates, where the precipitations are only in the shape of snow, humid climates, where the quantity of the precipitations exceeds that of the evaporation from the surface, and arid or dry climates, where the possible evaporation from the surface of the soil is substantially higher than the amount of atmospheric precipitations. Soils which were formed under different climatic conditions differ considerably one from the other in all their characteristics and properties.

The amount of humus in soil also depends to a great extent upon the climatic conditions: there is more humus in soils of a moderately warm climate and less in soils of a cold or hot, dry climate.

The importance of climate in soil formation is quite clearly revealed by a simple comparison between the maps of the climatic and soil zones. The zonality of climate and soils depends upon the distribution on the soil surface of the radiation and water balance and, tied with these, of the thermal and moisture exchanges between the earth's surface and the atmosphere.

The part played in soil weathering by the local climate as a whole and the soil climate in particular is of enormous importance. The soil climate has a definite bearing on the soil's properties (humus-content, temperature, humidity, aeration conditions and others) and, in turn, depends upon the soil, the vegetation growing on it and the elements of the relief. The biological, hydrological, geomorphological, lithological, social and economic factors may exert an influence upon the course of soil formation.

The most important climatic factors in soil formation are the atmospheric precipitations and the flow of radiant energy from the sun (heat and light). The atmospheric precipitations may be in the form of snow and rain, and water vapour may, upon condensation, fall as dew on the surface of the earth and in soil. Not only does the absolute amount of atmospheric precipitations play a role in soil formation but so does also the nature of the soil's humidity, particularly during the vegetative period. Part of the atmospheric precipitations percolates into the ground, the rest runs off or evaporates. The running off atmospheric precipitations effect surface washout and ablation of soils on slopes and deposition of sediments. Atmospheric precipitations may contribute to the establishment of optimum moisture conditions in soil and the formation of highly fertile soils, for instance the chernozems. On the other hand, insufficient moisture and drying out lead to the formation of sierozems of low-humus content, and excess of humidity to that of peaty-gley, half boggy and boggy soils. Irregular periodic precipitations create, in places, an unfavourable water regime in soil, characterised by periods of drought alternating with periods of excessive moisture. Under particularly dry conditions, the upper horizons of soil may dry out, which not infrequently brings about an accumulation of salts and the salinisation of soil. Atmospheric precipitations in the solid form, which form a snow cover of varying thickness and compactness, protect the soil from loss of heat and freezing and, in addition, exert an influence upon the water regime of soil. Atmospheric precipitations, which govern the moisture and the water regime of soil, exert a great influence upon the differentiation of the soil layer into genetic horizons. They also exert some influence upon the mechanical composition of soils. Thus, soils formed under conditions of limited moisture are characterised by a low clay content (soils of hot deserts and semideserts, soils of the cold deserts of the Arctic). At relatively constant temperatures, the clay content in soils increases with an increase of humidity. At the same time, upon constant humidity, the clay content goes up with a rise in temperature, the effect of which is more pronounced in humid and less pronounced in arid regions.

The soil-forming role of water is closely tied with the fall of atmospheric precipitations and the supply conditions of the ground water. The Soviet scientist A. N. Kostyakov used for the determination of the relative humidity of soils the water balance coefficient:

$$K = \frac{(1-\sigma)}{\varepsilon} p,$$

where p=mean annual amount of precipitations in mm;
ε=calculated amount of evaporation (in mm) during the same period;
σ=index of surface run-off.

Taking into account the conditional character of this formula, it should be noted that upon $K=1$, soils have a relatively unstable water balance, which is characteristic of the chernozemic zone; $K>1$ characterises soils of excessive moisture, i.e., podzolic soils; $K<1$, soils of insufficient moisture.

The second climatic factor of importance in soil formation is the temperature of the air and soil. Temperature conditions influence the rate of the chemical and biologic processes which take place in soil, which is in accordance with the well-known temperature law formulated by van't Hoff that the rate of chemical reactions is directly proportional to temperature rise (the rate is accelerated by approximately 2-3 times for a rise of 10°C).

The local temperature conditions and the duration of the vegetative period govern the duration of intensive seasonal soil formation. At temperatures below freezing point, soil formation, when it does not stop altogether, proceeds at a very slow pace. In the belt of the earth's surface comprised between approximately 30° of northern latitude and 30° of southern latitude (wider or narrower in places, depending upon local conditions) and in the tropical zone, soil formation goes on uninterruptedly. Intermittent soil formation proceeds to the north and south of the indicated belt, in connection with winter conditions. However, even there, on the greater part of the territory, soil formation is only profoundly modified in winter, in connection with the passage from summer positive temperatures to winter negative ones. Negative temperatures also exert some influence on soil formation. At low temperatures, irreversible coagulation (setting) of humic acids occurs, with the formation of insoluble substances; the same goes for certain other organic, organo-mineral and mineral colloids of the soil. The periodic action of positive and negative temperatures is attended by freezing, defrosting and thawing of soils. In connection with the crystallisation of water in the soil pores, there is a formation in it of cracks and unconsolidated units of a peculiar character which mark cryogenic structures. Freezing alternating with partial thawing of the soil, with excess of humidity on slopes, causes the ground to flow (solifluction), to distend and bulge out, and a semiliquid mud-like mass to pour out of the cracks.

Under permafrost conditions, the summer thaw affects only the uppermost horizons. The lower horizons of the soil remain frozen the whole year round.

Low temperatures combined with light evaporation of the water and excessive humidity of the soil lead to the formation of half-peaty half-boggy and peaty-gley bog soils, in which occurs a progressive accumulation of semi-decayed vegetable organic matter. High temperatures are attended by a pronounced evaporation of the soil's moisture, drying up of the soil and accelerated mineralisation of the organic matter. In such a case, soil undergoes a degradation, losing its structure, becoming packed, cracked, etc.

Wind may exert some influence on soil formation, by removing the dusty portion of the soil, transporting it far away and bringing material from other places. Wind favours the exchange of air between the atmosphere and soil, accelerating the evaporation of water from the earth's surface and from soil.

The climatic conditions of the native zones leave their mark on all physico-geographic processes as a whole and on soil formation in particular. Depending on the existing climatic conditions acting together with other factors, we get the formation of fairly varied soils in the native zones.

The Importance of Relief in Soil Formation

The surfaces of earth and soil develop in interdependence. Internal (endogenous) and external (exogenous) forces are at work on the earth's surface. One of the diverse exogenous forces at work on the earth's surface is soil formation, which leads to the consolidation and levelling of the surface, in contradistinction to the other exogenous and endogenous forces which lead to the formation of an uneven surface.

The main elements of the relief are watersheds, slopes running down from them and valleys which, in turn, are differentiated into their elements. A mature valley possesses terraces with slopes, a flood plain, a delta and a bed. All the elements of the relief possess their own inherent conditions of soil formation and soil. Soil formation proceeds in interdependence with the main categories of the relief: macro-, meso- and microrelief.

Macrorelief is the collective term designating the most prominent land forms in any particular area, which may be mountainous, hilly or form a plain. The term mesorelief applies to the medium land forms located on the elements of the macrorelief (secondary convex and concave land forms: hollows, hillocks and other irregularities).

The term microrelief applies to the smallest land forms, noticeable only in the immediate vicinity and formed on the elements of the macro- and mesorelief. They include the various microheights and depressions with an area of from one or several square metres to dozens and hundreds of square metres, with an amplitude of height which does not exceed some dozens of centimetres (cups, holes, grooves or slightly raised areas, heaps, ridges, mounds, etc.).

The formation of soils with a typical zonal profile proceeds on flat, broad watersheds.

In the upper parts of the slopes, there is a formation of washed-off and towards the bases of the slopes, of washed-in soils. Only on gentle slopes with a gradient of less than 5° is there a formation of soils with a profile near to normal, owing to the fact that there, erosion is only slight and has no influence upon the formation of a typical soil profile. On gentle slopes (5-10°), moderate (10-15°) and pronounced (15-20°) ones, erosion is fairly pronounced. On steep slopes (20-45°) erosion is so marked and the surface so denuded, that the soil cover subsists only in the shape of isolated patches preserved from washout. On precipitous slopes (45°) usually only rocks appear and there is no soil whatever. The inclination of slopes may, all other conditions remaining unchanged, influence the mechanical composition of parent rocks and soils, which, as the incline becomes more pronounced, gets more coarse-grained.

The development of soils proceeds in interrelation with the formation of ascending, neutral and descending relief. Ascending relief is characterised by convex slopes, formed upon the lowering of the base level of erosion and marked downcutting or river valleys. Owing to the fact that the denudation of watersheds proceeds at a slower pace than the cutting of valleys, there occurs, in consequence, an increase of the relative height of the divide above the thalwegs of the valleys. The relief acquires a sharper outline and is rejuvenated. On steep convex slopes, soils are subjected to washout. Here, we get what are termed eroded soils of an incomplete profile. A subsequent intensification of erosion and washout of watersheds leads to a general lowering of the elevation and maximum lowering of raised areas. This causes an attenuation of the ascending relief which changes to a so-called neutral relief with straight slopes and mature, fully formed soils with a complete profile. Neutral relief is characterised by a relative correspondence between the denudation of watersheds and the pace at which the river valleys are being cut down. But later, upon a relatively stable position of the base level of erosion, cutting down of the valleys stops and they start getting filled up with sediments. Straight slopes pass into gentle and then hollowed out ones, with well formed, thick, old soils. A neutral relief passes into a descending relief, which forms under conditions of continuous lowering of the height of watersheds and uplift of the thalwegs of the valleys. The relief acquires a levelled off (worn by old age) character with a smoothed outline. With a new displacement of the base level of erosion, the relief formation cycle may be repeated at a new stage of development.

Any land form and with it also soil has only a temporary existence and possesses a history. On the basis of some land forms and, consequently, also of soils, others arise.

Geomorphology recognises the following main principles:

1) In time, any ensemble of land forms passes into another one; for example, a mountainous region passes first into a hilly one and then into a flat one (reduced to a peneplain).

2) Similar land forms may arise in different ways, upon different conditions; hills, for example, may be of glacial, erosive, aeolian, volcanic and other origin.

3) Similar conditions may lead to the formation of different land forms; glacial phenomena, for example, lead to the formation of hills, ridges, sanders (outwash plains), valleys, etc.

4) All land forms develop in interdependence one with the other and with soils. A river valley, for example, which makes a substantial cut in the earth's crust, causes intensive erosion of ravines and ablation from slopes and watersheds. Abundant sediments transported by flows from watersheds on to slopes, from slopes into ravines and from ravines into river valleys, cannot fail to influence the development of the latter. Erosion, ablation and the accumulation of sediments, humidification, drainage, and, linked with them, various physical, chemical and biological processes, exert a corresponding influence on the course of soil formation and on changes in the soil. It can thus be seen that the role of the relief in soil formation is quite substantial, but it is far from being limited to that. Of particular importance is the role of the relief in the influence it exerts on the magnitude of the influx of the cosmic factor of plant life, viz., warmth and light, which

produce the local surface and soil climate, in particular the hydrothermal conditions which determine the pace and energy of the biochemical processes.

The influence of land forms on soil formation manifests itself in a redistribution of the elements of natural fertility, in connection with the varying supplies of water and warmth under different relief conditions. The flat areas of watersheds receive the total amount of atmospheric precipitations characteristic of the climate of the particular spot where the most typical soils of the area are being formed. In the temperate belt, in depressions into which an additional quantity of water flows, apart from atmospheric precipitations, there is a formation of soils corresponding to a more northerly position, whereas on slopes, particularly those with a southern aspect, there is a formation of soils similar to those of relatively more southern regions.

Soils on hills, which have a drier microclimate than soils of flattened areas, may be called "locally arid" (dry), whereas soils on depressions, which receive more water than flattened watersheds, may be called "locally humid" (wet). Soils of one and the same region, forming on different elements of the relief will, therefore, differ sharply in structure, composition and properties. Thus, the humic horizon may be substantially thicker in soils of depressions than in soils of hills or slopes, especially the southern ones. In the relatively less drained soils of depressions, the organic matter content and the total content of nitrogen is usually higher than in soils of the raised elements of the relief. Under conditions of greater ground and surface moisture content, the reaction of soil shifts towards acidification together with a decrease of base-saturation.

As distributor of surface water, the relief is divided into two categories: areas with a run-off and areas devoid of run-off, which may be flat or mountainous.

On a plain, as well as in mountains, of great importance are the absolute height of the place, the inclines of the slopes, their forms and aspect. The relief intervenes in the distribution of rain and water from melting snow. It has an influence on climatic conditions, on the distribution and growth of the vegetation and, consequently, on soil formation. Slopes of different aspects (southern and northern) receive unequal amounts of heat and moisture and are subjected to the action of different winds; hence, soils of slopes of different aspects have different water, thermal and air regimes and, consequently, different properties. Southern slopes are not only warmer and drier than northern ones, they are usually subjected to more pronounced washout than northern ones and, for this reason, are more inclined and soils on them are thinner and contain less humus. Not infrequently they contain carbonates and are even salined or eroded.

All other conditions being equal, soils of northern slopes in temperate latitudes are relatively more leached than soils of southern slopes. In connection with the uneven distribution of heat and moisture on the different elements of the relief, the character and rate of accumulation and decay of organic residues varies. This affects the quantity as well as the chemical composition of humus. The higher humidity of soils on concave slopes or hollows in the steppe zone results in a more vigorous growth of the vegetation and a higher humus content of the soil.

The relief also exerts an influence on the depth at which lies the water table and on the degree of mineralisation of the ground water. Underneath a concave surface, the ground water is often sweeter and lies closer to the surface. Not infrequently, soils with a high water table have a lower pH than soils with a good drainage and a deeper lying water table. The relief, thus, exerts also an indirect influence on soil formation, affecting, to begin with, the climatic and hydrologic and then also the biological conditions of soil formation. The microrelief exerts a marked influence upon soil formation notwithstanding the fact that the amplitude of its elevations is expressed in but dozens of centimetres and the diameter of the microheights or microdepressions is limited to a few dozens of metres. The microrelief conditions the complex distribution of soils on an irregular land surface. The soils of microdepressions, such as saucer-shaped hollows, small sinks, holes and pits, are characterised by a higher moisture and humus content than soils of microheights, such as mounds, ridges, hillocks.

The influence of the microrelief on the development of cultures is plainly seen from the appearance of the grass stand. The grass stand of microdepressions is usually vigorous, whereas that of microheights is stunted. Crop variations of this kind are most undesirable and it is therefore necessary to modify the surface so as to smooth out the irregularities of the relief and get uniform relief conditions of soil formation.

Most important in soil formation is the role of the macrorelief and its elements. This is particularly noticeable in mountainous regions, where the soil cover is strongly disunited and, in places, deformed, in connection with the intense denudation of the surface. Soils, here, are formed in accordance with the climatic mountain zones.

The Role of Biosphere in Soil Formation

Soil is a medium of a special kind, where the organic and inorganic worlds are in such close contact as to become practically undistinguishable. Soil is the seat of thoroughgoing processes brought about by the interaction and interdependence between living and inert nature.

In spite of the fact that there is a constant formation of newer and newer quantities of ash food elements of plants, the mineral mass of the soil is not responsible for their accumulation and concentration. The concentration of the elements of ash and nitrogenous food in soil is due to the biologic absorbing capacity of soils. Only in the living organism does assimilated mineral matter escape from getting dissolved in water; every time living organisms die, their mineral elements become dissolved and, if not taken up once again by new organisms, they are drawn into the major geological cycle of changes.

The concentration of the elements of ash food in soil is the function of living organisms, chiefly of the vegetation. The progressive accumulation of organo-mineral material in soil is due to the plants that grow and die in it. The dry matter of plants is composed, on the average (in percentages) of: C—45, O—42, H—6.5, N—1.5 and ash—approximately 5. Ash is composed of fairly numerous elements: their quantity varies greatly depending on the plant concerned.

The formation of soil is due, in the main, to the surface vegetation and the microorganisms which accompany it.

All green land plants are divided into 2 groups: woody and herbaceous plants. Woody plants are all perennial. Under coniferous forest vegetation poor in ash elements, are formed podzolised soils impoverished through leaching, whereas under deciduous forest vegetation, are formed soils relatively rich in humus and ash food elements.

Herbaceous plants, both annual and perennial, contain in their ash a large amount of alkaline and alkaline-earth elements. Hence, the influence on soil of a herbaceous vegetation differs considerably from that of a woody vegetation. The influence of herbaceous plants on soil formation varies, in view of the variations in the conditions of growth.

Under a meadow vegetation, we get a meadow type of soil-forming process which, in a number of cases, passes subsequently into a boggy process. Steppe vegetation gives rise to the steppe type of soil formation. Under a cover of desert vegetation soils of semideserts and deserts are formed.

The root system, which is sometimes more abundant than the part of the plant situated above ground (as is the case with many grasses) and whose development in soil depends on the distribution of moisture and nutritive elements, forms root layers variously distributed in the genetic soil horizons. Roots give out metabolic products (CO_2, soluble carbohydrates, organic acids such as lactic, malic, and other mineral compounds). The dead epidermal cells which are constantly thrown off by the roots are a source of organic matter on which microorganisms feed. The layer of organic matter forming around the roots is called the rhizosphere and the microflora—the bacteriorhiza. The rhizosphere and bacteriorhiza zone is the seat of a mutual nutrition between plants and microorganisms.

Dead roots form the main source of the organic matter leading to the formation of humus; humus gives to the soil its dark coloration down to the depth to which the main mass of the roots penetrates. Plants extract nutritive elements from the ground below and, as they die, there is an accumulation of nitrogenous and mineral nutritive elements, as well as of organic matter. The best results in this respect are achieved through herbaceous plants, which extract 10-15 times more mineral substances from the soil horizons than woody plants (see Table 6).

Table 6

Quantities of Ash Substances Extracted per Annum
(in kg per hectare) (after G. F. Morozov)

Plant	Ash	K	P_2O_5
Pine forest, wood	16	1	0.5
" " , needles	49	3	4.5
Wheat	246	41	24.0
Grassland vegetation	328	82	31.0

To each vegetable formation corresponds a complex of microorganisms composed of varying species, changing in time, in connection with changes in soil formation. There is a very close connection between the soil formation process and the soil organisms.

Plant roots are covered, as with a muff, by a live layer of microbial cells (bacteria and fungi), some of which are useful, others harmful.

The choice of suitable plants in the rotation affords the opportunity to check the development of undesirable microorganisms in soil, rendering it healthier.

The vital activity of microorganisms is amenable to regulation. The accumulation of humus in the soil can be influenced by suitable changes in the moisture conditions.

Under conditions of prolonged excessive humidity, anaerobic organisms may bring about a loss of nitrogen (denitrification). Correct optimum irrigation leads to an increase in bacterial accumulation of nitrogen (nitrification). Irrigation leads to an increase in the number and species of microorganisms and of their physiological activity. One should bear in mind, however, that watering may cause the leaching of readily soluble nitrates. In this respect, different watering systems lead to different results. Thus, irrigation through complete flooding depresses nitrification. Through the appropriate system of irrigation it is also possible to regulate the process of humus formation as a process of microbial synthesis and decay of microbial remains. The destruction of humus by microbes liberates nitrogen and ash nutrient elements, which become available to plants. The decomposition of the remains of green plants is brought about by the lower achlorophyllous plants: bacteria and fungi. Not only the organic matter but also the mineral part of soil are subjected to a vigorous transformation by microorganisms.

The decay of organic matter may be brought about (when there is free access of oxygen from the air) by what are called the aerobic microorganisms: by bacteria, when the soil reaction is close to neutral, by fungi when it is acid. In the absence of free oxygen from the air, the organic matter of the lower horizons and of boggy soils is decomposed by anaerobic bacteria. Anaerobes obtain oxygen from various substances. Part of the substances utilised by microorganisms comes from the soil solution.

Wood, which contains tannic and resinous substances, may be decomposed under aerobic conditions as a result of the vital activity of fungi thriving in an acid medium.

Soil is the seat of a biological cycle of a particular kind whereupon bacteria and fungi decompose the remains of animals and plants and serve as food for the protozoa: rhizopods, amoebae and infusoria which, in turn, serve as food for worms, the remains of worms and insect larvae serving as food for bacteria and so on.

Modern methods of bacteriological analysis permit the quantitative and qualitative study of soil organisms (edaphon). Average samples are selected and their microorganisms, stained by aniline dyes, are examined under the microscope and ultramicroscope. Bacteria appear as mobile and immobile rods, chains and colonies. The number of microorganisms in soil reaches enormous proportions. 1 gram of soil contains thousands of millions of microorganisms.

From 0.1 to 3% of the organic matter of soil consists of the cells of live organisms. The reserve of the microbial mass goes up substantially in cultivated soils.

The arable layer of cultivated soil contains up to 7 tons of live microbial cells per hectare, which die out and are constantly replaced by new ones. According to N. A. Krasilnikov, in one month, 2-3 generations of microorganisms succeed one another in soil, and up to 12-18 (27) generations in one vegetative period (more in the south, less in the north). This means that in summer the arable layer of soil is the seat of the activity of dozens of tons of a live microbial mass composed of bacteria, ultramicrobes and viruses, fungi, actinomyces, algae and protozoa.

The amount of microorganisms is also governed by the general physico-geographical natural conditions, especially climate, season, soil temperature, physico-chemical and chemical processes and the agricultural condition of the soil.

Microorganisms are not uniformly distributed in the soil. In aerated, warm places with optimum moisture content, their number goes up. The maximum amount of microorganisms is found in the upper horizons of the soil (10 cm) decreasing with depth (200 cm): at a depth of several metres, soil is relatively sterile.

The most favourable temperature for microbiological processes is 20-30-40°. Arable ploughland, well cultivated and manured, harbours more microorganisms than untilled soil, and sweet, neutral and limed soils more than salined ones.

The species of microorganisms found in soil vary according to the genetical horizons. In summer, the number of microorganisms in soil rises and the species are more numerous than in winter at a depth below the frozen layer. In winter, the upper soil horizons are subjected to partial sterilisation.

Here are the most important species and varieties of bacteria, one-celled anuclear organisms, living in soil. Particularly important are the nitrous bacteria which transform ammonia into nitrous acid. The nitrogen of organic matter turns to NH_3, which is converted to NO_2 and the nitric bacteria further transform it into NO_3. The enormous agricultural role played by these microorganisms will be fully realised if we take into account the fact that neither atmospheric nitrogen nor the nitrogen of organic compounds, which get into soil with plant and animal remains, are directly available to plants, which can only assimilate combined nitrogen in the form of soluble salts.

The bacteria which fix atmospheric nitrogen deserve a special mention:

a) free-living aerobic (*Azotobacter*) and anaerobic (*Clostridium Pasterianum*) bacteria;

b) nodule aerobic bacteria, which live on the roots of chiefly leguminous plants and assimilate molecular nitrogen. They are able to accumulate up to 160-300 kg of nitrogen per hectare of soil.

Upon the decomposition of proteins, sulphur bacteria oxidise H_2S to molecular sulphur and oxidise this further to H_2SO_4. The part played in soil by the sulphur bacteria is highly important if we take into account the fact that soil receives its main mass of sulphur in the form of albuminous compounds not directly available to plants.

Iron bacteria oxidise ferrous salts to hydrated ferric oxide (bog and lake iron ores) in small lakes, swamps and bog soils.

We now know of bacteria which break up silicates and assimilate potassium. Phosphorus-dissolving bacteria convert compounds of phosphoric acid to phosphorous, hypophosphorous ones and even to phosphorated hydrogen. Microbes can thus improve the supply of phosphorus to plants.

The decomposition of organic matter is brought about by aerobic and anaerobic bacteria.

Aerobic bacteria (*Bac. Micoides*) make up the bulk of what is called the rhizosphere. To this category also belong *Bac. vulgare* and *Bac. coli* which get into soil with muck. Ammoniacal fermentation of urea is brought about by the urobacteria which thrive only in an alkaline medium. The soil bacteria which decompose cellulose with the liberation of H, CH_4, CO_2 are also aerobic bacteria.

To the anaerobic bacteria belong the butyric bacteria which bring about buttery fermentation. The albuminous matter in soil is to a great extent decomposed by putrefactive bacteria.

Excessive irrigation or excess of fresh organic matter (sawdust, straw) in soil create favourable conditions for the vital activity of anaerobic denitrifying bacteria which reconvert nitrates to nitrites, nitrogen oxides, and finally, free nitrogen, which is lost from soil in the following way: $KNO_3 \rightarrow KNO_2 \rightarrow 2KNO \rightarrow N_2$.

Soil may also harbour pathogenic (causing diseases) bacteria which cause tetanus and other diseases.

Some part in the decomposition of organic matter in soil is also played by microorganisms belonging to the myxobacteria group. Soil contains, in addition, ultramicrobes, viruses and bacteriophages. Close to the surface of the soil live soil algae: green in a cold climate (*Chlorophyceae*), blue-green in a warm climate (*Cyanophyceae*) and diatoms (*Diatomeae*) with a silicified cell wall composed of two valves. They are most abundant in soil in spring, less abundant at the end of summer, in autumn and winter. Algae play an important part in the processes of weathering and primary soil formation. Diatoms are able to decompose kaolinite (in solodised soils) and exert a dissolving action on limestones. These algae bring about the accumulation of silicic acid in boggy soils and solods. Algae possess chlorophyll and are able to photosynthesise. They assimilate carbon dioxide and liberate oxygen, which gets dissolved in water, and in this way they improve, as it were, the aeration of boggy soils.

Diatoms are easily seen through a microscope of low magnifying power thanks to the queer shape of the tiny shells present in massive amounts in the so-called diatomaceous earths.

Soil shelters fungi, which are heterotrophic, saprophytic plants feeding on decaying organic matter of zoogenic and phytogenic origin. The most widely distributed are the mould fungi (belonging to the *Ascomycetes*) which develop chiefly in acid soils. With liming, the quantity of fungi in soil goes down. Farmyard manure and green manuring greatly increase, on the contrary, the quantity of soil fungi. Of the fungi distributed in forest soils, *Penicillium, Aspergillus, Trichoderma* and others play an important part in the decomposition of wood (cellulose). Fungi play a part in the decomposition of proteins with liberation of NH_3. Endowed with the faculty of utilising soluble nitrous compounds, fungi are responsible for maintaining the nitrogen cycle in podzolised soils. Certain fungi produce special substances, viz., antibiotics (phytoncids): penicillin and others.

Soil harbours various parasitic fungi and fungi which form what is known as the mycorrhiza, which is a particular symbiosis of higher plants with fungi. The fungi establish themselves on the roots of plants, forming on them a muff-like cover. Mycorrhiza may be superficial (ectotrophic) whereupon the hyphae of the fungi replace the root hairs of plants such as pine, beech, alder and others. Mycorrhiza provides plants with nitrogenous and other nutrients. Endotrophic, i.e., internal mycorrhiza penetrates within the roots and cells. It is found on plants of the orchid and heath families. Close to the fungi are the actinomycetes or ray fungi, whose presence in soil is revealed by the specific smell of freshly ploughed earth, due to the liberation of volatile substances.

In soil also live lichens growing in the tundra and on the weathering surface of compact rocks (lithophytes). The hyphae of lichens penetrate into crystalline rocks to a depth of several millimetres, forming a frond within the rock. A multicoloured crust (red, brown) representing a primary soil in which such elements as S, partly P, Ca, as well as a number of microelements are concentrated, arises on the surface of compact rocks; this underlines the fact that soil formation is, from the very beginning, due to the living organisms and leads to the accumulation of plant nutrients in accordance with the law of progressive increase of fertility. Active substances are formed in soil in the shape of vitamins and antibiotics, which exert a depressing bactericidal action including bacteriolysis (dissolving of cells). A number of bacteria produce substances like auxins, which stimulate plant growth. Sometimes, in connection with the production of special substances by some bacteria, conditions are created which slow down plant growth; this can be put right through the corresponding agromeliorative steps.

Soil harbours the simplest forms of animal life (*Protozoa*) the most numerous of which are the flagellates (*Flagellata*), amoebae (*Rhizopoda*) and infusoria (*Ciliata*). They are mostly aerobic and more seldom anaerobic organisms. Protozoa take some part in the decomposition of organic matter in soil.

In soil also live invertebrate animals, worms and insects. Worms are permanent habitants of the soil. Larvae, which represent a certain stage of development of insects, live only temporarily in soil. Insects (ants, termites) only build their dwelling in soil, spending the rest of the time on the soil's surface.

Invertebrate animals exert a great influence on the course of soil formation in view of the fact that, through their activity, they alter the physical conditions of soil, burrowing subterranean passages or passing soil through their digestive tract. They can exert a chemical action on soil, utilising it for food or discarding their waste matter into it. The invertebrates exert a direct and indirect influence on the soil microflora and microfauna and on soil-forming biochemical processes.

The invertebrates mix soil by bringing it upwards from down below and partly enrich it with organic matter. The earthworms, whose number on irrigated land reaches 1 million per hectare, play a particularly active part in soil formation and the building up of soil fertility. Soil passing through the bowels of earthworms becomes enriched with nitrogen, calcium, and acquires increased absorption capacity. Earthworms improve, therefore, both the chemical and physical properties of the soil, increasing its porosity, aeration and water capacity. Earthworms cannot live in markedly acid and alkaline or dry soils, but can tolerate a prolonged (15-20 days) stay in water.

Finally soil harbours also vertebrate animals, chiefly rodents (susliks, steppe marmots, hamsters, polecats, mice, mole-rats, moles) which, in places, dig numerous burrows.

The queer shaped burrows made by burrowing animals, which in section appear as oval patches of varying diameters, are known as krotowinas. Unsettling of the soil exerts an adverse influence on its properties, increasing the carbonate content and the permeability, which leads to excessive losses of water through percolation. Ploughing and other agrotechnical measures reduce the harm done by burrowing animals.

The Role of Time and Space in Soil Formation

Soils, which, like any natural body, develop in the course of time, undergo constant changes in composition and internal properties, and thereby also of external characteristics, in connection with the dynamics of all the factors of soil formation and intrasoil processes. In certain spots and areas of the earth's surface, due to it being renewed (washout, erosion, deposition), soil formation is intermittent. In places, soil is partly or totally destroyed and is formed again and continues its constant course through the stages of a progressive development from primitive to mature soils. In some cases soil develops (matures) faster, in others, slower.

On the more friable sedimentary rocks, rich in plant nutrient elements in available form, soils develop faster than on compact hard rocks. These differences between soils have a direct bearing on the level of fertility.

A distinction is made between the absolute and the relative age of modern soils. The absolute age of a soil is the absolute time which passed from the moment the formation of that soil began until the present time. If the beginning of the appearance of some soils goes further back in time in comparison with some others, then their age is absolutely greater. Soils of one and the same absolute age may sharply differ in their development. They are then of different relative ages: some of them, viz., those that develop faster, have a greater relative age than those which lag behind in their development. Some soils may be relatively less developed (new soils) than others, which develop (mature) faster, although they may be of the same age; or it may even happen that absolutely older soils may appear younger than some absolutely young soils.

Phenomena of this sort are tied not only with differences between soil-forming rocks, but also with differences between land forms, with physico-chemical and biological phenomena and a number of other physico-geographical conditions. The absolute age of the majority of soils is reckoned in many millenia, the exact number of which is lost in the distant past. It is easier to determine the age of tame soils, from the vestiges of human activity, such as pieces of bricks, glass, asphalt, concrete, ebonite, flinty and metallic weapons of man, even coins bearing dates.

On the whole, the formation of soils proceeds at a relatively fast pace. In road excavations and on embankments of unfinished and abandoned roads, on the dams of ponds and reservoirs, one can observe the formation of new soils which come quite close to the natural soils of the adjacent territory.

In some cases, soils disappear from the surface of the earth, having been blanketed by newest deposits, such as alluvium, deluvium, moraine, volcanic eruptions, mud flows, aeolian deposits, etc. Soils may vanish under the water of transgressing seas and reservoirs or may become covered by glaciers.

During the ice age, an enormous territory was covered by ice, which put a complete stop to soil formation. Glacial retreats and advances went on for many thousands of years. On land liberated from ice, there was a new growth of vegetation and new soils were formed. The spaces of southern regions, which were liberated centuries earlier than in the north are now covered with absolutely older soils than the younger soils of northern regions.

In the area which escaped glaciation, soils are considerably older but even there we find no soils of pre-quaternary age, in view of the fact that the earth's surface is constantly renewed as a result of denudation, erosion and the accumulation of deposits. For the same reason, in any region, we find older soils on flat watersheds and younger ones on slopes, which are being renewed by washout and deluvial deposits.

Many methods have been evolved for the determination of the age of soils, of which we should mention the geochronological, archeological, geological, geochemical and others. The age of an alluvial soil can be determined from the age of the alluvial silt depositions. From the structure of the soil humus, the amount of C^{14}, the disintegration products of elements and isotopes, we can find the age of the oldest soils on earth. The determination of the absolute and relative ages of soils is of great scientific and practical significance.

The significance of space in soil formation is viewed in the broad aspect of the geographic distribution of soils on the earth's surface within the framework of the native soil zones and, in a narrower sense, of the territorial distribution of soils in relation to the local borders, such as the sea coastline, the distance from the edge of a forest or a forest belt, from the limit of sand distribution, from the beginning to the end of a slope or from the summit down to the foot of mountains, etc.

The intermittent character of the distribution of soils is revealed by the differences between them, corresponding to the various native zones. Thus, podzols are distinguished from soddy-podzolic soils. The latter are distinguished from the grey forest soils and the chernozems. As for the continuous character of the distribution, it is revealed by the completely imperceptible passage of one soil into another. The limit between the various groups, subgroups, varieties, is, of necessity, only conventional.

In accordance with the geographical coordinates of longitude and latitude of the area, limits are drawn separating the native zones, one from the other, the soils of some provinces and facies from those of others.

The Production Activity of Man and Soil Formation

Existing soils are more and more influenced by the production activity of man, which, up to a recent past, was, on the whole, haphazard and destructive. It was sometimes positive, sometimes negative. Upon its utilisation with the purpose of obtaining maximum profits without bringing about a radical improvement or caring about its future fertility, soil became exhausted and was subjected to destruction. The soil of enormous areas was losing its structure and was, in places, eroded by water or wind action. Indiscriminate irrigation of land led, in the past, to mass salinisation and swamping of soils. The burning down of steppes and forests and deforestation aggravated the harmful effect of drought, which resulted in the soils suffering a depletion of humus and a worsening of their physical and chemical properties. It was only in the vicinity of build-up areas and in plots adjoining dwellings that soil was, in the past, subjected to insignificant taming and slow enrichment.

At the present stage of development of science in socialist conditions, soil is the main means of agricultural production and, at the same time, a product of labour. The production activity of man has now become the leading factor in cultural soil formation. If it takes centuries, in nature, to achieve a highly fertile and rich soil, the same result can now be attained with the minimum of delay under arable conditions. The process of taming of soils is now progressively being intensified. Further cultural soil formation will lead to the creation of soils far outstripping in potential and effective fertility the best soils of today.

Soil Formation

Soil is the result of a soil-forming process representing an intricate exchange of energy and matter, a complex of prolonged interactions between soil-forming rocks, the organic world and the environment (including relief, climate, geographical situation).

Taken in the broad sense, the term factor of soil formation applies to any of the natural interdependent phenomena or sources of matter and energy which take some part or other in the creation of soil, as well as to the circumstances (conditions) which govern the course of soil development in space and time. They can be divided into the following two groups:

First group (factors):

1) lithosphere (soil-forming rock);

2) biosphere (living organisms of vegetable and animal origin);

3) atmosphere (principally its lower part and in particular the layer of air lying close to the ground);

4) hydrosphere (surface and ground water).

Second group (conditions):

5) cosmic phenomena (mainly the influx of radiant energy, i.e., light and heat);

6) relief (land forms) and gravitational phenomena;

7) space (geographic situation);

8) time (age of soils);

9) production activity of man (agricultural and other).

The factors of soil formation develop in interdependence with the soil.

A most important condition in soil formation is the influx of radiant energy on earth (light, heat). This condition is only partly regulated by the soil. The development of the relief is still less influenced by soil formation and soils.

The production activity of man at an early cultural stage depended on the soil. The influence of soil upon man's culture and production activity is not direct but conditional, in so far as soil is an element of the geographical environment.

There is, between soil, time and space a unilateral and relative dependence, i.e., only the soil depends on the situation and age of the country.

A leading factor of soil formation, which is dependent on soil, is the green vegetation, which synthesises organic matter from CO_2 and H_2O with the help of sunlight (photosynthesis). Green plants play a cosmic role, by capturing an insignificant part (not exceeding 1%, in the main) of the solar energy which reaches the earth's surface in the form of units of light energy (quanta).

Soil is a peculiar mineral-biological system characterised by a complex of properties inherent to the living organisms and mineral bodies. All the properties of soil are functionally interdependent. A change in one of them leads to changes in the others. For example, the soil's temperature fluctuations necessarily influence the course of the mineral and biological processes. A change in the amount of water present in soil not only leads to an increase or decrease of humidity but to a change, in one direction or another, of all the other properties, such as the heat and water regimes, electroconductivity, composition and concentration of the soil solutions, compactness, porosity, etc. It should be noted here that the properties change in vectors, i.e., differently in different directions, owing to the fact that soil is an anisotropic body. In conformity with this fact, not only do the properties but also the composition and structure of soil undergo changes, which can be represented by complex curves, indicatrixes and block-diagrams (Fig. 8).

Soil formation on earth began with the appearance of the first simplest forms of life and the interdependence of biological and mineral factors, i.e., when the uninterrupted process began of the transition of one form of matter into another, viz., mineral into organic and vice versa.

In the long run, plants do not exhaust but enrich the soil year after year, raising its fertility. If this were not so, we should not find such extremely rich and fertile soils as the deep chernozems or the humus-carbonate soils (rendzinas) and others. In the course of the vegetative period, the leaves of a maple deposit on the surface of one hectare of soil more than 120 kg of CaO, 16 kg of K_2O, 5 kg of P_2O_5. The other broad-leaved species and herbaceous plants enrich the upper horizons of soil with ash elements in no lesser degree and, in this fashion, counteract washing out (leaching, podzolisation). The living organisms which grow on and in the soil, concentrate phosphorus within their bodies and as they

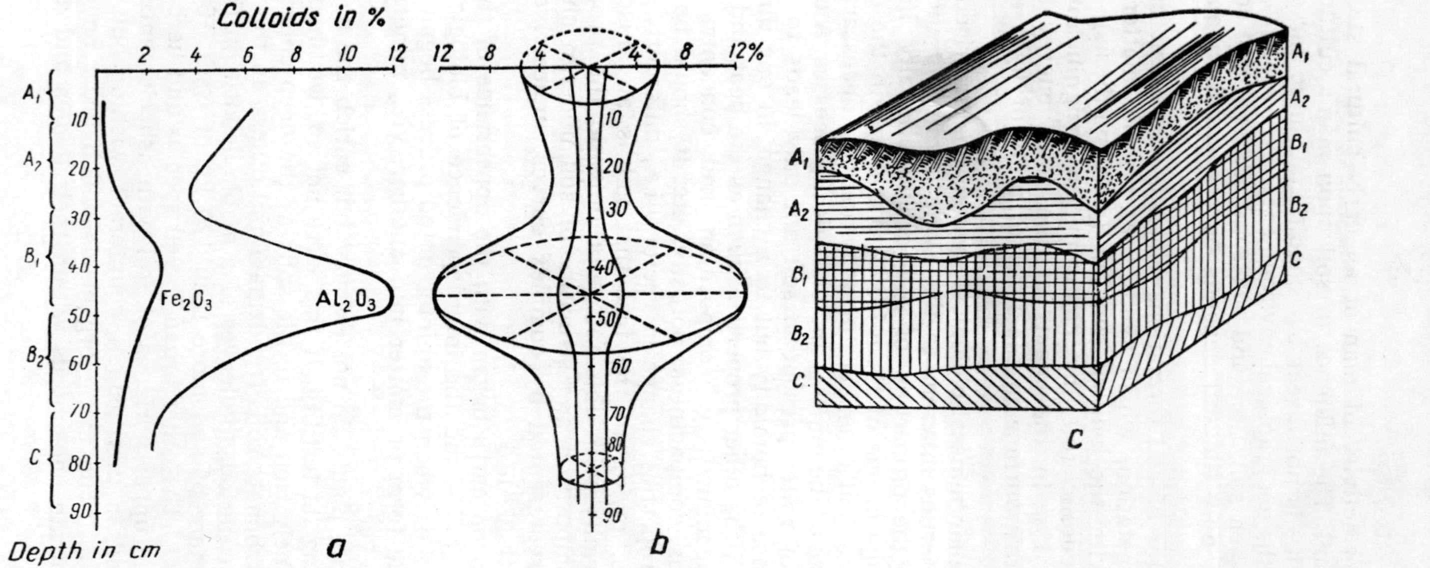

Fig. 8. Changes in the composition and structure of podzolic soil:

a—curve of the colloids content through the genetical horizons; *b*—same, indicatrix (after Ienni); *c*—block-diagram: relief and structure of the soil in perspective section (schematic) die, phosphorus accumulates in the upper horizons. But phosphorus, like the other biogenic elements, is removed from soil with water, taken up by crops and carried to the towns. These losses must necessarily be compensated by taking steps to prevent leaching and by applying mineral and organic manures.

The rise of the soil's fertility, which is conditioned by the development of plants, is not a mere quantitative process but a qualitative transformation, expressing the various stages of development of soil formation.

As the soil itself changes, it undergoes changes in its level of fertility, which normally rises. But at isolated stages in nature, an extremely slow and even regressive course of development of fertility can be observed. The soil may develop at a temporarily subsiding level of fertility (podzolisation, swamping, salinisation, solonetrisation, solothisation, steppe formation, etc.).

If the process of synthesis of organic matter went on endlessly on earth or exclusively prevailed over the process of its decomposition (mineralisation), then, as a result of soil formation, soil would, in the end, everywhere pass into the exact opposite, i.e., organic rock: peat, coal, graphite, which does in fact take place, but on a limited scale.

In nature, on a par with the synthesis of organic matter occurs also a decomposition of it as a result of the gigantic work performed mostly by the lowest organisms which transform and decompose vegetable matter. The synthesis and decomposition of organic matter constitute the essence of soil formation.

In contradistinction to weathering, soil formation is, on the whole, a creative, biologic process. This process is developing in a progressive fashion, reinforcing the procreative faculty of plants and raising the fertility of soils. There is absolutely no limit in nature to the rise of soil fertility. But the fertility of a soil cannot be identified with the presence in it of plant food elements. This complex property is amenable to regulation, not only through agrochemical or agrotechnical measures but also through the control of the soil formation process as a whole. This control is therefore particularly important from the production point of view.

All the complex physical, chemical and biological processes of soil formation may, in turn, be differentiated into the following elementary processes:

1) Physical processes of soil formation: a) formation of suspensions; b) formation of soil solutions; c) translocation of solutions and suspensions due to the force of gravity, suction force, osmosis, diffusion, etc.; d) precipitation as a result of conversion to insoluble salts, in connection with a change in the composition of the solutions, the concentration, the moisture content, the soil temperature, the absorption capacity, etc.; e) adsorption.

2) Chemical processes of soil formation: a) chemical reactions brought about by the action of H_2O, CO_2, O_2, HNO_3, H_2SO_4 (hydration, dehydration, carbonatisation, decarbonatisation, oxidation, disoxidation, silicatisation, desilicatisation and others); b) chemical reactions tied with the action of humic acids (displacement of silicic acid and carbonic acid from their compounds, intensification and acceleration of the processes of decomposition of silicates and alumosilicates, reduction under conditions of excessive moisture, etc.).

3) Biologic processes of soil formation: a) mineralisation of organic remains, i.e., modification of the composition of complex organic compounds to simple mineral compounds, upon the participation of microorganisms and of the enzymes which they excrete. To these processes belong: ammoniation, nitrification, denitrification, decay of albumens, methane and hydrogenous fermentation of cellulose, fermentation of starch and pectic substances, decomposition of fats, resins, waxy and tannic substances, destruction of wood and so on; b) humification, i.e., the formation of relatively stable humic substances from plant remains with the intervention of bacteria and microscopic fungi; c) synthesis of organic matter; d) assimilation by microorganisms and plants of certain elements of the atmosphere and volatile organic substances emitted into the air by soil.*

The main processes of soil formation as a whole are based on a complex of elementary soil-forming processes. To each type of soil formation corresponds a certain soil profile depending on the actual genetical horizons. The following processes are in progress during the formation of the humic-accumulative (upper) horizons of soil:

1) Formation of organic matter.

2) Decomposition of organic remains: a) mechanical comminution (physical change), b) dissolution, c) mineralisation (decomposition), d) humification (formation of humus).

3) Fixation and accumulation of humic substances.

4) Accumulation of ash elements (building up and fixation).

To the main processes of soil formation also belong the accumulation of organic matter, the formation of sod, peat formation, podzolisation, leaching, gleisation, solonchak formation, solonetrisation, solothisation, ortstein formation, ferruginisation, allitisation, formation of soil silicates and silicates of aluminium, formation of a colloidal complex, etc.

The formation and accumulation of organic matter of vegetable and animal origin proceed in interdependence with the development of soil fertility. The process of accumulation of organic matter in soil is due to the mechanical changes undergone by organic

* According to investigations conducted by Soviet Academician N. T. Kholodny.

remains, dissolution, mineralisation in the form of decomposition, humification, etc.

Aerobic decomposition is an accelerated process which goes on in the presence of free oxygen. Aerobic decomposition produces a large amount of heat and the temperature of the decaying matter may rise sharply. Anaerobic decomposition is slow and is not attended by a heating of the decaying matter.

Humification is the formation of humic organic matter; in the process of its fixation, a chemical reaction takes place between the humic acids and calcium carbonate, attended by the coagulation of the organic colloids by the $Ca^{\cdot\cdot}$ cations. The most pronounced fixation or enrichment with humus occurs in soils which contain much $CaCO_3$ and R_2O_3. A significant temperature fall and an increase of the moisture content of the soil have the same effect. The formation of soil humus proceeds through the following three types of disintegration of plant and animal remains, with the formation of products of synthesis:

1) Bacterial aerobic process, the breakdown products being, to begin with, NH_3, then H_2O, CO_2, carbonates, sulphates, nitrates, phosphates, chlorides and the products of synthesis being humic acids and compounds.

2) Bacterial anaerobic process with the formation of breakdown products such as CH_4, H, N, NH_3, H_2S, PH_3, FeS_2, $(NH_4)_2S$, and intermediate products such as organic acids: butyric acid $C_4H_8O_2$, lactic $C_3H_5O_8$ and others, the product of synthesis being ulmic acid.

3) Fungal process, the breakdown products being H_2O, CO_2, N, carbonates, sulphates, chlorides, NH_3, and the products of synthesis being crenic and apocrenic acids.

In the first case the process goes on in a relatively loose aerated mass of organic matter. In the second case the breakdown of plant and animal remains proceeds in a tightly packed mass. In the third case, the breakdown of organic matter is effected under aerobic conditions by fungi.

Humus accumulates in soil because in a given period of time, more organic remains get into soil than can be broken down into simpler products, such as CO_2, H_2O, NH_3 and others. This is sometimes reversed, however, depending on the soil formation conditions of that particular period of time and the general physico-geographical conditions at that place. Among the conditions which exert a direct influence on the humus balance of soil is the physical structure of soil, such as its looseness or density. In loose soils, decomposition is more intense than in packed or dense soils. By regulating soil structure we may therefore influence the accumulation of humus.

The decomposition of organic remains is slowed down in a very wet soil, with the formation of the raw humus of grassy bogs and peaty-gley soils. An acid reaction also slows down decomposition, which leads to a great accumulation of half-decayed humic substances in soil, etc. As a result of humus formation, soil acquires a dark colour, particularly in the upper horizon, with gradual fading out further down. The decaying organic remains which lie on the surface and penetrate downwards in the shape of humus tongues also play a part in the accumulation of humus. This oozing down of humus interferes sometimes with the gradual fading out of the dark coloration. Humic substances may penetrate down to the illuvial horizon along pores and cracks.

In connection with the mineralisation of the organic remains, the accumulation of humus in soil is attended by an accumulation of ash elements. The roots of plants, by penetrating deep into soil, extract the nutrient elements which they need and when they die and become mineralised, they leave them in the soil in the shape of various compounds. After mineralisation of the organic remains has taken place, the ash elements pass into a soluble form, are partly absorbed by the upper humic horizons and partly carried down by water to the zone of activity of plant roots where they are taken up once more and used for building up new organic matter and so on. In the course of time, a certain amount of ash plant food elements is accumulated in the upper horizons of soil and the more so, the richer the vegetation thriving on them.

The ash content of plants varies and is higher in herbaceous than in woody plants. Meadow and steppe grasses contain 7-7.5% of ash; beech litter 5.6%, spruce litter 4.5%, pine litter 1.5%. In accordance with this, different quantities of nitrogen and ash elements get into soil with the plant remains (Table 7).

Table 7

Amounts of Nitrogen and Ash Elements Received by Soil with Litter Fall (in kg per hectare) (after N. P. Remizov)*

Plants	N	SiO_2	R_2O_3	CaO	MgO	K_2O	P_2O_5	SO_3
Pine forests	7-8	3-7	3-7	11-25	2-5	2-5	1-4	2-5
Fir-woods	19-20	11-20	1-8	20-22	4-5	5-6	4-7	2-3
Birch groves	44-64	2-4	3-6	22-53	12	12-30	10-12	5-15
Oak woods	31-46	41-53	5-17	85-99	15-18	18-25	22-29	13-20
Meadow steppe grasses	100	450	—	40	8	200	35	—

Simultaneously with the humus-accumulative horizon, a lower lying horizon is formed called the eluvial horizon and still lower, a washed in horizon called the illuvial horizon (*eluere*=to wash

* See pp. 24-25.

out, *illuere*=to wash in). Washed out horizons are formed in connection with the solubility of a number of substances and their translocation in solution, due to the force of gravity, to deeper soil horizons, even as deep down as the ground water. Leaching in the podzolic zone is particularly pronounced. Soils of the podzolic type lose their soluble salts and the eluvial horizons lose even their sesquioxides. Even humus is partly washed out in soluble form. The washing out of humic substances in the podzolic zone is revealed by the brownish coloration of the local ground and bog waters. In places where evaporation exceeds precipitation, salts are not washed out of soil, they may, on the contrary, rise to the surface of the soil, not infrequently causing its salinisation.

Various compounds reach the illuvium. Gypsum settles in the illuvial horizon in the form of isolated faded patches, veins and concretions or grains. Lime, which is transported as bicarbonate, falls out of solution as carbonate. Lime falls out of solution in the small pores. It falls out in the shape of faded patches and tiny crystals, and in large cavities and cracks, lime may fall out in the shape of concretions and inclusions of various shapes and names. Sometimes there is a formation of lime tubes. Part of the lime permeates the whole washed in horizon more or less uniformly. Leaching of the sesquioxides is attended by a packing of the illuvial horizon and is marked by darkening, browning, clayisation, ferruginisation, with, sometimes, the formation of a ferruginous horizon (ortstein) or isolated veins, e.g., pseudofibres (ortsand). In anaerobic conditions, upon excessive moisture, occurs what is known as gley formation or gleisation, i.e., the formation of a particular kind of gleyey horizon in the form of a viscous mass of a bluish-grey colour with, sometimes, a greenish or bluish tinge. In interdependence with gleisation a process of peat formation goes on, giving rise to the peat horizon of bog soils. This process is favoured by an increase of the moisture content and an intensification of the anaerobiosis which, in turn, put a stop to the mineralisation of the organic remains and create suitable conditions for the conservation and accumulation of a poorly decomposed vegetable mass on the soil's surface. Later on, mineralisation is reduced to the very minimum whereas the accumulation of peat and increase of the peat layer continue. In this fashion, peat deposits arise, where soil formation passes into a geological process, in which the biological factor becomes completely predominant over the mineral factor.

A special kind of soil formation gives rise to soils known as solonchaks which contain an enormous quantity of soluble salts lying in the soil and on its surface in the form of a salt crust and impeding the growth of plants and microorganisms. The formation of solonchaks leads to a progressive attenuation of the effect of the biological factor and a reinforcement of the geological factor.

At the same time, soil, as the natural body for the production of green plants, loses its significant property, viz., fertility.

We do not know how soils developed in the distant past. We can only get an approximate and indirect idea of the ourse followed by soil formation, from geological, paleoclimatic, paleogeographic and paleontologic data. Thus, the direct evidence and vestiges of the coniferous forests which once grew in the Crimea, the skeletons of willow grouse and white foxes, the remains of dryads (*Dryas octopetala*) in the quaternary deposits in the Crimea are an indication of possible conditions for podzol formation where now chestnut soils are being formed. The presence of fossil tree trunks in tundra soils is evidence of the existence there, in the

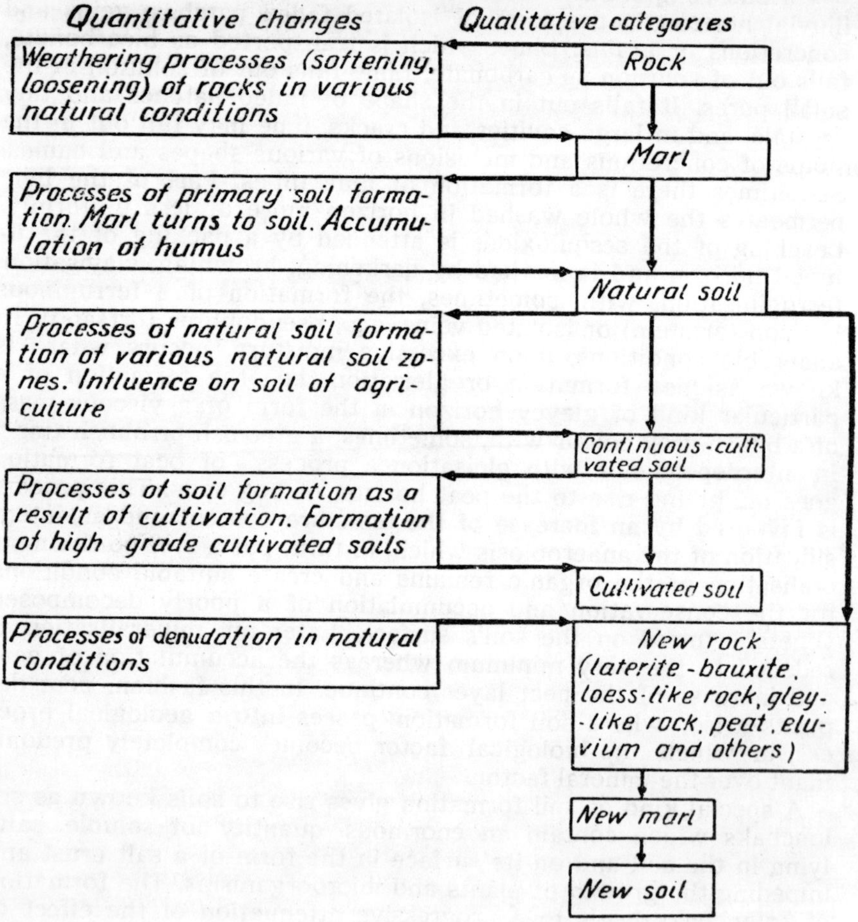

Fig. 9. Diagram illustrating the development of soils

past, of forests and taiga soils. All this is evidence that in the past the native zones were displaced in diametrically opposite directions.

It can be seen from the general diagram of the development of soils (Fig. 9) that the processes of cultural soil formation are closely linked with and can be traced to natural soil formation. A knowledge of the properties of soil and the ability to control soil-forming processes will allow us to work out a scheme of meliorative measures whereby the desirable aspects of soil formation will be enhanced and the undesirable ones checked, with, as a result, a progressive increase of fertility.

Chapter III

COMPOSITION OF SOIL

Soil is made up of three phases: solid, liquid and gaseous. The solid phase consists mainly of mineral particles (up to 90% of the soil's weight) and to a lesser extent of organic substances. This phase is part of the mechanical, microaggregatory and structural composition of soil. The liquid phase in soil is represented by water and an insignificant amount of liquid organic substances. The gaseous phase is made up of soil air, and contains water vapour and a certain amount of CO_2, NH_3 and other gases.

Mineralogical Composition

In mineralogical composition, soil is close to the parent rock from which it was formed. The mineral part of soil consists, in the main, of primary minerals: quartz, felspars, micas, hornblende, augite, magnetite, hematite, apatite and others. Of the secondary minerals distributed in soils we have: opal, kaolinite, halloysite, montmorillonite, beidelite, nontronite, hydrargillite, diaspore, bauxite, limonite, hydrohetite, hetite, turite, pyrolusite, psylomelane, calcite, gypsum, mirabilite, halite and others.

The weathering of parent rocks is inevitably attended by a sorting out of the products of disintegration according to their mechanical and mineralogical composition. In sands, on the whole, primary minerals predominate, viz., quartz and felspars, whereas in clays the predominating minerals are the secondary ones with a certain amount of quartz. More variegated in mineralogical composition are the loams, which represent a mixture of primary and secondary minerals. The primary minerals which form the residue of magmatic rocks passed into soils without changes in their composition. They are admixed with secondary minerals which form the main mineral mass of soils. Primary minerals are the sources of formation of secondary minerals and of the elements of ash food of plants. Secondary minerals are found in soil chiefly in the colloido-dispersive condition. They determine the main properties of the soil (hydrophily, swelling, filtration) including its fertility.

Chemical Composition

The chemical composition of the soil depends on the chemical composition of the parent rocks, the organic matter and other constituents of the soil. Of the substances making up the soil, SiO_2 and R_2O_3 are the most abundant, then come the organogenous elements: C, H, O, N, P, S, K, Ca, Mg and others. Soil analyses reveal, in addition, the presence of Na, Al, Mn, Fe, Ti, Cl and, in small quantities, of Rb, Li, Ba, Zr, Th, As, U, F, Cu, I, Co, Se, Cs, Sr and other rare elements, the contents of which in soil vary. Thus, cobalt (Co) is abundant in clayey chernozems and sparse in acid peaty and sandy podzolic soils. It has been established that cesium (Cs^{137}) accumulates in soils as a result of atomic tests (Japan). Nitrogen is a constituent of humus, of nitrates and ammonium salts. Nitrogen is found in the soil air.

Nitrogen finds its way into soil as a result of electric discharges in the atmosphere. It is synthesised to compounds of nitrogen by ultraviolet rays and is assimilated by bacteria.

In view of its high mobility, nitrogen is easily lost from soils as a result of the activity of denitrifying bacteria and by chemical means, via the reactions that take place between nitrous acid salts and amino compounds when the soils are poorly aerated.

Phosphorus is a most important organogenous element. Its content in soil changes in connection with its utilisation by plants. In a number of cases, soils may suffer from a shortage of phosphorus, causing sickness in plants.

The contents of humus, nitrogen and phosphorus in chernozems, are as follows (in percentages):

Depth, cm	Humus	Nitrogen	Phosphorus
0-18	6.6	0.21	0.24
18-26	5.5	0.20	0.23
26-55	1.6	0.06	0.11
55-90	0.3	0.02	0.06

Potassium is quite an important element of the ash food of plants. It is usually present in soil in adequate amount but cases of deficiency are frequent. For the obtention of high yields, its application in the form of mineral fertilisers may become necessary.

The soil's contents of N, P, K, which are nutrients of prime importance, do not exceed 1% of the total amount of the ash elements present in soil. The soil contains also many soluble salts: chlorides, sulphates, nitrates and nitrous compounds, phosphates, carbonates.

Changes in the chemical and mechanical composition of soil take place simultaneously. Passing from coarse particles to fine ones, the amount of silica decreases and the amount of Al, Fe, Mg, and K increases. In the course of soil formation, the chemical composition of soil-forming rocks and of soil itself undergoes substantial changes in connection with the soil formation processes of the native soil zones. The chemical composition of soil is determined by various chemical analyses: bulk, acid, and water extracts, analyses of soil solutions, etc. The data of these analyses are cited in the description of the native soil zones.

Mechanical and Microaggregatory Composition

Soil is a polydispersive body consisting of particles of extremely varied sizes, from large fragments (stones) down to the minutest colloidal particles which can only be seen through an electron microscope when magnified several thousands of times. It is therefore necessary to resort to a grouping of the particles according to their diameters within the conventional limits of what are termed the fractions of the mechanical composition. Such a grouping of the particles is referred to as the classification of the mechanical elements of soil (Table 8).

Table 8

Classification of the Mechanical Elements of Soil after Williams-Kachinsky

Diameter of particles, mm	Names of the mechanical fractions	
Less than 0.0001	Colloids	Physical clay
0.0001-0.001	Silt	
0.001-0.005	Fine dust	
0.005-0.01	Medium dust	
0.01-0.05	Coarse dust	Physical sand
0.05-0.25	Fine sand	
0.25-1.00	Medium sand	
1.00-3.00	Coarse sand (grit)	
3.00-10.00	Gravel	
10.00-larger	Stones	

Stones and gravel (more than 3 mm) are fragments of rocks whose structure and properties they have retained and which exert an influence on the physical properties of soil.

Sand (3-0.05 mm), an immobile constituent of marl and soils, consists mainly of crystalline silicic acid, partly of flakes of mica and fragments of other minerals, takes almost no part in chemical processes and is not a plant nutrient, but it has an influence on the soil's physical properties. Sand represents the end product of the mechanical weathering of silicic rocks.

Dust (0.05-0.001 mm) consists mainly of crystalline, stable and to a lesser extent amorphous silicic acid. The dust fraction represents a relatively inert part of the marl. Chemical and physical processes in it are relatively weak.

Silt (particles smaller than 0.001 mm, including the colloids or particles smaller than 0.0001 mm) is the most active and mobile clayey part of marl and soils. It is composed of mineral and some organic matter, mainly alumina (kaolin and other compounds), free iron oxide, hydrated iron and aluminium oxides, iron and calcium phosphates. This fraction is endowed with Brownian movement and other properties possessed by colloids, coherence, plasticity and impermeability, is coagulated by electrolytes. Silt is one of the factors of structure formation.

It is customary to express the mechanical composition of soil by giving the relative amounts of the fractions in percentages. The division of the mixture of particles into fractions is referred to as its mechanical analysis. There are various methods for determining the mechanical composition of soils, based on the elutriation and subsequent settling of the particles according to size and weight (methods of V. R. Williams, Sabanin, Shöene, the pipette method), or on the separation of the mechanical elements of soil using centrifugal force (centrifugal method). The determination of the mechanical composition by the microscopic, photographic, radioscopic and aerometric methods have not, so far, received any application in view of the fact that they have not yet been perfected. The classification of soils according to their mechanical composition which is generally accepted is that of the Russian scientist N. A. Kachinsky (Table 9).

Using the data of the mechanical analysis of soil, based on the relative fractions of sand, dust and silt, we can further subdivide soils according to their mechanical composition. In time, the mechanical composition of soils usually changes and becomes heavier and in some horizons (A_2 podzolic) it may become somewhat coarser grained owing to the removal of the fine-dispersive part. Some physical and physico-mechanical properties of soils are governed by the mechanical composition. Thus, sandy soils and sandy loams are usually devoid of structure but are endowed with favourable air and heat regimes, possess good water permeability but low water capacity, offer little resistance to agricultural implements. At the same time they are poor in humus, nitrogen, ash plant food and, as a consequence, they rapidly become exhausted.

Table 9

Classification of Soils According to their Mechanical Composition
(after N. A. Kachinsky)

Content of physical clay (particles <0.01 mm) in percentages of weight of dry soil			Classes (names of soils according to mechanical composition)	Varieties (distinguished according to relative amounts and predominance of fractions)
for soils of podzolic type	for soils of steppe type	for solonetzes and strongly solonetzic soils		
>80	>85	>65	Heavy clay	Silt-like, dust-like
65-80	75-85	50-65	Medium clay	Ditto
50-65	60-75	40-50	Light clay	Silt-like, dust-like, coarse dust-like
40-50	45-60	30-40	Heavy loam	Ditto
30-40	30-45	20-30	Medium loam	Silt-like, dust-like, coarse dust-like
20-30	20-30	15-20	Light loam	Coarse dust-like, sandy
10-20	10-20	10-15	Sandy loam	Ditto
5-10	5-10	5-10	Coherent sand	Ditto
0-5	0-5	0-5	Loose sand	Fine, medium, coarse grained, gritty

Note: The remaining particles making up the difference up to 100 consist of physical sand. The terms "heavy", "light" refer to greater or lesser resistance upon cultivation. Clayey soils are difficult to work, sandy ones are lighter. The latter have a lower apparent density.

Clayey soils, on the contrary, are rich in ash plant food, possess high water capacity and low water permeability, do not easily part with their water, possess poor aeration and unfavourable heat properties (cold soils), offer much resistance to implements. Not infrequently, on such soils, a crust is formed of low water and air (gases) permeability. Loams, which are intermediate in properties and composition, are considered the best soils. Useful as it is, a knowledge of the mechanical composition of soils allows us to gain but an approximate picture of their physical properties, owing to the fact that the mere adding up of the mechanical elements (fractions) does not provide the sum-total of their properties. The physical properties of soils are also influenced by the chemism of the mechanical elements and the course of the biological processes. Silt, for example, possesses low water permeability and the movement of water along the capillaries, even upon maximum capillarity, is slow. The same silt, but saturated with Ca and Fe represents an important factor of soil structure, governing the formation of the most favourable physical properties of soil.

Differences in the chemical composition of otherwise equivalent mechanical soil elements condition the differences in effective (pervious, transit) diameter of the pores and differences in their water-air regime.

A great variety is observed in the mechanical composition of soils in nature in connection with differences in the combination and arrangement (mosaic) of the mechanical elements.

It has also been found that soils of close or identical mechanical composition differ considerably in their physical properties. The great diversity in soil properties is conditioned by the diversity of the microaggregatory composition.

In nature, the elementary particles of the soil's mechanical composition are usually coalesced in microaggregates, i.e., small clods of various sizes, from several millimetres down to thousandths of a millimetre in diameter. Just as the mechanical particles, these aggregates become joined together in fractions according to size. The division of the microaggregatory composition of soil into fractions is effected in the same manner as for the mechanical analysis, i.e., by a microaggregatory analysis. There are sometimes substantial differences between the data of the mechanical and microaggregatory composition of one and the same soil sample (Table 10).

Table 10

Mechanical and Microaggregatory Composition of One Sample of the Upper Horizon of Ordinary Chernozem
(in percentages of weight of dry soil) (after N. A. Kachinsky)

Composition of soil sample	Size of particles, mm					
	1-0.25	0.25-0.05	0.05-0.01	0.01-0.005	0.005-0.001	<0.001
Mechanical	0.0	1.8	35.2	10.9	16.7	35.4
Microaggregatory .	0.0	35.5	45.3	9.1	7.6	2.5

A mechanical analysis of the sample (Table 10) gave the percentage of silty particles as 35.4, whereas the microaggregatory analysis of the same sample put it at only 2.5, due to the fact that the remainder was used up in bringing about aggregation of the other particles into microaggregates, i.e., the remaining part of the silt was drawn into the composition of larger fractions of complex particles. Errors are possible in mechanical analyses: they occur when microaggregates, which for some reason were not broken up in the preparatory stage for the mechanical analysis, are classed as mechanical elements.

The ratio of silt in the microaggregatory composition to that in the mechanical composition determines the capacity of the soil to form aggregates. This ratio is known under the name of index or coefficient (*K*) of dispersion (after N. A. Kachinsky) which is computed in percentages according to the following formula:

$$K = \frac{a}{b} 100,$$

where a is the amount of silt obtained by microaggregatory analysis, and b the amount of silt found by mechanical analysis. The higher the K index, the lower the aggregatory capacity, the less steady the microstructure and structure of the soil. The K coefficient of chernozems does not exceed 10%, that of solonetzes may exceed 60-80%.

The microaggregatory composition of soil is a dynamic quantity and, in consequence, so are the main physical properties of soil.

Organic Matter

The organic matter of soil is rich in chemical energy which, in the process of decay and breaking up, is liberated in the form of thermal energy, given off upon oxidation. The organic matter of soil passes through a lengthy course of transformations at certain stages of which it acquires a specific character, being retained in soil in the form of humus. In humus, the solar energy reaching the earth becomes transformed and humus contains carbon and nitrogen, which are indispensable for the development of life.

Humus contributes to a better assimilation of the mineral substances, increasing the permeability of plant tissues (membranes). It makes soil friable, lowering its apparent density, increases the water capacity, the absorption of radiations, etc. Humus increases the coherence of light soils and decreases that of heavy soils. It contributes to the creation of optimum moisture conditions of soils. Humus partly reacts with the mineral part of the soil to give particular organo-mineral compounds. As a result of polymerisation, part of the humic substance becomes packed and accumulates in soil, remaining there for a more or less lengthy period.

Humus is the main source of plant foodstuffs. The organic matter of soils consists of the organic remains of animals and plants. Nearly 7 tons per hectare of air-dried vegetable material situated above ground and up to 25 tons per hectare of root material remain every year in chernozemic soil. In soils of desert steppes (sierozems) there remain annually during the vegetative period nearly 1 ton of air-dried material situated above ground and nearly 15 tons of roots per hectare. The transformation of these remains in soil proceeds in two ways: a) breaking up (mineralisation) with the formation of simple compounds (CO_2, H_2O, NH_3 and others); b) formation of new, more stable organic compounds (humification process). These transformations take place as a result of the following processes:

1) Physico-chemical changes of the organic matter, which begin immediately following the death of the living organisms, with the participation of the atmospheric agents: water, air, heat, light, plant enzymes.

2) Changes of the organic remains under the influence of the activity of animals (worms, insect larvae), aerobic and anaerobic fungi and bacteria. Aerobic decay proceeds rapidly and can be likened to burning. Anaerobic decay proceeds slowly and does not go to the end; this incomplete decomposition leads to the accumulation of organic matter of the semi-peaty and peaty type.

3) Change of the soil's organic matter under the influence of the chemical action of its acids and salts on the organic remains. The same goes for the physical effect of the soil on the process of transformation in it of organic matter.

As a result of the decomposition of organic matter and the life activity of microorganisms, there is a formation of humus, and carbon dioxide is given off in soil and into the air layer close to the soil. The carbon dioxide utilised by green plants goes towards the building up of organic matter. In this way, a constant circulation of carbon dioxide is set up between soil and vegetation.

Humus is not a product of a simple decomposition (mineralisation) of organic matter. Natural humus must also be regarded as a product of synthesis by the lowest achlorophyllous plants (microorganisms) plus the decayed organic mass of the microorganisms themselves and the products of their life activity (waste products, toxins, exoenzymes). Dry bacterial matter in a 20 cm layer of soil represents from 0.12 to 0.75 tons per hectare and for the whole of the vegetative period nearly 2.5-3.3 tons per hectare (after I. V. Tyurin). Decay of the organic matter begins and partly proceeds afterwards, outside the cells of the living organisms. But the main processes of decay and transformation of plant remains proceed with the participation of the micro- and macroorganisms living in soil. The varied processes of transformation of organic remains in soil may be represented graphically in a diagram after I. V. Tyurin (Fig. 10.).

The pace of the processes of decomposition of remains in soil is closely linked with its moisture-content. Air-dried plant remains are hardly at all decomposed in soil. In the absence of adequate moisture there is a preponderance of physico-chemical phenomena leading to humification. This takes place in chernozemic soils, where the decomposition of organic matter is of brief duration and proceeds only in spring and autumn. In summer, the soil dries up, decomposition slows down, but humification goes on. Upon increase of soil humidity, the decomposition of organic matter is accelerated but up to a certain limit, in connection with the fact

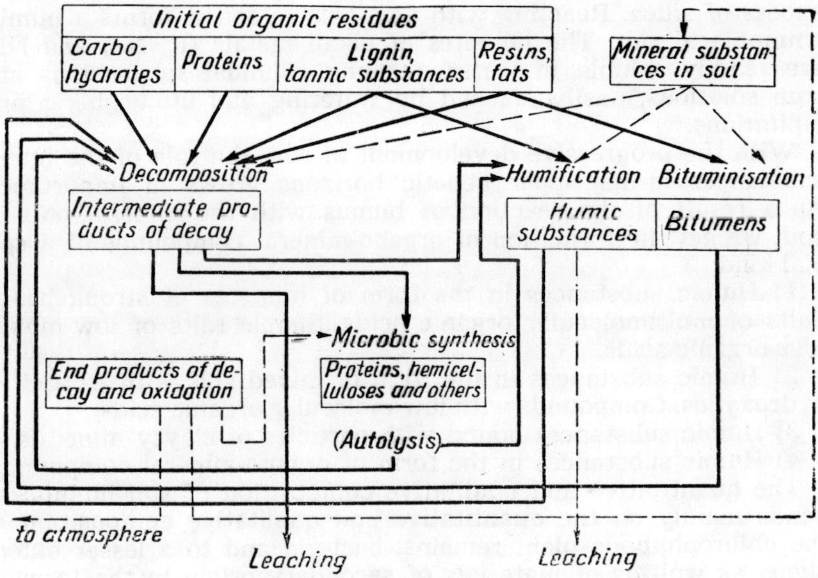

Fig. 10. Diagram illustrating the transformation of organic remains in soil (after I. V. Tyurin)

that the access of oxygen is hampered and aerobic processes are replaced by anaerobic ones. Upon aerobic decomposition, the end products of mineralisation are CO_2 and H_2O. Among the products of anaerobic decomposition preponderate unoxidised and incompletely oxidised substances: CH_4, H_2S and a number of organic acids, which inhibit the life activity of microorganisms and put a stop to decomposition. The process of humification is suppressed, there occurs only a bituminisation of fats, waxy substances and resins. Under anaerobic conditions there is an accumulation of semidecayed, semipeaty and peaty material. Plant remains, poor in bases, give more free organic acids. The richer in bases the soil-forming rock and soil, the less acid the humus formed.

Humus contains fulvic acids (crenic and apocrenic), ulmic and humic acids. Of the fulvic acids, colourless crenic acid is a product of the fungal decay of wood, it is soluble in water in unlimited amount and is not denatured, giving a distinct acid reaction to the solution. Ulmic acid is brown, it is a product of a bacterial anaerobic decay, and is readily soluble. Becoming denatured with low temperatures in winter, it gives ulmin, which has a neutral reaction. Ulmin accumulates in soil and gives it its brown colour. Humic acid is black, it is a product of an aerobic bacterial process. It is practically insoluble in water but is soluble in weak alkalis. Upon denaturation it gives humin, intensifies the hydrolysis of the silicates of aluminium, promoting the dissociation of the excess of silica. Reacting with ammonia in soil it forms a humic-ammoniacal salt. The humates of alkali metals (K, Na) and NH_4 are readily soluble in water, forming colloidal suspensions and true solutions, easily leached by watering and atmospheric precipitations.*

With the progressive development of soil, the role of the humic substances in its upper genetic horizons grows in importance. As a result of the reaction of humus with the mineral part of soil, we get the formation of organo-mineral compounds of a special kind:

1) Humic substances in the form of humates of strong bases. Salts of multimolecular organic acids. Simple salts of low-molecular organic acids.

2) Humic substances in the form of mixed gels with Al and Fe hydroxydes. Compounds with low-molecular organic acids.

3) Humic substances bound with particles of clayey minerals.

4) Humic substances in the form of organo-mineral compounds.

The quantitative and qualitative composition of soil humus depends mainly on the quantitative and qualitative composition of the chlorophyllous plant remains, bacteria and to a lesser extent algae, as well as of materials of secondary origin in the form of animal remains.

The amount of humus depends on the physical, chemical and biological properties of soils, which govern some or other favourable conditions for the development of plants and the subsequent accumulation of organic remains. According to data obtained by D. V. Khan, some 60-70% of the humus is bound with the fraction of the mechanical composition smaller than 1 μ (clayey minerals of the montmorillonite group), from 5 to 23% with the amorphous silica fraction and some 10-14% is in the free state.

The organic matter of soil consists of the following compounds:

I. Nitrogen-free substances—carbohydrates (sugars, peptose, hexose, cellulose), lignin and others.

II. Nitrogenous compounds (proteins, fats, resins, waxy and tannic substances).

Decomposition of nitrogen-free substances, like cellulose, pectic substances and hexose, begins with their hydrolysis, which is brought about by the bacteria responsible for buttery, methane and

---

* The composition of the acids is not exactly known. The question of the nature of humic substances remains to this day somewhat obscure.

The tentative composition of humic acids is: C—52-62%, O—31-39%, H—2.5-5.5% and N—2.6-5.1%. They contain some carboxyl (—COOH) and phenolhydroxyl (—OH) groups. The composition of the group of crenic acids is characterised by a low C content and higher O and H contents (C—44-50%, O—42-48%, H—4.6-6%, N—2.5-5.5%). According to certain data, humic acids should be regarded as organo-mineral compounds.

hydrogenous fermentations. Wood components, like lignin, are decomposed by fungi. The conversion of lignin to humic substances proceeds further, through a chemical process, this time upon the participation of oxidising enzymes. The sugars are decomposed by bacteria with the formation of organic acids and the liberation of CH_4 and H, followed by CO_2 and H_2O.

Nitrogenous compounds and fats are decomposed with the help of bacteria, mould fungi and actinomycetes. The hydrolysis which takes place thereupon is accompanied by the formation of glycerine and fatty acids. The decomposition of wax, resins and tannic substances proceeds slowly, due to the fact that they are not easily attacked by microbes. The decomposition of protein is also accompanied by hydrolysis; it is broken down into simpler compounds with the formation of amino acids and partly of NH_3. At the same time, there occurs a partial desamination. As a result of synthesis, some enrichment in nitrógen (microbial albumen) and the accumulation of phosphorus, sulphur, potassium and others are possible.

After they die, microbes decay just as any other organic matter. The process of decomposition of organic matter in soil does not stop so long as there is a continuous supply of organic remains, provided it is not hampered by special adverse conditions (temperature, salts, moisture). When the supply of fresh organic remains stops, there is a progressive slowing down of the decomposition.

On a par with microbial processes of decomposition and synthesis, other processes of transformation of organic matter liable to oxidation and polymerisation reactions are in progress; more stable multimolecular compounds are formed, having the character of humic substances. In time, humic acids turn to more compact humic substances, akin to coal and graphite. Attempts are being made to determine the age of these dense humic substances from their structure and the amount of C^{14}.

Humus itself decays but more slowly than the original organic remains from which it was formed. Decomposition is all the more intense as the organic remains contain more nitrogen. When there is a shortage of nitrogen, decomposition is weaker but losses of nitrogen are higher, owing to the fact that microbial synthesis is weaker when there is a shortage of nitrogen. The vigour of the decomposition is governed by the conditions and degree of aeration and moisture, i.e., depends on the correlation of aerobic and anaerobic processes. Aerobic decomposition may lead to an almost complete mineralisation of the organic matter. Anaerobic decay leads to the accumulation of undecayed organic matter (peat, sapropel).

Decay usually proceeds under both aerobic and anaerobic conditions, which alternate according to the season. Decomposition increases with a rise in temperature up to an optimum of 30-40°; upon further rise of temperature, the intensity of the decomposition falls and finally stops altogether. The intensity of the decomposition is higher in soils which are lighter in mechanical composition and in structural soils. But in light soils, which are more aerated and water permeable, the products of decay are more rapidly lost. That is why light soils are usually less rich in humus. These soils are poor in Ca and, in this connection, there is no stabilisation of the humic substances.

Different soils possess qualitatively different humus, which depends on climate, relief and soil organisms. Humus of podzolic soils of coniferous forests is relatively light in colour. It contains 2-3 times more fulvic than humic and ulmic acids. Humus formed in steppes is of a dark colour. It contains many stable neutral humic and ulmic substances rich in oxygen, which confer a dark colour to soil.

Humus is an important factor of the exchange absorption capacity of soil and of structure formation. It does not directly provide nutrients for plants and living organisms owing to the fact that it is a colloidal substance, suspensions of which are unable to diffuse and penetrate through plant membranes into the plant roots. The plant foodstuffs contained in humus become available upon disintegration of the organic matter. But humus is, in the long run, a source of mineral and nitrogenous foodstuffs.

The amount of humus in soil changes, depending on the system of cultivation practised, the plants cultivated and the microorganisms present in soil, i.e., its amount in soil is amenable to regulation and accumulation. Accumulation of humus (*S*) can be brought about by controlling the aerobic and anaerobic processes of the decomposition of the original organic matter as well as of humus itself. This accumulation can be approximately expressed by the formula:

$$S = \frac{(1-\alpha)}{x} A,$$

where S—amount of accumulation of humus;
A—annual amount of organic mass reaching soil;
α—coefficient of decay of organic matter;
x—coefficient of decay of humus.

Progressive agrotechnical practices and a high level of agriculture lead to an increase of the humus content. Applications of peat and lime ($CaCO_3$) to soil promote a slow, steady accumulation of humus in them. The importance of humus in the development of soil properties is exceptionally high.

Chapter IV

SOIL COLLOIDS AND ABSORBING POWER OF SOIL

The absorbing power of soil is its ability to retain or absorb various substances which react and are in contact with its solid phase.

The theory concerning the absorbing power of soils, which has gained world-wide recognition, was worked out by a leading Soviet soil scientist K. K. Gedroitz.

Soil is an energetic chemical reagent and absorbent. It is capable of retaining or absorbing gases, various compounds from their solutions, as well as particles of mineral or organic composition, live microorganisms and suspensions. The main plant nutrients, as K, NH_4, Ca, Mg, P, are energetically absorbed and preserved.

The absorbing power of soil is tied with the highly dispersive, finest, mainly colloidal part of the soil, which possesses an extensive specific surface area for a small amount of substances and a high energy of hydration of the ions.

Soil Colloids

Among the constituents of soils are the colloids, which play an exceptional role in the life of soil and in connection with the possibility of controlled changes. In particles' size they occupy an intermediate position between suspensions and true solutions. Colloidal particles have sizes ranging from 0.0001 to 0.000001 mm (from 0.1 μ to 1 m μ), they pass through ordinary filters but not through ultrafilters, do not diffuse and are not dialysed, can be seen through an ultramicroscope.

The size of colloidal particles is thus at the limit between what can and what cannot be seen with the naked eye.

Colloidal solutions are identified by what is known as the Tyndall effect, which consists in that a ray of light passing through a colloidal solution becomes visible, as distinct from true solutions through which it passes without being revealed. A ray of light passing through a colloidal solution becomes visible because colloidal particles diffract light, owing to the fact that their size is larger than the length of light waves. The degree of diffraction varies due to the length of light waves and the sizes of the colloidal particles being different.

Even up to the beginning of the present century, all substances were divided into colloids and crystalloids. It is now known that any substance may assume either a crystalloidal or a colloidal state, if we create the conditions that will bring about a finer division of its particles or, on the contrary, their enlargement. The first way of obtaining a colloid through a finer division of the particles in suspension is the dispersive way, through the use of chemical reagents or mechanically in what are called colloidal mills, where a high degree of division of the substance can be achieved. The colloidal properties appear progressively, as the particles decrease in size. The second, opposite way of obtaining colloids from molecular solutions by enlargement of the particles—with the formation of sparsely soluble substances—is the condensative way.

Soil contains mineral, organic and organo-mineral colloids.

All colloids form systems consisting of a dispersive medium and dispersed in it a colloidal dispersoid phase. The medium and the phase may be solid (S), liquid (L) and gaseous (G) in various combinations (Table 11). As systems, colloids consist either of particles of various sizes (polydispersive systems) or of particles of relatively equal sizes (monodispersive systems).

Table 11

Possible Cases of Heterogeneous Systems

Dispersive medium	Dispersoid phase	Name of system. Examples
S		Metallic sodium in common salt. Soil
S	L	Moist soil. Clayey "solutions"
S	G	Dry soil. Pumice
L	L	Emulsions and emulsoids
L	S	Suspensions and suspensoids. Clayey suspensions. Colloidal solutions
L	G	Froth. Air bubbles in soil water
G	G	Mix in all combinations
G	L	Mist (aerosols). Water droplets in soil pores
G	S	Smoke, dust

Any colloidal particle situated in the plane of separation of two phases (for instance L and G, S and L) always possesses some reserve of free surface energy, measured in dynes per 1 cm^2, which conditions the energy of absorption (adsorption), which is all the higher as the specific surface is larger, i.e., as the surface of the separation, possessing a reserve of free surface energy, is larger.

The total surface area of the soil particles (specific surface) rises progressively as they become smaller (Table 12).

Table 12

Increase of Surface Area of a Body upon Division

Length of edge of cube	Number of cubes	Total area	Specific area	Perimeter (P) cm (length of edges)	Number of tetrahedral angles
1 cm	1	6 cm^2	6	12	8
1 mm-1,000μ	10^3	60 cm^2	6×10	12×10^2	8×10^3
100μ	10^6	600 cm^2	6×10^2	12×10^4	8×10^6
10μ	10^9	6,000 cm^2	6×10^3	12×10^6	8×10^9
1μ	10^{12}	60,000cm^2=6m^2	6×10^4	12×10^8	8×10^{12}
0.1μ-100mμ	10^{15}	60 m^2	6×10^5	12×10^{10}	8×10^{15}
0.1mμ	10^{24}	60,000 m^2=6 ha	6×10^8	12×10^{16}	8×10^{24}

The term specific surface (S_0) designates a particular quantity, which is determined by the ratio of the total surface of the dispersive particles (S) to one unit of volume (V=1 cm^3); $S_0 = \frac{S}{V}$; $S = S_0 V$, i.e., the total surface area of the particles is equal to the product of the specific area by the volume.

The main mass of the soil colloids consists of humic particles and clayey minerals of crystalline structure in the form of discs of from 0.01 to 0.05 microns in thickness. The total surface of these particles amounts to from 15 to 100 m^2 to 1 g (after A. A. Rode).

Colloids exhibit Brownian movement, i.e., a continuous spontaneous irregular motion of the particles in water, which occurs as a result of the colloidal particles being repeatedly struck by the moving molecules of the dispersive medium (examine a suspension of gamboge under the microscope). The tiny particles of the dispersive phase, subjected to countless blows of varying strength, are set into motion and all the more intensively as they are smaller. Brownian movement prevents the particles of the dispersive phase from coming to the surface or sinking.

Not being dissolved, colloids may take part in chemical reactions with compounds of the contiguous solutions, with those molecules or ions which compose the upper layer.

In a reaction involving electrically neutral substances coming into contact with charged ions, colloids acquire a corresponding charge. But ions of one sign cannot exist without an equivalent number of ions of the opposite sign.

And so, around the nucleus of any substance two layers of ions are formed: an inner layer (determining the potential) and a double outer layer which together constitute the colloidal particle or micelle.

The nucleus and inner layer of ions constitute a granule and the latter together with the outer layer of ions constitutes a colloidal particle. The inner ionic sphere is formed as a result of the dissociation of the nucleus.

According to the electric charge of the particles, all colloids are divided into: a) acidoides—negatively charged colloidal particles, having cations or H^+ in the diffusion layer; b) basoides—positively charged, having anions or OH^- in the diffusion layer; c) ampholytoides, which change their charge according to the reaction of the medium. The positive charge in acid medium changes to a negative charge in alkaline medium and vice versa.

Under the influence of the neutralising effect of electrolytes, colloidal particles lose their charge and rapidly grow larger, as if curdling, i.e., from a state of colloidal solutions or sols, they pass into jelly-like gels. This phenomenon is referred to as coagulation. In coagulation there is a formation of particles of the first, second, third order, etc. The particles grow larger due to unresilient blows or agglutinate. The amount of electrolyte necessary to cause coagulation is called the edge of coagulation and the moment when the particles lose their charge corresponds to what is referred to as the isoelectric point.

Upon the coagulation of a colloidal solution, a certain amount of ions of the coagulator passes into the sediment. These captured ions are not washed out by water, they can be separated by washing with the solution of another electrolyte, replacing them in equivalent number. This constitutes the basis of the absorbing power of soils. An equilibrium is set up between the ions of the medium and the diffusion layer of the colloid. In a diffusion layer absorption and displacement of varying energy occur. The further the absorbed ion from the nucleus of the particle, the more easily it becomes displaced, and vice versa. The nearer to the nucleus, the more firmly the absorbed ion is held. The energy of the absorption is all the greater as the sign of the cation is greater. Coagulation can be reversible and irreversible. The process of irreversible coagulation is a potent factor of cementation in soil-ground. In the case of a reversible coagulation, colloids form compounds with the solvent, drawing it into the composition of the gel. But within the coagulated complexes of colloidal particles or gels, they retain their individuality. That is why coagulated particles, upon absorption of ions, for example, or dissociation of ionogenic groups, recover their charge and change back to sol. This phenomenon of passage from gel to sol, which is the opposite of coagulation, is called peptisation. Peptisation occurs all the easier as the gel is more aqueous and younger. In the first stage of peptisation, the colloids swell.

Colloids can be hydrophobic, in which case the nuclei of their micelles have no affinity for water molecules, and hydrophilous, when their aggregatory stability depends not only on the ionogenic group but also on the hydrated pellicle surrounding the colloidal particle, which is little affected by electrolytes.

The edge of coagulation is substantially higher for hydrophilous than for hydrophobic colloids. A mixture of hydrophobic and hydrophilous colloids with preponderance of the latter exhibits low sensitivity to electrolytes. The first is protected, as it were, from the effect of electrolytes by the second, as a result of the formation of complex sols. Thus, the formation of podzolic soils, gives rise to complex organo-ferruginous and organo-clayey sols, in which the protecting colloid is made up of fulvic acids and the protected colloids, of hydroxydes of iron and aluminium. When the amount of the electrolytes is small, gel formation is slow. In certain cases even a mechanical influence (shaking) can change it back into sol (thixotropy).

The separation from the gel of part of the water it contains, referred to as syneresis, occurs as a result of the reciprocal attraction between the molecules in a liquid and reciprocal attraction between the solid colloid particles, i.e., as a result of the overcoming of the forces of attraction between the micelles of the colloidal and the liquid. Syneresis (discharge of water) in soil is characterised by the fact that the reciprocal attraction between water particles and soil particles continues to manifest itself even after the formation of a gel. As a result of a contraction of the volume of the solid phase of the soil, there occurs a discharge, squeezing out, exudation of water.

In soil occurs a periodic, often recurring (rhythmical) penetration and deposition of some colloids into the mass of others, often with the formation of multicoloured streaks, spirals or rings. This phenomenon is referred to as the formation of Liesegang rings. It accounts for the formation of patterned soil horizons, pseudofibres, tongues, pockets, streaks and patches.

Colloids possess the property of coacervation, occurring as the result of the effect of high concentration of electrolytes. Instead of coagulation we get the formation of drops of coacervate (phase containing the colloids) which progressively grow larger (Fig. 11). Coacervation may pass into coagulation and vice versa. Coacervates are considered as the initial stage of the formation of archaeobions in soil.

The varied physico-chemical phenomena occurring in soil are tied, in the main, with the manifestation of the properties of the various soil colloids. The most widespread colloid in soil is SiO_2 of mineral origin. Silicic acid passes into solution upon weathering and soil formation, after which, upon polymerisation, micelles of the colloid are formed. The colloid SiO_2 is a typical acidoide. Reacting with proteide substances it forms complex proteide-silicic compounds. The distinctly aqueous hydrophilous colloid SiO_2 plays a protecting role for the colloids of sesquioxides but it may change its negative charge to a positive one in the presence of an excess of R_2O_3. Among the negatively charged mineral colloids are

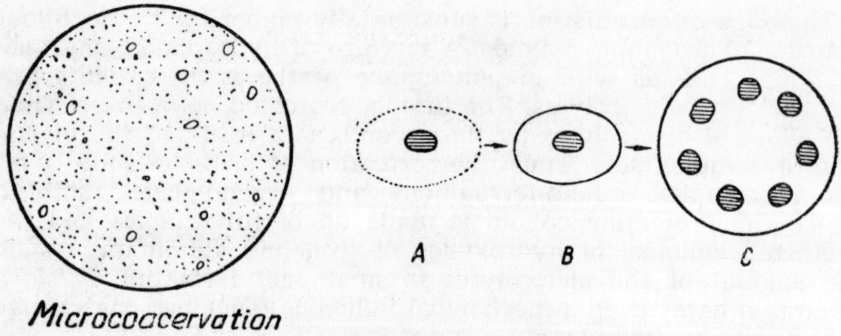

Fig. 11. Process of coacervation.
Left—microcoacervation; *right*—diagram illustrating the coacervation process:
A—colloidal particle surrounded by a diffuse water pellicle (dotted line); *B*—the same particle devoid of diffuse hydrated layer, with one compact layer of bound water (completely surrounding it); C—particles whose compact hydrated pellicles have coalesced into one (coacervation)

compounds of manganese, found in the illuvial horizon of soil, sometimes in large quantities (up to 15%). These compounds precipitated together with Fe_2O_3 colloids give iron-manganese concretions.

Among the positively charged mineral colloids, we should, in the first place, name the hydrates of sesquioxides whose isoelectric point is pH-7. In an acid medium R_2O_3 colloids carry a positive charge and in an alkaline one, a negative charge. Positively charged colloids absorb anions, for instance P_2O_5.

R_2O_3 hydrates, reacting with negatively charged mineral and organic colloids, form elaborate complex compounds.

On a par with the mineral colloids, the soil contains organic and organo-mineral colloids, i.e., negatively charged humic substances. In the ionogenic layer, the COOH of humic acids splits off into COO^-, attached to the nucleus and H^+ in the diffusion layer which, reacting with electrolytes, is displaced by cations. Organic and organo-mineral colloids possess hydrophilous properties and impart a high stability to clayey colloids, causing their translocation from the upper horizons into the illuvial ones where they coagulate and become fixed. Proteins, i.e., albumens reaching soil with plant remains, are organic colloids of the ampholytoide type, their organic radical being:

$$R\begin{cases} NH_2\text{—weak base} \\ COO\text{—}H\text{—weak acid} \end{cases}$$

the weakly acid properties being more pronounced.

Of the negatively charged organic colloids, cellulose is a typical hydrophilous colloid, as well as lignin to which COO^- gives a negative charge. Organic colloids, containing COO^-, are, for the most part, negatively charged. The organo-mineral colloids which have particles of unequal surface, represent complex ampholytoides in which we find SiO_2 and humic substances as acidoides and R_2O_3 in the form of $\frac{Al}{Fe}(OH)_3$ as basidoides.

The properties of the soil colloids and their charge is determined by the ratio $\frac{-SiO_2}{+R_2O_3}$.

All the soil colloids taken together represent a colloidal complex known under the name of soil's absorbing colloidal complex, a highly dispersive part of the soil, conditioning the absorption phenomena.

The colloidal complex is richest in clayey, humic soils and chernozems and poorest in sandy soils. In the latter it may be found only as films colouring the sand particles. In some soils, the colloidal complex forms up to 75% of the weight of the soil, in others it hardly reaches a tiny percentage. The colloidal complex is a most important constituent of the soil's absorbing complex, in the composition of which coarse dispersions are also found.

The soil's absorbing colloidal complex consists of a mixture of compounds in a colloidal condition:

1) of particles of finely divided minerals, of the products of their disintegration in the form of SiO_2, $Al(OH)_3$, $Fe(OH)_3$, MnO_2, compounds of silicates of aluminium and ferruginous silicates of aluminium and other mineral colloids (the zeolitic part of the complex);

2) of the products of the decomposition and synthesis of organic remains: humic substances, organic colloids (humate part of the complex);

3) of the products resulting from the reaction of organic and mineral colloids and complexes arising as a result of the reciprocal coagulation of organo-mineral colloids (zeolite-humate part of the complex).

All soils possess an absorbing colloidal complex. It is inseparable from soil. This complex changes in composition and quantity according to the genetic horizon of the soil in which it is found.

Absorbing Power of Soils

The absorbing power of soil is differentiated into several types: the mechanical, physical, physico-chemical, chemical and biological.

The mechanical absorbing power of a soil, i.e., its property to retain mechanically substances suspended in water, depends on its mechanical composition, structure, texture, porosity and capillarity. As a filter, soil is capable of retaining particles filtering through it, depending on their size, the diameter of the capillary pores and their disposition. Mechanical retention of suspensions occurs mainly in the upper layers of the soil horizon, at some depth from the surface. Mechanical absorption contributes to a great extent to the formation of the illuvial (B) genetic horizon of soil.

The physical absorbing power of a soil is its power to absorb from solutions molecules of electrolytes, the products of the hydrolytic dissociation of the salts of weak acids and strong bases, and colloids, as a result of their coagulation. The physical absorbing power of a soil is tied with its surface energy, which is conditioned by the total area of the finest particles of the soil. It is manifested on the surface of the fine dispersive particles of soil in the increase of the concentration of substances coming from the soil solution. Physical absorption is attended by polar adsorption, i.e., a condensing of molecules on the surface of separation of two phases: solid and liquid, solid and gaseous, conditioned by the existence of unsaturated surface energy on the surface of the dispersive soil particles. This energy is all the greater as the specific area is larger, i.e., as the mechanical composition of the soil is finer. The physical absorbing power is therefore higher in clayey and weaker in sandy soils.

Physical absorption protects water soluble substances from leaching. Physical absorption is not infrequently attended by the coagulation of colloidal substances under the influence of electrolytes, which also prevents the leaching of soluble compounds. That is why it is indispensable, through chemical melioration, to promote the coagulation of colloids and to prevent their peptisation.

Physico-chemical or exchange absorbing power in the proper sense of the term (absorption, adsorption) is the property of soil to exchange part of the cations and to a lesser extent of the anions of the solid phase for an equivalent amount of cations or anions from contiguous solutions, in which case occur physical and chemical absorptions.

The chemical absorbing power of a soil is its property to retain ions as a result of the formation of insoluble or slightly soluble salts. It consists in the falling out of the soil solutions of sediments and their fixation in soil. Upon a reaction between soluble and medium soluble salts there is a formation of salts of low solubility, which join the solid phase. For example: $Na_2CO_3+CaSO_4=CaCO_3+$ $+Na_2SO_4$; $3CaSO_4+2Na_3PO_4=Ca_3(PO_4)_2+3Na_2SO_4$. Readily soluble salts, like Na_2SO_4 are removed from the sphere of reaction. Chemical absorption occurs also when the anions of the solution give insoluble compounds with the ions located upon the surface of the solid particles of the soil, K and Mg ions may be absorbed without exchange, becoming incorporated in the composition of clayey minerals of soil, etc.

The biological absorbing power (worked out by V. R. Williams) is conditioned by the vital activity of the soil's organisms (mainly the microflora), which, by absorbing food, fix in their organisms various substances with which, upon their death, they enrich the soil. Soluble substances, coming from the solutions, as well as substances which organisms assimilate from the solid and gaseous phases of soil, pass into an insoluble form in the living bodies of organisms. Thanks to its biological absorbing power there is an accumulation in soil of the ash and nitrogenous nutrients indispensable to plants. This absorbing power is therefore a selective absorbing power with regard to the nutrient elements of plants. Selective absorbing power is a biological phenomenon, i.e., a quality of living organisms (biosphere) and not a physical or physico-chemical property of soil. The biological absorbing power of soil is of particular importance in soils easily leached and poor in nutrient substances.

Soil retains bacteria and adsorbs them as a physical medium. This property is more pronounced in chernozems and less so in sandy soils. The adsorbing power of soil varies with respect to the different species of bacteria.

The absorbing power of soils is at its best under conditions of optimum soil moisture, when there is a progressive accumulation of humus and plant nutrient elements and soil fertility rises. The absorbing power of soils and fertility develop in interdependence.

The soil's absorbing power is conditioned by its absorbing complex, mainly its highly dispersive part, consisting of mineral, organic and organo-mineral colloids. But all fractions smaller than 0.001 mm (silt) possess considerable absorbing power. All this part of the soil is referred to as the absorbing complex. The colloidal part of the absorbing complex is present in soil mostly in a coagulated condition, forming microaggregates or coalescing mechanical particles.

The soil's absorbing complex retains foodstuffs and replenishes their reserve. Tied with this complex are structure formation and the character of the structure. The complex can be destroyed as a result of physico-chemical and biological processes. It is subjected to destruction by water and all the more so as it is more finely divided. Anything that contributes to the coagulation of colloidal particles will reduce the destructive effect of water on the absorbing complex. The absorption of sodium by the absorbing complex contributes to its destruction, whereas the entry of calcium, on the contrary, contributes to its preservation.

In the composition of the absorbing complex of soil we find humus, silt and clay. Their removal upon treatment by acids, through leaching, or their destruction through drying out, lead to a lowering of the absorbing power.

The absorbing power of a soil is likewise lowered upon lowering of the pH.

Mechanical division of the soil's microaggregates likewise increases the absorbing power in connection with the appearance of additional units capable of active absorption. New surfaces become available for exchanges, surfaces which were previously concealed within the aggregates.

The absorbing complex of soil is not uniformly distributed in soils but a considerable portion of it is contained in the upper horizons of soil in the shape of organic, organo-mineral and mineral colloids. In connection with the course of soil formation and exogenous factors the absorbing complex changes in time. It may rise and with it rises also fertility. By exerting an influence on the absorbing complex, it is possible to influence the fertility of soil.

The absorbing power of sandy soils is due to the colloidal pellicles on the grains of sand, whose destruction leads to a sharp decrease of the absorbing power of sandy soils and a lowering of the fertility. On the surface of the fine dispersive soil particles occurs an energetic adsorption of chemical active gases, which become condensed and compressed. The diffusion of gases within soil aggregates proceeds substantially more slowly. The nature of the absorption by soil of gases is fairly complicated if we take into account the fact that soil is the seat of adsorption and absorption, chemosorption, capillary, molecular and thermal condensation of vapours, etc. The absorption of gas by the surface of a solid body is due to forces of attraction of an electric nature. Absorption of ammonia is energetic, that of oxygen is weaker and that of nitrogen weaker still. The more colloids there are in soil and the less water, the more intensely is air absorbed. Methane, ozone, hydrogen, chloropicrin are only adsorbed by a dry soil. The point is that the coating of dropping-liquid water present on the surface of soil particles hampers the adsorption of gases and air. But once the air is adsorbed it is difficult to oust it with water.

Of particular importance in agricultural practice is the absorption by soil of various substances from solution. This kind of absorbing power counteracts the leaching from soil of water-soluble substances indispensable to plants. In particular it increases the efficiency of water-soluble fertilisers, protecting them from leaching. Absorbing power plays the same role with regard to the prevention of leaching upon watering.

A Soviet scientist, Academician A. N. Sokolovsky was the first to show that the absorbing power of the soil changes through the soil profile in conformity with natural laws but differently for each type of soil, in connection with the composition peculiarities of soils and their genesis.

Soil possesses the capacity to exchange in equivalent amounts the cations of the absorbing complex for any cations of the solutions. An insignificant absorption of cations is effected without exchange, particularly upon periodic overdrying of soils.

The exchange absorbing power of soil consists not in the absorption of all the salt but of isolated ions (cations or anions) depending on the charge of the colloid. In most cases, the colloids of the soil absorb not one but several cations. The sum of all the cations which can be absorbed and exchanged is referred to as the absorbing capacity. The extent to which this capacity of soils is realised depends upon the natural conditions. To the various types of soil correspond relatively definite degrees of saturation of the exchange capacity with cations. The degree of introduction of cations depends upon the concentration of the solution, as well as upon the energy with which the cation is bound with the colloidal particles. This energy is proportional to the valency of the cations. Within the limits of one valency the energy depends on the value of the molecular weight. The higher the weight, the more energetic the absorption (Fe>Ca>Mg>K>Na), an exception being H which is energetically absorbed. Hydrogen ions also possess a higher coagulative capacity than monovalent cations. Hydroxyl ions, in concentrations below the electrolytic edge, hamper coagulation, even in the presence of another electrolyte. The absorbing capacity is higher in alkaline medium and lower in an acid one. It is also linked with the chemical composition of the absorbing colloidal soil complex. The greater the amount in soil of negatively charged colloids (SiO_2, MnO_2, humus), the greater the number of absorbed cations.

The absorbing capacity is expressed either in gravimetric percentages of the sum of all the cations, reduced to Ca, or in milligram equivalents. The capacity in percentages, reduced to calcium is expressed in accordance with the following correlations: Ca $\frac{40}{24}$ for Mg, Ca $\frac{20}{23}$ for Na.

From the percentage contents of a given cation we may arrive at the expression in milligram equivalents, by multiplying the percentage expression by 1,000 and dividing it by the equivalent weight. For example, a Ca content of 0.5% will in milligram equivalents be $\frac{0.5\times 1,000}{20} = 25$ mg. equiv. to 100 g of soil. The content in milligram equivalents can be expressed in percentages by multiplying the milligram equivalent by the equivalent weight and dividing it by 1,000.

Soil can be saturated with bases (Ca, Mg, K, Na) or unsaturated when H and Al cations in the absorbed condition are present, in addition. The degree of unsaturation with bases is determined from the amount of unreplaced hydrogen ions. The following combinations of cations are usually found: Ca+Mg, in soils of the steppe type of soil formation, Ca+Mg+H, in soils of the podzolic, boggy and lateritic types of soil formation and Ca+Mg+Na, in soils of a solonetzic type of soil formation.

The actual composition of the absorbed bases in soil depends upon the conditions of the soil formation, the origin of the soil and the composition of the soil's solution. The composition of the exchangeable cations reflects the soil formation (Table 13).

Table 13

Composition of the Exchangeable Cations

Soil	Absorbed bases in mg-equiv. to 100 g of soil				
	Ca	Mg	Na	H	Capacity
Chernozem (A + B)	30-55	5-15	—	—	35-70
Podzol (A_1+A_2)	1-25	0.5-10	—	0.1-5	2-30
Solonetz B_2	17	20	19	—	56

The composition of the exchangeable cations changes under the influence of the crops, agrotechnical and meliorative measures. Taming of the soil brings about an increase of the exchange capacity and a substantial rise of the sum of absorbed bases (Table 14).

Table 14

Changes in the Absorption Capacity in Accordance with the Degree of Taming of Soils

Degree of taming of soils	Sum of absorbed bases in mg.-equiv.	
	soddy-podzolic soil	chernozem
Weakly tame	9.5	27
Medium tame	12.0	34
Highly tame	18.0	—

The absorbed cations determine the properties of the soil. Soil saturated with Na is easily dispersed, dissociating back into its colloidal particles, swells, increases in stickiness and tenacity. A soil of this kind exhibits a worsening of its physical properties. It becomes impermeable to water and the seat of anaerobic-reducing processes. In soil which contains absorbed sodium, the reaction becomes alkaline and not infrequently there is a formation of soda, which is quite harmful for plants. In the presence of water a hydrolysis of the sodium salts takes place. Soda is formed in the presence of CO_2, first as hydrated carbonate, then as carbonate. This salt may also arise in soil as a result of biochemical process.

The soil's properties are different when its absorbing complex is saturated with calcium. In that case, the colloids are coagulated; when humus is present, the soil acquires structure. Favourable physical properties are created. In this connection, the degree to which soil is saturated with calcium is quite important.

The saturation of soils with calcium decreases to the north and south of the chernozemic zone.

Upon saturation of the soil with H, its properties occupy a somewhat intermediate position between the properties of soils saturated with calcium and those of soils containing a high amount of sodium.

When a podzolic soil contains absorbed hydrogen or a solonetzic soil contains absorbed sodium, there is a leaching of the humus, which in both cases passes into a soluble state. In spring, in the podzolic zone, water in rivers and wells acquires a dark coloration, which indicates the loss of soluble organic substances.

Apart from the fact that humus is a source of ash and nitrogenous nutrients, it plays quite an important role in the absorption of cations.

It was found by A. N. Sokolovsky that in the upper layer of the Poltava region chernozem, humus was responsible for nearly 67% of the total absorption (the ratio of humus to silt being 1:4), that in the podzolic soil of the Moscow region it amounted to 65%, etc. The absorbing power of humus is 8 times higher than that of the mineral colloids of soil. Peat possesses a high absorbing capacity (>200 mg.-equiv.).

An accumulation of decayed organic remains leads to an increase of the absorbing capacity of any soil (I. V. Tyurin). It has been shown by practice that up to 80-90% of the soil's total absorbing capacity should necessarily consist of absorbed Ca and that for various cultures a certain amount of Mg, K, H, NH_4 and Na was also necessary.

The absorbed bases present in soil are amenable to regulation. It is possible to replace certain cations by others or to alter their quantitative correlation through the application of fertilisers of direct or indirect action. The unsaturation of soil with bases is removed by liming and alkalinity is removed by gypsuming. Lime exerts an indirect effect on the accumulation of humus in soil and on its fertility.

The calcium requirements of soils (liming of soils of the podzolic type and gypsuming of soils of the solonetzic type) are determined not only by their degree of acidity or alkalinity but also by the extent to which they are deficient in absorbed calcium. The same goes for podzolised soils.

Chapter V

SOIL MORPHOLOGY

The morphology of a soil is the aggregate of its external morphological characteristics, or its external appearance. The external morphological characteristics of a soil are in unity with its internal phenomena. The morphology of a soil reflects the major moments of its development.

The morphological characteristics are best studied in deep trenches specially dug for the purpose; soil sections of this kind reveal the whole of the soil's thickness down to the soil-forming rock. On the walls of such sections-excavations, the genetical soil horizons which form what is referred to as the soil profile stand out quite clearly. The horizons can be differentiated one from the other owing to the changes in coloration caused by the introduction into soil of new substances from the biosphere (humus) and the atmosphere (N, H, O, H_2O, NH_3) or the falling out from solution within the soil of various substances and their redistribution through the soil profile. The formation of the soil's genetic horizons is also tied with the transference and distribution of colloids in the soil layer.

The most obvious morphological characteristic of a soil is its coloration. The coloration may be uniform through the genetical horizons or it may be irregular, appearing as bands and patches. It is closely tied with the soil's physical properties. One and the same soil of one and the same horizon may present different colorations depending on whether it is in a structural (aggregated) or in a powdery condition. The shade of the coloration changes with the moisture content. The soil's coloration depends also on the colour of the soil-forming rock, on new formations and on the formation of humus, salts, etc. The shade of the coloration is influenced by the character of the lighting under which the colour

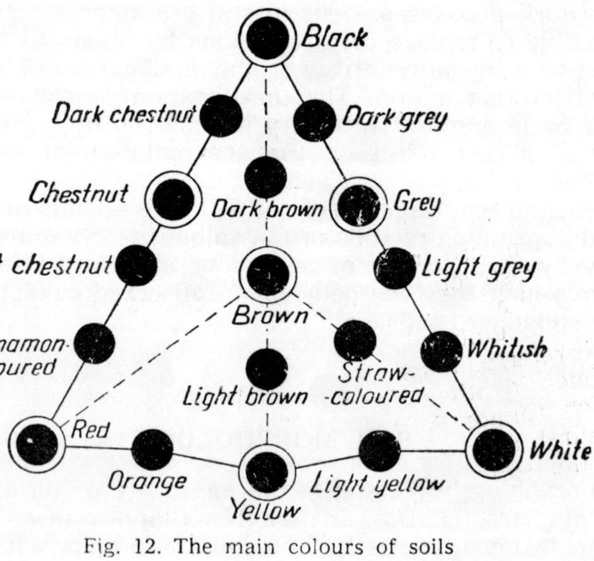

Fig. 12. The main colours of soils

of the soil is determined. That is why the determination of the coloration effected on the spot is corrected by examining air-dried samples of the soil under normal daylight.

The soil's colour is defined by reference to a number of main colours and shades. The main colours of soils and their combinations when mixed can be represented by a triangle of colours after S. A. Zakharov (Fig. 12). A black coloration is usually due to the presence in soil of organic matter and partly to the presence of manganese oxides and others. A red coloration is due to the presence of haematite, brown haematite ($2Fe_2O_3 \cdot 3H_2O$) and others. A white coloration can be explained by the presence in soil of silica, lime, hydrated alumina, kaolinite, marl, tripoli, etc. One and the same section of soil presents a combination of many colours and shades.

In the section of a typical chernozemic soil, the following approximate succession of colours can be observed: the top part (A horizon) is uniformly coloured in a relatively deep dark-grey or black colour. Lower down, the coloration gets lighter, passing to a brownish-dark grey. At a depth of more than 1 m (in the B horizon), the coloration loses its uniformity and becomes brownish with dark streaks and patches. Still deeper, the brownish coloration passes into a brownish pale yellow one with white rounded off patches of lime, nodules and lower down (in the C horizon) it gradually turns to a uniform pale yellow coloration.

Soil Structure. Structure Formation and Its Significance

Structure is an important morphological characteristic of the soil. The term soil structure designates the macroaggregate arrangement of the solid phase or the complex of soil units (aggregates) of various sizes and shapes, arranged in various proportions through the genetical horizons of the soil. The soil aggregates are formed as a result of the coalescing (sticking together) of elementary soil particles and microaggregates into larger units of a diameter of up to 1 to 10 mm and more. This property of the solid phase of soil to form complex macrostructural units from smaller mechanical elements and microstructural elements can be called its structure-forming capacity or aggregative capacity. A soil which is capable of breaking up into units ("aggregates") of different shapes, sizes and hardness is referred to as structural.

Soil structure is sometimes identified with the aggregates or soil units themselves, which cannot be considered correct, due to the fact that the structural aggregates are the elements of soil structure into which the soil mass breaks up or may be divided.

A distinction should be made between structure in the morphologo-genetical sense and structure in the agronomical sense. Soil structure in the first sense is that which arises under natural conditions in the process of soil formation. Such a structure, together with the other factors, conditions a clear differentiation between the genetical horizons of the soil. To each type of soil formation corresponds its preponderating structure, such as cloddy-granular for chernozems, laminated-foliate for the A_2 horizon of podzolic soils, columnar-prismatic for the B horizon of solonetzic soils, etc. The morphologo-genetical structure can exhibit great diversity. The agronomical structure appears under conditions of agronomic production. It is predominantly a cloddy-granular structure, which is more uniform for different zones, but which also varies substantially with regard to the shapes and sizes of the aggregates.

The morphologo-genetical structure is most clearly displayed by the chernozems. It is due to this factor that these soils exhibit an optimum agronomic structure. It would in fact be more correct in this case to talk of the coincidence of the agronomical with the genetical structure.

Soil in the agronomical and genetical senses may be structural (cloddy-granular) or structureless, i.e., in a single-grain state (Fig. 13). A cloddy-granular structure confers to the soil the necessary friability, a favourable water, air and heat regime, ensures the best conditions for the development of useful microorganisms and vegetation. Such a structure, by accelerating the percolation of water, confers to the arable layer great stability against ablation and washout. Structureless soil, i.e., soil in a singlegrain state may be in a pulverised or in a lumpy compact state. Structureless soil is incomparably worse, it loses considerably more moisture through evaporation.

Structure arises under a cover of annual and perennial plants under conditions of accumulation of organic remains and humus. Structure formation begins with the fragmentation of the soil-forming rock into facetted fragments (lumps). This is attained through working the soil when it contains the optimum amount of moisture, as well as through growing crops. With their powerful root system, plants exert a splitting action, separating the compact mass of soil into fragments of various sizes. Soil structure is attained not only through the fragmentation of large aggregates but also through the enlargement of primary microaggregates and elementary soil particles into complex units. The processes of humus formation, the accumulation of soil colloids, the vital activity of microorganisms, protozoa and worms (particularly earthworms) all contribute to the development of structure. Tillage and manuring, especially liming of acid soils and gypsuming of solonetzic ones, also assist in the development of structure. Structure is partially formed as a result of periodic freezing, defreezing,

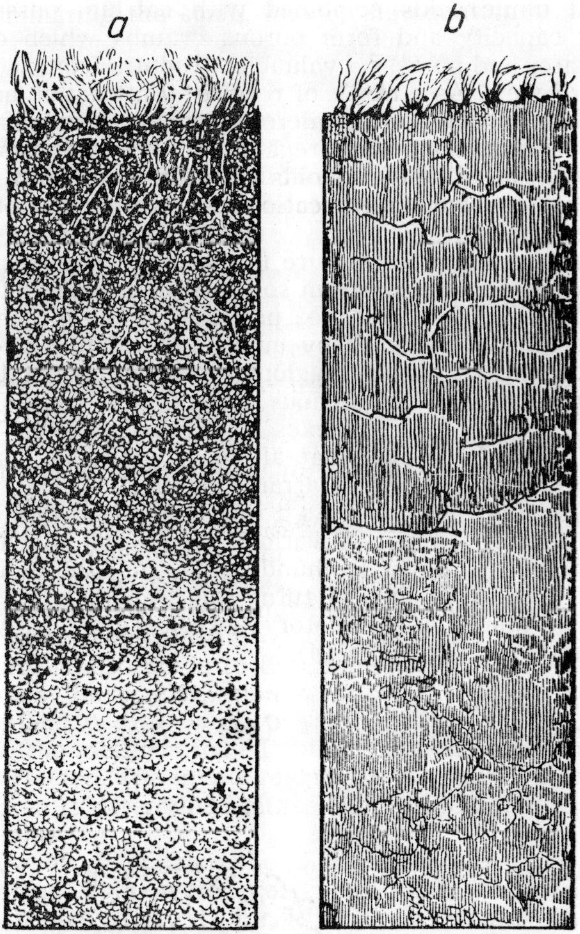

Fig. 13. Structural (*a*) and structureless (*b*) soils

thawing, moistening and swelling alternating with drying up and cracking, etc.

The process of structure formation consists in the main in that separate mechanical elements of the soil and groups of particles become covered by colloidal coatings, which promote their subsequent clumping together into complexes. Quite an important role in structure formation is played by organic colloids, in particular soluble (active) humus. Humus saturates the primary units, binding them together. Thereafter, upon low temperatures and under the influence of electrolytes, humus, coagulating and ageing, passes into an insoluble form, firmly cementing the component elements of the clod. The organic substances of acids such as humic and ulmic acids combined with calcium possess a high cementing capacity and form porous crumbs which do not fall apart in water and possess a valuable structure from the agronomical point of view. As a result of reciprocal attraction and coalescing, there is a formation of microaggregates which subsequently agglomerate to give macroaggregates of various orders and sizes.

The cementing role of colloids is reinforced by high valency cations (Fe, Ca). Monovalent cations (K, Na) do not promote the coalescing of particles.

The main element of structure is the clod. It possesses a complex arrangement differing from soil to soil. The clod is a complex of elementary soil particles, i.e., particles of the mechanical composition, cemented together by amorphous humus into microaggregates of various orders agglomerated, in turn, into complex macroaggregates, also of various orders. Macroaggregates of a higher order represent complexes of macroaggregates of a lower order. In view of the fact that all the constituent parts possess different volumes, forms and arrangements, the clod cannot present a uniform arrangement. As a rule it is penetrated by interaggregatory capillary pores of various forms and diameters, from several millimetres down to hundredths and thousandths of a millimetre. The clods are, in turn, so assembled together as to possess a certain interaggregatory porosity, principally noncapillary but partly capillary (Fig. 14).

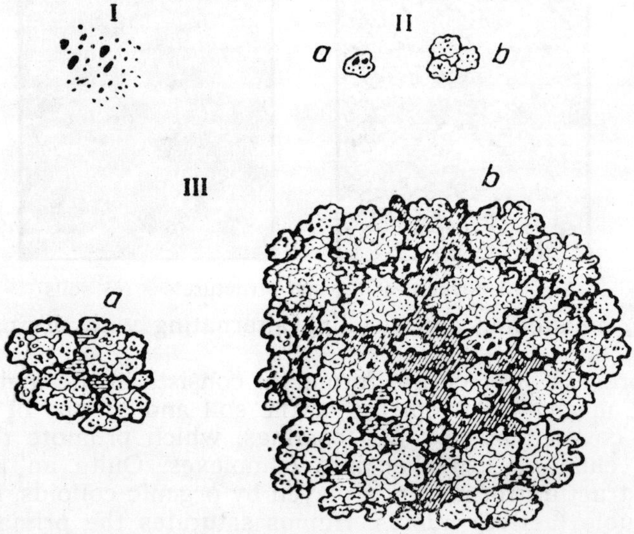

Fig. 14. Diagram illustrating the structure of soil aggregates (magnified several times):
I—elementary soil particles; *II*—microaggregates: *a*—of the first order, *b*—of the second order; *III*—macroaggregates. *a*—of the first order (complex of microaggregates, "grain"), *b*—of the second order ("clod")

The surface of a wet clod is the seat of aerobic processes involving oxidation and dissociation of organic matter with liberation of ash plant food elements. This ensures a continuous supply of plant food elements in an available form, as they are needed by plants. Within the clod occur anaerobic processes involving the slow (incomplete) decay of organic matter. In this fashion, the organic matter inside the clod is, as it were, conserved, so long as anaerobiosis preponderates, i.e., so long as the aggregate is not broken up and is sufficiently packed, and its pores filled with water.

The clod, which conditions a certain combination of aerobic and anaerobic processes as if automatically regulates the requirements of plants in food elements and water. Each clod undergoes complex physico-chemical changes and, in time, it dissociates. The disintegration of the clods is tied, to a considerable extent, with the behaviour of water and air in the interaggregatory pores.

In time, the structure of a soil changes and an incorrect agricultural treatment may bring about its destruction and loss. The causes for the loss of structure are: mechanical breaking up of the clods, physico-chemical and biological processes. Unnecessary tillage operations or a decrease of the soil's moisture content promote aerobic processes and lead to an accelerated disintegration of the humus which binds together the soil particles into aggregates. In time, humus also loses its coalescing capacity as a result of its molecular rearrangement, i.e., packing upon ageing. That is why active fresh humus is needed for the formation of structure.

Stated in very general terms, the points of difference between a structural and a structureless soil are as follows: in a structural soil, the water regime is at its most favourable level. Water of irrigation or water of atmospheric precipitations penetrates freely into soil via the noncapillary pores, i.e., the interaggregatory spaces, and then via the capillaries inside the clods where it is retained. Interaggregatory pores remain free from water and ice due to the fact that all the water in excess percolates downwards, finding its way to the ground water. In structureless compact soil, the interspaces are, in the main, capillary and filled with water, which slowly moves from its source as a continuous body. Upon drying out of the soil, the ascending flow slowly wets the soil and stops when the source is exhausted. The greatest part of the atmospheric precipitations falling on a structureless compact soil runs off along its surface. The reserve of water in a soil of this kind is not high but transitory (ephemeral) and is hard to regulate.

The advantages of structural soil are obvious. The soils which most need structure are, in the first place, the compact, heavier clayey and loamy soils. Soils of medium and light mechanical composition, light loams and sandy loams possess more favourable inherent physical properties and what they need, above all, is a supply of organic matter and mineral plant food elements. A high degree of fertility and high yields are not always necessarily associated with a cloddy-granular structure. Certain soils of southern regions akin to structureless soils possess but a microstructure (the sierozems, the brown soils) and yet they are characterised by high fertility. Where high yields and progressive agricultural methods are the rule, whereupon a great amount of organic residue remains in the soil, it cannot fail to acquire structure, i.e., this structure can be obtained with all crops. Soil structure is a most important condition of fertility.

A classification of structural units after S. A. Zakharov is given in Table 15.

Table 15

Classification of Structural Units (after S. A. Zakharov)

Class	Genus	Variety	Size
I. Cuboidal	1. Lumpy	Coarse-lumpy	>10 cm
		Fine-lumpy	10-5 cm
	2. Cloddy	Coarse-cloddy	5-3 cm
		Cloddy	3-1 cm
		Fine-cloddy	1-0.5 cm
	3. Cuboido-nut	Cuboidal	>2 cm
		Coarse-nut	2-1 cm
		Nutty	1.5-1 cm
		Fine-nutty	0.5-1 cm
	4. Granular	Coarse-granular (pea-like)	7-5 mm
		Granular	5-3 mm
		Fine-granular	3-0.5 mm
	5. Dusty	Dust-like-floury	<0.5 mm
II. Prismoidal	6. Column-like (not well formed)	Coarse-column-like	>5 cm
		Column-like	5-3 cm
		Fine-column-like	<3 cm
	7. Columnar	Coarse-columnar	>5 cm
		Columnar	5-3 cm
		Fine-columnar	<3 cm
	8. Prismatic	Coarse-prismatic	5-3 cm
		Prismatic	3-1 cm
		Small-prismatic	1-0.5 cm
		Fine-prismatic	<0.5 cm
		Pencil-shaped	<1 cm
III. Platy	9. Flaggy	Cleavable	>5 mm
		Flaggy	5-3 mm
		Lamellar	3-1 mm
		Foliated	<1 mm
	10. Imbricate	Shelly	>3 mm
		Coarse-imbricate	3-1 mm
		Fine-imbricate	<1 mm

Texture. Inclusions and Neogeneses

The structure and mechanical composition determine the character of the soil's texture, i.e., its consistency, its degree of compactness and friability.

According to the compactness of the soil we may distinguish the following textures: highly compact, compact, consolidated, weakly loose, loose and crumbly. According to the dimensions of the pores and the character of the porosity, texture can be finely porous (dia. up to 1 mm), porous (1-3 mm), spongy (3-5 mm), spongy-foraminated (5-10 mm), alveolate (cavities>10 mm), tubular, etc.

When soils exhibit marked cracks, this cracked texture should be taken into consideration and a distinction is made between finely fissured (distance between the walls up to 3 mm), fissured (3-10 mm) and cracked (>10 mm). The cracks may be temporary or lasting, more or less permanent. Not infrequently, on the walls of such cracks are found small crusts of plaster, iron hydroxides or silicic and other formations. Transient, temporary cracks occur periodically, for example when the soil dries out. Such cracks disappear when the soil is moistened and swells, without disturbing the physical homogeneity of the mass. The existence of cracks can cause drainage of the soil and lead to harmful deformations, viz., funnels of leaching and caving in. The latter occurrence is not infrequently tied with the removal of the fine dusty and silty fractions of the mechanical composition from soils lying on a gravel subsoil.

A friable and alveolate texture, irrespective of the structural condition and mechanical composition of the soil, may be governed by the degree to which the soil is dug up by rodents, worms, grubs, by the penetration of plant roots, etc. Texture, like the other morphological features, has no permanent character, it changes in time, depending upon the course of physico-chemical and biological processes in soil. In time, under the influence, for example, of annual and perennial grasses, worms and microorganisms, soil, acquiring structure, becomes more friable, in other cases, losing its structure, it becomes packed. Hydromelioration leads to marked changes in texture: correct regulation of the moisture content and improvement of the structure cause the texture to become friable.

Other external features of soil are inclusions and neogeneses. The term inclusion designates various bodies which are mechanically drawn into the soil mass. Their origin is not tied with soil formation but they may serve as an indication of certain conditions under which the soils were formed. Soil sometimes includes: shells, bones, bark, fruit, wood, rock fragments, as well as various small bodies and streaks. Inclusions can consist of chance objects such as coal, broken crockery, pieces of bricks, glass and archeological finds (coins). Fragments of rocks and minerals larger than 2-3 mm in diameter form part of what is called the soil skeleton. Are distinguished soils devoid of a skeleton (fine earth soils), weakly skeletal (skeleton$<$10%), skeletal (skeleton 10-30%) and strongly skeletal (skeleton$>$30%). A classification of the soil skeletons is given in Table 16.

Table 16

Classification of Soil Skeletons

Sizes of fragments	Shape of fragments	
	Rounded	Angular
2-3 mm	Grit	Sharp gravel, spalls
3-10 mm	Gravel	
1-7 cm	Pebbles	Broken stone
$>$7 cm	Cobbles	Stones

The term neogenesis designates various secretions and accumulations in soil, tied with soil formation. Neogeneses can be of chemical and biological origin. Chemical neogeneses comprise: stains of salts or films on the surface of soil units, small crusts, dabs, tongues along cracks and on the surface of the units, veins and tubules along the passage made by roots and worm-holes, concretions, streaks, etc.

Neogeneses may consist of stains of soluble salts, such as chlorides and sulphates (Glauber salt, of a whitish colour, gypsum, of a white colour); lime, of a floury white colour, appearing as pseudomycellium, loess dolls, lime nodules; compounds of iron hydroxide, of reddish brown, rusty and ochre colours; ferrous iron compounds of a dirty greenish and bluish colour; manganese compounds of a black colour, R_2O_3, silicic acid of light and whitish colours, etc. The sesquioxides (Al_2O_3 and Fe_2O_3) give ochre films and stains, patches, dabs, streaks, tongues and patterns, the so-called dendrites, pseudofibres and others. Sometimes they appear as small brown pipes, rusty veins, brown ore grains, beans and streaks of ferruginous formations or so-called ortstein. Compounds of iron and manganese are found in soil in the shape of concretions and beans. Silicic acid conditions the appearance of a silicic grey powder along cracks and pores, white and whitish patches and tongues, whitish veins, etc. Humic substances give brown and dark brown glossy patches, tongues and small fine crusts, brownish black incrustations on the surface of structural units. We also find them in brown ore grains and humic streaks of ortsand.

Neogeneses of biologic origin may be zoogenic, in the shape of caprolites, balls and nodules, excreta of worms and insect larvae, small structural lumps cast by ants when building their nests, worm burrows and worm-holes, krotowinas, i.e., burrows of rodents and other animals, and phytogenic in the shape of dendrites (patterns made by fine rootlets on the surface of structural aggregates), large cavities left where roots decayed, etc. The cavities left by roots, when filled with friable earth, sometimes look somewhat like krotowinas but can easily be distinguished from them. Krotowinas in soil sections appear clearly as rounded elliptical or elongated patches of a dark colour on the light background of the lower horizons if filled with earth from horizons higher up or of a lighter colour on the dark background of humic horizons if the passages are filled with earth brought upwards by the burrowing animals from down below. The sizes and shapes of krotowinas depend on the size of the burrowing animals (babocs, moles, marmots, jerboas) and on the angle at which the section is cut in relation to the direction of the burrows. Soil also contains roots of trees and grasses, rhizomes, tubers, bulbs and other bodies in the shape of organic remains preserved from decay. In the taiga zone, the surface of the soil is covered with duff, which is sometimes felt-like, forming the A_0 horizon. In bog soils, above the mineral horizons lies a peaty A horizon of varying thickness.

Structure of Soil Profile

The general aspect or structure of the soil profile, i.e., the vertical section of the soil through the genetical horizons is conditioned by all the above mentioned morphological characters taken together. The uppermost soil horizon is usually coloured to a certain depth by humus, sometimes down to 1 m and more but in most cases not exceeding a few centimetres.

The limit separating the humus horizon from the horizon lying below it is sometimes straight but more often sinuous, wavy, wedged in in the shape of pockets, tongues and corners. In the middle part of the soil profile often occurs an accumulation of sesquioxides, carbonates and sulphates in the shape of patches which, merging together, may condition the formation of corresponding subhorizons.

The following genetic horizons appear in the soil profile:

A—humus-accumulative, horizon of biological accumulation of humus, nitrogen and ash plant food elements. The most structural, dark coloured horizon of mass root distribution. It is subdivided into sub-horizons:

A_0—organic remains, usually without admixture of mineral matter—duff of forest soils, "peat dust" of peat-boggy soils, undecayed plant remains of other soils;

A_1—humus-accumulative proper, usually structural subhorizon;

A_2—eluvial—washed out horizon.

B—illuvial horizon containing most of the washed in material; it comprises sub-horizons:

B_1—transition horizon;

B_2—illuvial horizon proper, in steppe soils it contains carbonates; sometimes there is a sulphate sub-horizon (B_3), etc.

C—considerably changed by soil formation but easily recognisable soil-forming or parent material. This horizon comprises sub-horizons:

C_1—rock considerably changed by soil formation;

C_2—rock less changed by soil formation.

D—parent rock, hardly changed by soil formation.

In peaty and swamped soils a gleyey horizon can be distinguished, designated by the letter G. A loess horizon is designated by the letter L.

The above named soil genetic horizons are not always equally well marked and do not always clearly show in the soil profile. Sometimes they pass one into the other so gradually that it may be difficult to distinguish between them. Owing to the fact that soil is an anisotropic natural body, its genetic horizons and all its morphological characteristics together with its composition and properties change differently but according to certain laws, in the vertical as well as in the horizontal direction. The thickness of the soil horizons varies within a short distance in a horizontal direction, even on a relatively plane surface, not to mention the fact that their thickness varies considerably in connection with irregularities of the relief; furthermore, some horizons grow thicker, whereas others grow thinner, in full accordance with the manifestation of the soil-forming factors. The changes of morphological characteristics and properties of soil in three directions can be represented by a complex block-diagram (see Fig. 8).

When describing the horizons of the soil profile, figures are given indicating the various depths at which begin and end the various horizons; it is therefore quite easy to deduct from these figures the actual thickness of each particular horizon. If, for example, we are given the following figures: A—0-15 cm, A_2—15-40, B_1—40-80, B_2—80-130 cm, etc., it follows that the respective thicknesses of the horizons are A_1=15 cm, A_2=25 cm, B_1=40 cm, B_2=50 cm, etc. Changes in thickness in a horizontal plane can be followed in trenches dug in a vertical plane. When describing a soil profile, each genetic horizon must be described separately. Are indicated the colour of the soil, its structure, texture, the character of the distribution of separate substances, inclusions, neogeneses, moisture content, course of the soil formation, spatial distribution of soils, changes in soil under the influence of the various soil-forming factors and in particular under the influence of man's productive activity.

Chapter VI

CHEMICAL AND PHYSICAL PROPERTIES OF SOIL

The essential property of soil is fertility. It is bound up with the chemical, physico-chemical, physical and biological characteristics of soil. Differences between these properties determine the qualitative differences between the various soils.

The soil characteristics are, in turn, interrelated with the soil-forming processes, the composition and structure of the soil. If, for example, we change the composition of the soil through the application of various substances (mineral fertilisers, organic matter, clay, sand, microorganisms, water, etc.), we also cause the properties to change. The same thing occurs in soil when changes are brought about in its composition and structure through cultivation and meliorative measures. Soil can be improved and made highly productive by changing its properties.

Soils of different origins possess different properties. In order to be able to control these properties, we must get to know them. As soil develops, it changes and passes from one qualitative state into another, with increased fertility. But not all changes in soil lead to an increase of its productivity. Some changes, such as leaching or salinisation, swamping or drying out, excessive accumulation or destruction of organic matter, etc., lead to a worsening of the soil's water-physical and chemical properties and a lowering of its fertility.

Chemical Properties

The numerous chemical transformations—changes in the solid, liquid and gaseous phases, colloidal and cristalloidal systems, mineral and organic substances—which take place in soil, constitute the essence of its chemical properties.

The chemical properties of a soil are governed by the chemical phenomena which take place in the main between its solid and liquid phases. According to the law of mass action, various substances are formed which go into solution. A state of equilibrium is set up in soil between the solid part and the soil solution whereupon, in case of a diminution of the concentration of the solution, part of the substances in the solid phase passes into solution and, conversely, when the concentration goes up, part of the substances falls out of solution and joins the solid phase of the soil.

Soil solution. The water present in soil, which contains substances in solution (various salts and acids), constitutes what is called the soil solution. This solution is formed in the process of soil formation in the course of a fairly long time, as a result of the movement of water in soil, moistening of the soil and dissolution of its salts. This occurs under the effect of acids, kaolinisation, oxidation-reduction processes, under the influence of the hydrolysis of substances, etc. The composition of the soil solution is governed by the interaction of the soil constituents, water and the living organisms. This interaction results in the dissolution of mineral and organic substances, in peptisation, coagulation and exchanges between the ions of the solutions and the soil colloids, in the manifestation of the influence of the vegetation and microorganisms on the composition of the soil solution.

The soil solution contains molecular as well as colloidal dissolved mineral, organic and organo-mineral substances. It also contains almost all the readily soluble simple salts in various amounts, as well as the medium and slightly soluble salts of many elements. The soil solution is a highly active and dynamic part of the soil.

The nutrient regime of soil is tied with the dynamics of the soil solution. Plants and microorganisms alter the soil solution, drawing from it their ash food elements. The substances removed from the solution by plants and microorganisms are constantly being replaced by new ones. By utilising the bases from the physiologically acid salts [$(NH_4)_2SO_4$, K_2SO_4, KCl and others], plants liberate the combined acid and thereby reinforce the solving power of the soil solution. By removing the anion part from the physiologically alkaline salts ($NaNO_3$ and others), plants raise the alkalinity of the soil solution.

Water and soluble substances pass from the soil into the plant through the root hairs owing to the osmotic pressure exerted by the plant's cellular sap or suction force, which reaches 5-10 atm. (in halophytic plants up to 25 atm.). The composition and concentration of the soil solutions exhibit seasonal fluctuations. Dry periods of the year correspond to increases, wet periods to decreases of the concentration.

The soil solution is not homogeneous throughout the soil profile, due to the heterogeneity of the soil's composition and differences in the physical condition of the water and the degree of taming of the soil. The distribution of water of different forms through the soil profile is highly irregular and this leads to corresponding variations in the amount of the dissolved substances.

Soil solution can be extracted from soil by subjecting wet soil (paste) to a pressure of 50 atm. per 1 cm^2, or by centrifuging it, or by displacement methods using indifferent liquids (benzine, paraffin oil), etc. Natural soil solutions can be obtained by making use of what are known as lysimeters, which are isolated soil masses or large monoliths, whose percolating solutions are collected. The soil solution extracted from the soil possesses a characteristic coloration, reaction and concentration.

The salt regime of the soils of different native zones is correspondingly reflected in the composition of salts and the concentration of the ground water solution. The ground water lying under salined soils possesses a relatively high concentration, whereas the ground water lying under leached podzolised soils is usually sweet. These rules are not infrequently complicated by interference from local mineralised water or water from deep lying sources.

The peculiarities of the soil's chemism are manifested in the reaction of the soil solution which is set up upon the interaction of the soil with water or with solutions of salts. The reaction of the soil solution is determined by the predominating concentration of the hydrogen and hydroxyl ions. It may be acid, alkaline or neutral. The reaction is neutral when the concentrations of H and OH are equivalent. The composition of the absorbed cations and the calcium carbonates present in the soil condition the potential reaction of the soil's solid phase and the reaction of its solution. The reaction of the soil solution is expressed in terms of pH, which is the negative decimal logarithm indicating the degree of concentration of H in the soil solution or the number of H-ions in one litre of solution.

Acidity can be active (actual) and potential (exchange). Active acidity is due to the presence of weak acids (chiefly carbonic acid, organic acids), as well as acid salts and mineral acids, particularly H_2SO_4. Active acidity is revealed by the effect of water on soil, whose absorbing colloidal complex is base-unsaturated. Potential acidity is conditioned by the presence in the absorbing complex of soils and in solution of hydrogen and aluminium ions capable of being replaced by metals. Potential acidity can be exchangeable and hydrolytic.

Potential acidity proper is the capacity of soil to acidify solutions of neutral salts when interacting with them. This acidity is due to the hydrogen ions formed in the presence of a neutral salt, for example $BaCl_2$, KCl, $AlCl_3$, $FeCl_3$ and others. The exchange reaction proceeds according to the formula:

$$\text{s oil}\Big)\Big)\begin{matrix}H\\H\\Al\end{matrix} + 4KCl \rightleftarrows \text{soil}\Big)\Big) 4K + HCl + AlCl_3.$$

Upon the interaction of the soil with solutions of neutral salts, the resulting acid reaction is due to the presence of free acid and $AlCl_3$. The latter, being a hydrolytically acid salt, is hydrolysed* in water solution, forming free acid:

$$AlCl_3 + 3H_2O \rightleftarrows Al(OH)_3 + 3HCl.$$

* Hydrolysis is an exchange reaction between various substances and water. Essentially, hydrolysis consists in that the hydrogen ions of dissociated water displace the cations of the bases on the surface of soil particles. The

The quantity of HCl formed gives a measure of the amount of exchangeable acidity.

In mineral soils, the factor considered responsible for the exchangeable acidity is $Al^{\cdot\cdot\cdot}$ and $Al(OH)^{\cdot\cdot}$. A. N. Sokolovsky has suggested that $Al^{\cdot\cdot\cdot}$ and $Al(OH)^{\cdot\cdot}$ appear in salt extracts from unsaturated soils as secondary products, as a consequence of the dissolving action of the acid formed upon displacement of hydrogen ions. Exchange reactions proceed in solution after cations from the solid phase of soil gain access to it as a result of solution and hydrolysis. Exchange acidity is conditioned by the hydrolysis of aluminium salts, and of salts of organic acids with strong bases and hydrogen ions.

The capacity of salts for hydrolytic dissociation in soil is called the hydrolytic acidity of soils. Hydrolytic acidity manifests itself upon the interaction of the solid phase of soil with solutions of hydrolytically alkaline salts, for example, sodium acetate. This salt hydrolyses in a water solution:

$$CH_3COONa + H_2O = CH_3COOH + NaOH.$$

Maximum dissociation of the ionogenic groups takes place when the soil's solid phase reacts with the solution of a hydrolytically alkaline salt. Hydrogen ions appear in the diffuse layer, which are capable of exchange reactions. The cations of a hydrolytically alkaline salt displace the hydrogen ions of soil which are capable of exchange and those hydrogen ions which are further dissociated upon an alkaline reaction. The reaction is expressed as follows:

$$\text{soil}\left)\!\!\right)\begin{matrix} H \\ \\ H_2 \end{matrix} + 2CH_3COONa \rightleftarrows \text{soil}\left)\!\!\right)\begin{matrix} Na \\ \\ Na \end{matrix} + 2CH_3COOH$$

(H_2—hydrogen ions capable of exchange in an alkaline reaction, H—in neutral reaction). Hydrolytic acidity is therefore usually higher than exchange acidity.

Soil promotes the intensification of the hydrolysis of salts, owing to the fact that part of the products is removed from solution. Acid hydrolysis leads to the dissociation of R_2O_3 and enrichment of residual products with SiO_2. Upon an alkaline hydrolysis, SiO_2 is dissociated and removed and soil is enriched with R_2O_3.

Hydrolytic and exchange acidity are not forms but only degrees of potential acidity. Soils can be extremely acid (pH=4.5), moderately acid (pH=4.6-5.0) and slightly acid (pH=5.1-5.5).

An alkaline reaction in soil is due to absorbed sodium. The degree of alkalinity depends on the amount of exchangeable

---

hydrolysis of salts of a weak acid and a strong base leads to an alkaline reaction, whereas the hydrolysis of salts of a strong acid and a weak base leads to an acid reaction of the solution.

sodium. When an alkaline soil reacts with irrigation water or with water of atmospheric precipitations which contain, as a rule, a certain amount of carbonic acid, there is a formation of soda:

$$\text{soil}\Bigg)\Bigg)\begin{matrix}Na\\ \\ Na\end{matrix} + H_2O + CO_2 = \text{soil}\Bigg)\Bigg)\begin{matrix}H\\ \\ H\end{matrix} + Na_2CO_3.$$

Soda being a hydrolytically alkaline salt is hydrolysed in solution: $Na_2CO_3+2H_2O=2NaOH+H_2CO_3$. Some of the alkalinity is imparted to soil by lime which, reacting with water containing carbonic acid, is partly converted to hydrocarbonate: $CaCO_3+H_2O+CO_2 \rightarrow Ca(HCO_3)_2$.

The reaction of the soil solution is to some extent changed by microorganisms, manures, cultivation (agrotechnical measures) and cropping. The reaction of the soil solution has a corresponding influence on the development of plants. The majority of plants do not develop in a pH lower than 3.5 and higher than 9. Normal conditions for the development of plants are found within narrower limits: from slightly acid to slightly alkaline.

The reaction of the soil solution and its concentration change with changes in the environment. This is sometimes timed with seasonal changes but more often it spreads over lengthy periods of time, usually a number of years and decades. The dynamics of the fluctuations of the soil solution's concentration may be revealed by changes in the $\frac{Cl}{SO_3}$ ratio, in connection with the differences in the solubility of the chlorides and sulphates. An increase of chlorine will be an indication of salinisation, a decrease, of desalinisation, etc.

Buffering. The capacity of a soil suspension to resist changes in its active reaction (pH) upon application to soil of acids or bases is known as the buffer effect. As a result of buffering, the soil solution possesses a relatively stable reaction. The buffer effect is exhibited by the solid phase of soil and depends upon the composition of the soil colloids and the saturation of soil with bases, or, to be more precise, on the chemical, colloidal and mechanical composition of the soil.

That part of the soil which passes through a fine sieve possesses a more pronounced buffer effect. Clayey soils and those which are rich in humus (chernozems) possess a higher buffer effect than sandy and leached soils of low humus content (podzolic). The more humus and colloids are present in soil, the higher the buffer effect.

Gley-podzolic, acid humous and half-peaty soils withstand applications of alkaline fertilisers and, up to a limit, retain their acid reaction which can only be overcome by a considerable amount of alkali. Carbonate soils withstand changes in their alkaline reaction upon the application of acid fertilisers. The acid applied to a carbonate soil is neutralised until the carbonates are exhausted. After that, neutralisation is maintained on account of the absorbed bases and, finally, due to the presence of more firmly bound cations. The hydrogen ions formed thereupon attach themselves to the OH groups.

The application of alkali to an acid soil is attended by a neutralisation of the soil acids, with an exchange of hydrogen ions for the cations of the alkali.

The buffer effect is all the more pronounced as the absorbing complex is more powerful and its absorbing capacity higher. Under such conditions, plants suffer less from excessive applications of acid or alkaline fertilisers. That is why sandy soils, which possess a weak colloidal complex, are highly sensitive to excessive applications of fertilisers.

The solid phase of the soil is fairly inert. Relatively small amounts of soluble substances get into the soil solution. Their amount is not always sufficient to provide plants with adequate nutrients and so ensure high yields. It is often necessary to replenish and mobilise the soil resources by increasing the admission of substances into the soil solution.

It is fairly often found during the vegetative period that the chemical properties of soil do not correspond to the requirements of the plants in nutritive elements. For this reason, chemical soil analyses have but a relative significance, in view of the fact that they do not give a full picture of the chemical process of the soil in all its changes. Analyses can only reveal the presence in soil of some or other injurious substances. They may establish the richness of the soil in some or other compound or indicate what plant nutrient elements the soil is short of at a particular time.

The nutrient regime of soil could be studied in a more satisfactory fashion by improving the system of chemical analyses. But a rise in the fertility of soils can be achieved through the ability to forecast and regulate the changes in the chemical properties of soil. This constitutes quite an important task for agromelioration.

Physical Properties

The physical properties of soil comprise: apparent and true densities, pore-space, swelling, plasticity, stickiness, hardness, coherence, tilth, the water, air and thermal regimes, etc. The physical properties are closely tied with the other properties of soil; they change in accordance with the course followed by soil formation and changes in properties are, in turn, followed by changes in soil formation.

Soils of one and the same mechanical composition may possess different physical properties depending on the origin of the colloids which they contain, the ratio of SiO_2 to R_2O_3, the humus content and the composition of the exchange cations. A soil in which Ca constitutes the preponderant part of the absorbed cations is characterised by favourable physical properties (good aeration, water permeability).

Replacement of Ca by Na is attended by a worsening of the physical properties: collapse of the structure, increased resistance to agricultural implements and plant roots, deterioration of aeration and water permeability.

Soil is greatly affected by the accumulation of humus, the effect of which varies from one soil to another. Thus, by covering the separate grains of sand with a film and sticking them together, humus causes sandy soils to become more viscous. On the other hand, possessing lesser viscosity than clayey soils, it contributes to their loosening.

Water greatly affects the physical properties of soil. In the presence of excessive moisture, light loams and even sandy loams become viscous upon gleisation, sticky when wet and stone hard upon drying out. Friction in a wet soil is greater than in a dry one, but only up to a certain degree of moistening, after which friction goes down, due to the fact that water acts like a lubricant.

A further increase of moisture in a clayey soil brings about a sticky condition and, in consequence, friction increases, but, here again, only up to a certain degree of moistening.

The soil physical properties are divided into the main (such as apparent density, true density, pore-space, coherence and others) and the functional ones (water, air and thermal properties). To the latter belong the capacity of soil to absorb atmospheric precipitations or irrigation water, to allow the percolation of water, to retain or conserve it, to admit it to the top layers from deep-lying horizons, to supply plants with water, etc. The same applies to the thermal and air properties.

Apparent and true densities. The apparent density of a soil is the weight of one unit of volume of the soil in its native porous texture. It is determined by simply weighing a sample of the soil with an undisturbed structure, taking a strictly definite volume. The apparent density of a soil depends on the weight of the substances constituting the loose mass of soil as it is found in nature. The more porous and friable the soil, the lower its apparent density and vice versa. The more organic matter it contains (light in weight), the less the apparent density. The product of the thickness of the soil layer by the apparent density is a relatively constant quantity, which does not depend upon the degree of friability. The apparent density of a soil changes in connection with changes of its properties, particularly under conditions of taming. On the average it fluctuates between 0.9 and 1.8.

In light loams, the apparent density reaches 1.4, in medium loams 1.42, in heavy loams 1.45, etc. As a rule, the apparent density of the upper humic loose horizons is lower and that of the deeper horizons, usually more packed and devoid of sizable quantities of organic matter, higher.

The true density of a soil is determined by the ratio of the weight of an absolutely dry soil to the weight of a volume of water exactly equivalent to the volume of the hard particles of the soil without the pores. The true density of a soil depends on its mineralogical and petrographic composition, as well as its contents of organic matter. The true density of clayey soils is 2.6-2.7, of loams 2.5-2.6, of sandy soils 2.4-2.5, of humus-boggy soils 2.0-2.2.

The true density is all the higher as it contains more iron and other heavy elements, and all the lower as the soil is richer in organic matter. For example, the specific gravity of gypsum is 2.3, of kaolin 2.5, orthoclase 2.55, kaolinite 2.6, quartz 2.65, granite 2.7, basalt 3.0, magnetite 5.3, pyrite 5.0, humus 1.25, etc.

Pore-space. The total volume of the pores (volume of all the spaces) in a volumetric unit of the soil is commonly called the pore-space, in contradistinction to the porous texture of the soil or to the porosity of rocks and other bodies.

The sizes of the pores, their shapes and combinations vary considerably, owing to the fact that they are the result of the chance arrangement of the polydispersive particles of mechanical composition, elementary soil particles, microaggregates and structural units, which are quite different in sizes, shapes and surface characteristics. Many of them join to form oddly shaped curved elongated labyrinths or interstices between the soil particles and aggregates. The sizes and shapes of the interstices depend on the mechanical and aggregatory composition of the soil. Clay particles are like scales or densely packed planks and form numerous microspores of various sizes. Sandy rounded particles form more uniform, larger pores, but in smaller number.

In time, these interstices undergo considerable changes in shape and size in connection with the physico-mechanical and biological processes taking place in soil. As a result of partial or complete blocking, some pores disappear, others are formed anew.

Pore-space (p) is determined by the ratio of the apparent density (d_1) to the true density (d) of the soil. It is measured by the difference between the unit of volume of the soil and the volume of the solid (compact) phase (V):

$$p = 1 - V; \quad V = \frac{d_1}{d}; \qquad p = 1 - \frac{d_1}{d};$$

the pore-space in percentages being:

$$p = \left(1 - \frac{d_1}{d}\right) 100.$$

The volume of the pores (P) in m^3/ha is measured according to the formula: $P=ph$, where p is the pore-space in percentages of soil volume; h=thickness of the layer in cm.

If the aggregates or particles of mechanical composition of soil were spherical and of equal diameters, the pore-space would only depend upon their arrangement. The diameter size of the spheres would then have no bearing on the quantitative expression (coefficient) of the pore-space. For spherical particles of equal sizes exist only two possible types of packing: rhombic or cubic. In the first case, viewed in a vertical section, the spheres above come to rest between the spheres immediately below. In the second case, the spheres are arranged one above the other; the pore space is then equal to 47.64%, which can be ascertained from the following data: the volume of each sphere is equal to $\frac{\pi d^3}{6}$; the volume of the cube shown in the figure $=d^3$; the empty space in the corner of such a cube is:

$$p = d^3 - \frac{\pi d^3}{6} = d^3\,(1 - 0.5236) = 0.4764\ d^3,$$

in other words, whether d be large or small, the pore-space will always be 0.4764 of the volume of the cube, or 47.64%. The pore-space in a rhombic arrangement goes down to 39.54% (Fig. 15).

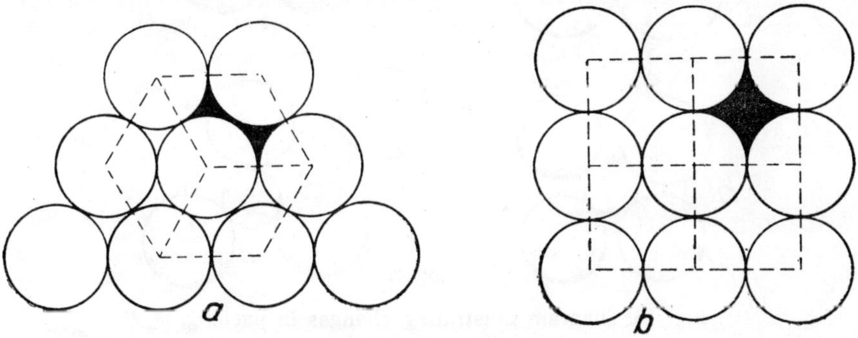

Fig. 15. Diagrammatic cross sections through the centres of spheres in differently packed arrangements:

a—rhombic packing; *b*—cubic packing

If the spheres were disposed so that the lines joining their centres formed a tetrahedron, we would obtain the highest possible compactness of packing in which the pore-space would go down to 25.95%.

The types of spaces (pores) in spherical arrangements can be tetrahedral, where the spaces are limited by four spheres, or octahedral, where the spaces are limited by six spheres. The radius of the spheres which can be contained within these pores in relation to the radius of the spheres forming the pores, if we take it as the unit, will in the first case (in tetrahedral interstices) be equal to 0.22 and in the second case (in octahedral interstices), to 0.41.

The tightest possible packing can be obtained in the case of equal particles or units. Tendency towards the most compact possible arrangement is a crystallographical law. It is impossible to get such maximum compactness in soil in view of the fact that its units and particles of mechanical composition possess quite different shapes and sizes. Due to this great variety in the sizes of the particles, they cannot occupy the positions corresponding to the most compact packing. Only the packing (pore-space) of some sands, consisting of rounded quartz grains, comes near to the type of maximum compactness. The pore-space of sands, the size of sand grains remaining unchanged, will be the larger the more they differ from the spherical shape. The soils may exhibit a compact arrangement when the spaces between the large particles are occupied by particles or aggregates whose diameter corresponds to the dimensions of the spaces (Fig. 16). We get the oppo-

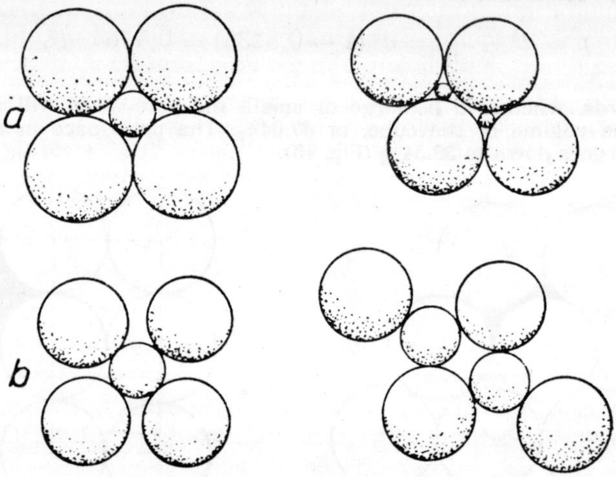

Fig. 16. Diagram illustrating changes in packing:
a—maximum packing; *b*—loosening of packing due to the separating effect of interstitial spheres

site effect when the sizes of the particles exceed the dimensions of the interstices, which does take place upon the growth of crystals of salts or ice, upon swelling, impregnation with gluey substances, enlargement of the microaggregates, etc. As a result of the growth of such interstitial particles or aggregates, which play the role of wedges, a compact arrangement is replaced by a loose one. Such phenomena take place upon carbonate accumulation in soil-grounds and their subsidence.

Pores can be large (noncapillary) and small (capillary); relatively to the soil aggregates they can be interaggregatory and intra-aggregatory, of any size, in view of the fact that each large aggregate is, in turn, composed of smaller aggregates, which are penetrated by pores of smaller and smaller diameter. Soils of a good structure (chernozems) have an overall pore-space of 55-65%. Their average pore-space is 40-50%.

The pore-space of soil-forming rocks and soils of various mechanical compositions varies considerably but the following somewhat approximate figures can be given: pebbles—35%, medium sand—39, sandy loam—45, loam—47-50, clay—52% and higher.

The pore-space is all the larger as the mechanical composition of the soil is finer. When the particles of the mechanical composition are large, the pores, although larger, have an incomparably smaller overall volume than the total volume of the numerous pores formed by fine particles. The relationship between overall, intra-aggregatory and interaggregatory pore-spaces is given in Table 17.

Table 17

Pore-Space, in Percentages (after N. A. Kachinsky)

Soil	Horizon	Depth, cm	Apparent density	True density of solid phase	Pore-space		
					overall	of separate aggregates	interaggregatory
Medium-podzolised loamy soil	A	0-12	1.33	2.61	49.04	39.02	10.43
	A_1	12-20	1.35	2.60	48.08	—	—
	A_2	20-32	1.39	2.65	47.55	38.09	15.28
	B_1	32-55	1.56	2.68	41.79	—	—
	B_2	55-85	1.72	2.72	36.76	—	—
	B_3	85-110	1.78	2.70	34.10	26.52	10.32
Leached slightly clayey chernozem. Steppe	A	0-4	0.90	2.49	63.86	52.87	23.32
	A_1	10-14	0.99	2.55	61.17	50.30	21.87
	A_2	40-44	1.06	2.57	58.75	—	—
	B_1	55-59	1.08	2.63	58.93	47.01	22.50
	B_2	80-84	1.10	2.61	57.85	46.09	21.82
	B_3	100-104	1.14	2.68	57.46	—	—
Nutty-lumpy sandy loamy solonetz. Virgin soil	A	0-4	1.07	2.46	56.50	38.31	29.49
	A_1	10-14	1.32	2.64	50.00	37.15	20.45
	B_1	15-19	1.36	2.73	50.18	28.92	29.91
	C	60-64	1.54	2.77	44.40	—	—

Pore-space plays an enormous role in the water, air, thermal and even nutrient regimes. It governs water capacity, filtration, water-raising capacity, aeration, etc.

Interstices between the particles which have a diameter narrower than 0.003 mm do not allow the passage of certain bacteria and still less of plant root hairs, which cannot penetrate even in interstices of 0.02-0.03 mm. Only pores whose diameter is not inferior to 0.01 mm allow the passage of root hairs. On the other hand, excessive looseness or porosity are not desirable, due to the fact that an important condition for the utilisation of absorbed ions is a close contact between the root hairs and the solid phase.

The physical properties of a soil depend on the correlation of capillary and noncapillary porosity; the optimum ratio being 1:1.

The porosity of soil undergoes constant changes in connection with the disintegration or formation of soil aggregates. Water circulating in soil modifies the sizes and shapes of the interstices, often decreasing their diameters and volumes, conditioning soil shrinkage and subsidence, which is particularly marked in loesses and loess-like materials. Subsidence of soils and grounds is partly connected with the dissolution of salts, their removal and changes in the colloidal properties of the particles which make up the soil-ground. Thus, on a plateau, loesses have a pore-space of 42.5-46.5%, and of 30% under steppe saucer-shaped depressions. When loesses are moistened, their pore-space goes down. In time, pore-space goes down due to silting (colmatage) and irrigation, or goes up, in connection with the loosening effect of soil cultivation, defreezing, gypsuming, etc.

Changes in the sizes of the pores lead to changes in the soil's properties, viz., absorption, wetting, filtration, water-raising capacity. A change of porosity is attended by a deformation of the soil either towards compactness or shrinkage, or, on the contrary, towards loosening up. Shrinkage can be linear and volumetric. Linear shrinkage is expressed in percentages of the original length and volumetric shrinkage in percentages of the original volume of the sample of soil.

The volume of particles and pores changes with the drying out of soil. At the same time, there is a formation of cracks in soil, which damage plant roots. Drying out alternating with swelling lead to a collapse of the structure which, in turn, changes all the physical properties of the soil.

Subsidence is a widely distributed phenomenon. It is tied with changes in the volume of the colloids, their downward displacement from the top layers, with filling up of the pores, translocation of water-soluble salts, transfer of suspensions, etc. Subsidence is attended by cracks and banks of subsidence, disposed in rows along channels. Some subsidences are disposed in across the channel. The ellipses of the subsidences reach 10-15 m and more along the axis. Sometimes, several ellipses of subsidences are formed on a distance of 1 km. In places, deformation of the ground affects the whole of the irrigated territory. Along the cracks we fairly often get, as a secondary phenomenon, washout and clayey karst. The formation of subsidences should be energetically counteracted through the improvement of soil structure, coagulation of the colloids, preliminary wetting of soil, gypsuming, etc.

The porosity of soil changes upon mechanical disturbance or packing, whereupon the volume of the soil may rise or fall. The volume of the soil changes also upon changes in moisture content, upon freezing and thawing. The porosity and volume of a moistened soil layer may change upon swelling of the colloids. In the latter case, there is an increase in the volume of the particles (up to 40%) and a decrease of the diameter of the pores. Swelling is, in the main, the result of the osmotic absorption of water, whereupon water and soil get into a complex physico-chemical interrelationship. The volume of the constituent parts of the soil changes also in connection with freezing, defreezing and thawing. In that case, as A. N. Sokolovsky points out, to the increase in volume from swelling is added the effect from the dilatation of water upon freezing.

When quicksands freeze, the surface of the ground becomes deformed, with the formation of chasms.

In the permafrost areas occur hillocks 2-6 m and more in height. Their appearance is tied with the manifestation of crystallisation forces upon the formation of ice and drawing in of newer and newer neighbouring water. This phenomenon is also met with, on a small scale, in areas outside the permafrost region. It occurs upon stagnation of water on ploughland, taking the form of bulging out, swelling and separation of the ploughed horizon from the lower lying layers of the soil. Bulging out of the ploughed horizon can be counteracted by preventive measures, by draining the soil, protecting it from freezing down to a great depth, etc.

The amplitude of the fluctuation of the soil's volume is greatly reduced by a lowering of its dispersiveness. In building and road building, to prevent volumetric changes, use is made of various materials such as broken stone, pebbles, coarse sand, gravel, which form a hard skeletal structure in the soil mass and reduce ground deformation.

Plasticity. The plasticity of a soil is its capacity, within a certain range of moisture content, to change its form when subjected to outside forces and to retain the new form (capacity for moulding). The most plastic soils are the rich, heavy clays consisting of scaly particles, disposed in the fashion of compact stacks of planks.

Plasticity is determined by the Atterberg method, based on the measurement of the soil's humidity, corresponding to various conditions of the soil mass. For this purpose one determines: a) the moment of crumbling upon minimum water content; b) the moment the soil forms a fluid upon maximum water content. The difference between the moisture contents at the lower and upper limits is called the plasticity number. The higher this number, the more plastic the soil and vice versa. The heavier the mechanical composition, the more plastic the soil. The plasticity number of clays is >36, loams 12-36, sandy loams 0-12, sand 0. Maximum plasticity is exhibited by clay. Plasticity is governed by the hydrated dense films covering the clay particles.

L. P. Rozov proposes to distinguish a lower and higher limit of fluidity. The former is determined by merging of a cut sample (a flat piece 1 cm thick); the latter is characterised by the condition of the soil when a groove made by a palette-knife on the surface of a moistened mass quickly disappears. This corresponds to the maximum quantity of water which can be retained by the upper layer of the soil. At maximum limit of fluidity, the furrows of a ploughed field run.

Dusty sands, sandy loams and light loams exhibit fluidity as a consequence of their lack of plasticity. Such soils and grounds can be called running grounds, capable, under a mechanical influence, of running and flowing with the water they contain. This property is also exhibited by relatively coarse-grained sands containing hydrophilous colloids. Soils possessing this capacity to run form chasms, bulge out, flow on the surface, as a result of the tightening forces acting upon freezing on the surface of the soil, etc. In this fashion, there is sometimes a formation of irregularities in the microrelief.

Adhesiveness (stickiness). Adhesiveness is the capacity of soil in a wet condition to stick to objects which are introduced into it or with which it comes into contact. Adhesiveness depends upon moisture content, mechanical and chemical composition and other properties. Adhesiveness begins to appear in a structural soil when the water content reaches 60-70% and in structureless ones when it reaches 40-54% (Vadyunina). Beyond that, adhesiveness increases up to the moisture corresponding to the lower limit of fluidity. Very fine-grained soils are almost twice as sticky as structural soils.

Adhesiveness is determined by the amount of moisture corresponding to the moment when the soil mass begins to adhere, upon a certain minimum moisture content. Up to a limit, adhesiveness goes up with increase of moisture; beyond that, it begins to fall and, finally, stops altogether when the overmoistened mass assumes the consistency of a paste.

Adherence is due to the so-called adhesive forces set up by the reciprocal attraction of molecules on surfaces coming into contact (Van der Vaals forces). Maximum interaction of these forces occurs when water displaces adsorbed air from the surface of adhering soil particles and the soil moisture becomes somewhat higher than the lower level of plasticity. Adhesiveness depends on the presence in soil of a silty fraction, on humus and on absorbed bases. Saturation of soil with sodium causes a rise in adhesiveness. A structural soil possesses lower adhesiveness than a structureless one.

Adhesiveness can be expressed by the force in grams necessary to remove a metallic plate 1 cm^2 in area from the soil to which it adheres. This force can be measured using simple scales, with a measuring plate several cm^2 in area fixed to one of the sides.

For solid bodies, adhesiveness is directly proportional to the dispersiveness, it depends upon the mechanical composition and the degree of moisture. Adhesiveness, just as plasticity, is higher in clays and considerably less in sands. According to the degree of adhesiveness, we get highly adhesive soils (over 15 g/cm^2), adhesive (2-5 g/cm^2), slightly adhesive and nonadhesive (crumbly) (below 0.5 g/cm^2).

Cohesion. The cohesion of soils is their capacity to resist forces which tend to disunite the soil particles. It is expressed by the value of the forces which hold the soil particles close to one another.

Cohesion is due to the forces of reciprocal cohesion, i.e., the attraction set up between the soil particles. The factors intervening in the manifestation of these forces are the soil colloids, cementing substances, moisture and other material factors. Cohesion is all the greater as the mechanical composition is heavier, as the adhesiveness or stickiness is higher and as the moisture content is lower. Cohesion depends on the form of the soil particles. It is higher when they have a disk-like or scaly shape. Cohesion is measured in kilograms, according to the load necessary to break up, crush, split samples of soil on special apparatus. The cohesion of sandy soils does not exceed 5 kg/cm^2, whereas in clayey soils it reaches 50-60 kg/cm^2.

Changes in the composition and amount of colloids, humus, absorbed bases and water are reflected in cohesion. In clayey soils, cohesion is greater when they are in a dry condition, in sandy soils upon medium moisture. It reaches a considerable degree in the illuvial horizons of podzolic and solonetzic soils. Active humus, which increases the cohesion of soils of light mechanical composition, brings down the cohesion of heavy clayey soils. This is explained by the fact that humus possesses medium cohesion in relation to clay and sands, i.e., its cohesion is lower than that of clay and higher than that of sands. When covering clay particles, humus hampers their agglutination but when covering sandy particles, it promotes it. Solutions of salts (electrolytes), which promote the coagulation of colloids, decrease cohesion. Substances which peptise colloids, increase cohesion. Cohesion goes down when the soil is frozen and for other reasons. The cohesion of a soil is all the greater as its dispersiveness is higher and as the particles are more tightly packed. Absorbed bases exert an influence on cohesion. Replacement of Ca by Mg almost doubles the resistance of soil to crushing; replacement of Ca by ammonium increases it by 4.5 times and replacement by sodium increases cohesion by more than 10 times. Upon replacement of Ca by Fe ions, cohesion falls sharply. The cohesion of certain clays, when they dry out, approaches that of concrete.

Another property of soil is its hardness, which is the capacity to resist in some or other degree the penetration into it of cutting bodies under pressure. The degree of hardness of soils is expressed in dynamometric terms or in kilograms per 1 cm^2 (kg/cm^2) and is measured in the laboratory and in the field using special apparatus called durometers (devised by V. V. Goryachkin, N. A. Kachinsky and others).

The hardness of a soil in the field can be approximately determined using a soil knife; the following degrees of soil compactness are distinguished:

a) friable—crumbles from the walls of the cut when touched by the knife, which penetrates easily into the soil (sands and sandy loams);

b) fairly friable—crumbles less than the previous type, easily dug with a spade, the knife penetrates with ease into the soil. Sandy loams, light loams and structural loams. Typical of the upper soil horizons;

c) packed (fairly compact)—fairly easily cut by spade and knife. The walls of the cut are firm. Loamy and clayey structurised soils or markedly cemented sandy loams and sands;

d) hard—difficult to cut with a spade. The walls of the cut are very firm, the knife penetrates with difficulty into the soil. Typical of the lower horizons of podzolic, solonetzic and other soils;

e) very hard—hardly yields to the spade. A knife only leaves a mark on the soil without penetrating into it. This degree of hardness is characteristic of illuvial horizons, strongly solonetzic soils, solonetzes and, in a number of cases, of podzols (ortstein, ortsands), etc.

The hardness of a soil depends not only on its mechanical composition and structure but also on the character of its texture and moisture content.

The mechanical composition being the same, some soil horizons are more compact (B_1, B_2) or, on the contrary looser (A_1, A_2) than others, which is tied with processes of soil formation. Moistened horizons of soil have a lesser degree of hardness. On the other hand, the degree of moistening can itself bring about a lower or higher degree of compactness.

The soil's degree of hardness is a dynamic quantity, which depends upon many factors. It is regulated by soil cultivation methods, the growing of crops and other operations (agromelioration, forest reclamation, hydromelioration, etc.).

Depending upon its cohesion and hardness, soil offers different degrees of resistance to agricultural implements and digging machines. This resistance is tied with the force of friction. Friction in soil depends on its moisture content and pressure on the rubbing surfaces. With increase in moisture content, it goes up, at first, then falls. The friction in a moist soil (loam) is greater than that in a dry soil. Upon further increase of the moisture content, it goes down again, due to the fact that water acts as a lubricant. Clay exhibits a second increase of friction tied with the adherence of the soil to metal, when there is friction between the soil particles. A further increase of the moisture content is attended by a sharp decrease of friction, because soil acquires the consistency of a fluid body (like a muddy mass). The heavier the mechanical composition of the soil, the greater the friction: minimum friction is exhibited by sandy soils and maximum friction by clayey ones.

The resistance which soil offers when worked (S) is determined by the formula:

$$S = Kab,$$

where K—specific resistance on the working surface of the implement expressed in kg/cm^2;
a—depth of furrow, in cm;
b—width taken, in cm.

At a speed of ploughing of 1 m/sec the specific resistance K varies between 0.2 and 0.9 kg/cm^2 and substantially more for solonetzes.

The resistance of soil to implements changes in accordance with its agricultural condition and its moisture content. It depends on the design and weight of the implement (plough) as well as on the speed of the tractor during ploughing, the depth and width of the furrow, the coefficient of friction, etc.

Gypsuming reduces the resistance during ploughing by 30-40%. The poorer the structure of the soil, the greater its resistance to implements and plant roots which penetrate into it.

When worked, the soil breaks up into units of various sizes and shapes, depending on its structure, texture and moisture content, i.e., its physical fitness, or "ripeness" for working. One of the important properties of soil is a capacity to crumble when worked, to break up into clods and grains, which is determined by optimum moisture content at the time of working.

The optimum period for working soil is determined by its physical "ripeness", whereupon it offers the least resistance to implements and breaks up into aggregates. A moisture content of 40% of the total pore-space corresponds in most cases to the ripeness of the soil, whereupon it is necessary to begin working it. The optimum moisture content of a structural soil varies noticeably: it is close to the moisture content corresponding to the limit of adhesiveness, but always somewhat below it.

Premature working of the soil, or delayed, i.e., after the moisture content has fallen, give unfavourable results. In the first case, the ploughed field becomes covered with a hard crust which adversely affects its physical and biological properties, and in the second case, the ploughed surface breaks up into big lumps. When a wet soil in a plastic condition is being ploughed, it is inverted and rests back unbroken, forming slightly torn bands (furrow-slices).

Soil cap and "pan" due to ploughing. On the surface of loamy and clayey soils, after they have been wetted, we very often note the formation of a puddled layer on the horizon of ploughing, showing vertical cracks, which is called the soil cap.

By increasing the losses of moisture from the ploughed land, the soil cap lowers the field germinating rate, worsens the growth conditions of plants and lowers the yields of all crops. It hampers the access of air to plant roots and exerts an unfavourable influence on the biological processes: gaseous exchanges are slowed down and the formation of plant nutrients in an available form is halted. The utilisation by soils of atmospheric precipitations and irrigation water is adversely affected. The upper soil cap should be broken up after each water application, which increases the labour and other costs involved in the management of the field.

Puddling of the soil when it becomes wetted and the formation of a cap on sown fields are tied with a relatively heavy mechanical composition of the soil, when the aggregate amount of clayey particles (0.005-0.001 mm) and silt (<0.001 mm) exceeds 25%. Soils which contain up to 60% of dusty and sandy particles and more, form no cap. When soil contains more than 20% of silt, it is liable to puddle after rain or watering and may form a cap after drying. The high specific surface of clayey and silty particles conditions their capacity to retain absorbed water on their surface. These particles serve as material for coalescing the particles of coarser soil fractions, with the formation of a hard crust upon drying up.

The formation of a pan on powdery soils is often tied with the leaching of the absorbed calcium. The latter is attended by disaggregation and puddling of the soil's surface. The soil gets "baked" upon drying out. The displacement of absorbed Ca from the soil absorbing complex by the Na, K and NH_4 cations worsens the physical properties of the soil, creating conditions leading to puddling and the formation of a cap. The application to soil of calcium salts acts in an opposite direction. But on structureless heavy soils with a high content of colloids, a soil cap is formed regardless of the character of the absorbed bases. On soils which have a water resistant structure, there is no cap formation as a rule.

The ploughing of unripe clayey soil, frequent harrowing and unnecessary cultivation pulverise the ploughed horizon of soil, lowering its water permeability. As a consequence, there is a risk of puddling after rain and watering and a cap forms upon drying up. Unnecessary tillage operations may lead to the destruction of the aggregates and intensification of the mineralisation of the organic matter in the plough horizon. Liming, gypsuming, green manuring, the sowing of grasses, furrow irrigation and other methods are used as preventive and active measures for counteracting the formation of a soil cap.

Immediately below the plough horizon of loamy and clayey soils (A_p), not infrequently occurs a packed subplough horizon called the plough "pan" (Fig. 17). Its origin is due to the mechanical consolidation of the subplough horizon owing to the fact that upon ploughing the plough slade presses down and pastes the unripe overmoistened soil. A consolidated layer may, however, also arise as a result of the translocation of colloids with water and their retention in the subplough layer (A. N. Sokolovsky).

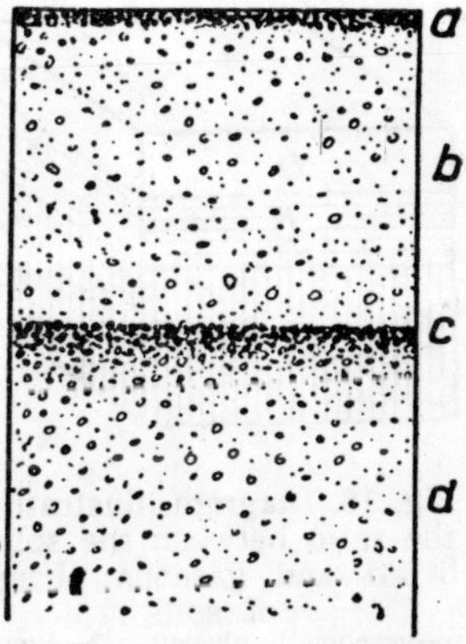

Fig. 17. Diagram illustrating the seat of formation of surface cap and plough pan: *a*—surface cap; *b*—plough horizon (A_p): *c*—plough pan; *d*—subplough horizon

The plough pan is a kind of microilluvium, tied sometimes with the displacement of calcium by the potassium, sodium and ammonium of the fertilisers applied to the soil. For this reason, in order to prevent the formation of a pan, it is advisable to combine deep ploughing with liming.

Above the plough pan, the plough layer (A_p) gets overmoistened, which leads to the soaking, rotting and death of autumn sown crops. The latter occurrence is often tied also with the formation of an icy crust on the soil surface, under which the plants perish, due to oxygen starvation. In addition, the plough pan hampers the infiltration of water to deeper lying soil horizons to which the roots can penetrate.

The formation of a plough pan should be counteracted by varying the depth of ploughing from year to year, by breaking the pan with a subsoiler, by liming acid soils and gypsuming alkaline ones, etc.

Chapter VII

WATER PROPERTIES OF SOIL

Water is a most important condition for the development and fertility of soils. By translocating the products formed upon the interaction of organisms and soil, it contributes to the formation of genetic soil horizons. From a meliorative point of view, water acquires a particular significance as a physical system closely interrelated with the solid and gaseous phases of the soil and with plants.

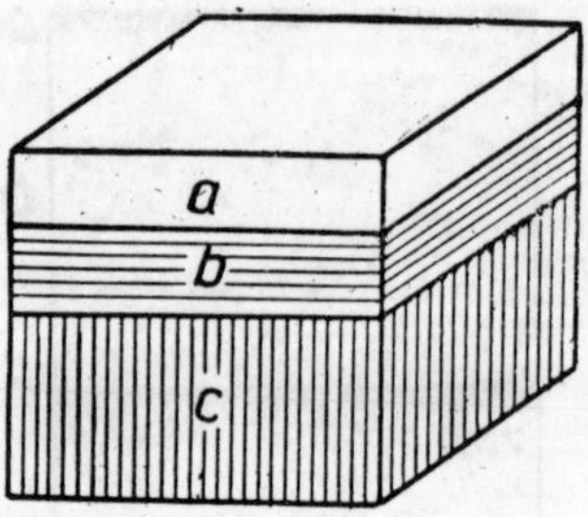

Fig. 18. Diagram illustrating the ratio between the solid, liquid and gaseous phases of soil:
a—gaseous phase; *b*—liquid phase; *c*—solid phase

The water present in soil is studied as a mobile system, in which the combination of the constituents, viz., gases and dissolved substances and with them also the biological phenomena, is subjected to changes which depend upon the thermodynamic condition of the soil medium.

In order that a soil may serve as a medium for the development of plants, it must contain liquid water in a certain ratio with its other phases, viz., gaseous and solid (Fig. 18).

The soil's water properties are closely connected with the properties of water itself, which consists of dipolar molecules. The H^+ and O^- ions of water molecules are disposed asymmetrically in triangular fashion (Fig. 19). The asymmetrical disposition conditions the retention of a negative charge at the apex of the triangle owing to O^-, and ensures a certain surplus of positive electricity at the other two angles owing to H_2^+. When electrically neutral, the molecule is dipolar, i.e., has two poles. The water molecules become bound with the other soil molecules, due to the attraction between poles of opposite charges. An insignificant part of the water molecules (2 in one thousand million) dissociates into H^+ and OH^- ions. The number of dissociated molecules goes up (up to 40) in water containing a certain amount of H^+ and HCO_3^-.

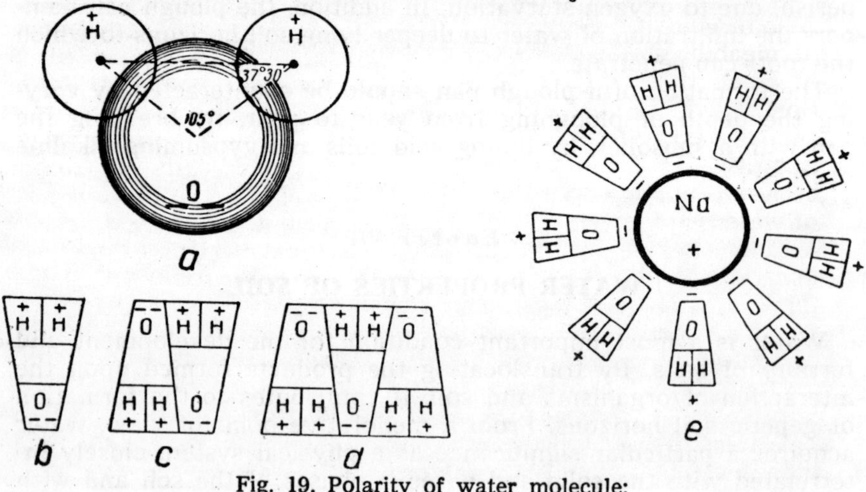

Fig. 19. Polarity of water molecule:
a—diagram of the molecule; *b*—hydrole; *c*—dihydrole; *d*—trihydrole; *e*—hydration of the ions (Na-ion, surrounded by water dipoles)

The dipolar molecules of water are attracted by the energetically unsaturated ions of the surface zones of crystalline substances, around which are formed water pellicles. Such is the way in which the process of hydration of soluble substances goes on in soil. The dipoles of water, which exert an attraction, as if wrest the surface ions out of the spatial crystalline lattice of the substances present in the soil mass. Tied with this is the destruction of minerals and other components of the soil. The process of hydration in soil proceeds as part of a complex, which involves other processes, such as oxidation, hydrolysis, etc.

Liquid water possesses thermal energy which supplies kinetic energy responsible for the movement of water molecules. Thus, upon its absorption by soil, part of the water passes into an immobile state, losing part of its kinetic energy (nearly 80 small calories of heat are given off per 1 g of water). The latter occurrence is attended by a lowering of the surface energy of the soil particles. The mobility and availability of water to organisms is governed, in the main, by the physico-chemical properties of the solid phase of the soil (absorbed bases, absorption capacity and others). In the presence of other favourable conditions, an optimum liquid water content of the soil ensures high yields. Part of the water in soil is held with a force of several thousands of atmospheres and is therefore unavailable to plants. It forms a sort of dead reserve, which in sandy soils reaches 1.5-2% of the soil's weight, and for clayey soils goes up to 10-15% and more. But soil possessing pore-space, is capable of containing and holding a considerable amount of water over and above the dead reserve. This capacity is referred to as the water capacity of soil.

A highly important property of soil is its water-raising capacity. Also of great agronomical and meliorative significance is the water permeability of the soil, i.e., its capacity to let water pass through it. Upon the water permeability and cultural (structural) condition of the soil depend the surface runoff and the degree of resistance to erosion. A most important property of soil is, in addition, its evaporating capacity. Of great importance is the exchange between soil moisture and the atmosphere and the sublimation of water vapour, the so-called subterranean dew.

The water regime of soils is of particular significance in melioration. It governs the direction followed by soil formation and fertility. The regulation of the water regime and water balance of soils constitutes therefore a highly important task confronting agricultural melioration.

Forms of Water in Soil

Water is found in soil in various aggregatory conditions and in different forms. The various forms of water are present in soil simultaneously and exhibit a complex interrelationship.

The moisture contained in soil is not uniform and in different points of the water and soil mass, it possesses different properties. Passing from one condition into another and from one soil horizon into another, water is, at different times, differently distributed in soil.

Water in soil can be in the following forms:

1) chemically combined water: a) of constitution, b) of crystallisation; 2) physically bound water: a) tightly bound (hygroscopic), b) loosely bound (film water, molecular); 3) capillary water: a) abutment water (pendular), b) funicular, c) perched, d) supported, e) mobile; 4) gravitational water: a) leakage, b) supported, c) percolating, d) ground water (soil-ground water); 5) water vapour; 6) water in the solid state.

Chemically combined and physically bound water. Chemically combined water is a component of the hydrated substances contained in soil. The amount of this water is generally small but it sometimes reaches 5-7 (12)% and even more. This water is an indication of the degree of weathering of the rocks. The higher the contents of silicates and silicates of aluminium in soil, the higher the content of chemically combined water in it. But this water in soil does not take a direct part in the physical processes and does not evaporate at a temperature of 100°C.

Chemically combined water is, in turn, divided into water of *constitution* and water of *crystallisation*. Water of constitution is a chemical component of minerals. Water of crystallisation is a constituent part of crystallohydrates and zeolites in which it is contained as separate molecules bound with the crystalline structure of the substance. Water of constitution is liberated in the process of the destruction of the chemical composition of minerals at high temperatures. The second is less strongly bound and is set free at lower temperatures (for example, the water of kaolinite, goethite, gypsum, minerals of the $R_2O_3+nH_2O$ type and others).

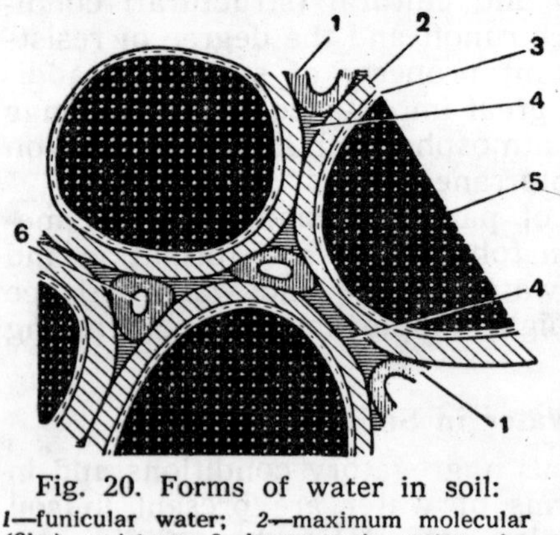

Fig. 20. Forms of water in soil: 1—funicular water; 2—maximum molecular (film) moisture; 3—hygroscopic water; 4—pendular water; 5—soil particle; 6—air bubble

Physically bound water in soil can be present in all three aggregatory states, viz., gaseous, liquid and solid and be subjected to the effect of forces varying in nature and value. A substantial part of this water is held on the surface of the soil particles with varying force, as a result of molecular reciprocal attraction between the molecules of water and soil (Fig. 20).

Water physically more strongly bound is referred to as hygroscopic moisture; less strongly, loosely bound water is referred to as molecular or film water.

Like any porous body, soil adsorbs vapour and water molecules that are tightly held on the surface of its particles. The soil's capacity to adsorb water vapour from the air is called hygroscopicity and the moisture adsorbed by the surface of the particles, *hygroscopic water*. The degree of hygroscopicity depends on the dispersiveness (total surface of the particles) of the soil and the amount of water vapour in the air with which the soil is in contact. Hygroscopicity is all the higher as the air is more saturated with water vapour and as the soil is more dispersed. When the temperature goes up, the hygroscopicity of soil goes down. In an atmosphere saturated with water vapour, the soil acquires maximum hygroscopicity. Hence, the maximum amount of water absorbed by soil from a space saturated with water vapour is called maximum hygroscopic moisture. The hygroscopicity of soil in any concrete conditions varies in accordance with the relative moisture of soil air and atmospheric air* and the properties of soil, in particular the capacity of the particles to attract molecules of gaseous water. Hygroscopic water is absorbed in virtue of the molecular interattraction of soil and moisture particles. Dipolar water molecules are attracted and become fixed on the surface of soil particles. Hygroscopicity is therefore tied not with thermal but with molecular condensation. It may manifest itself when the soil and the air containing water vapour have the same temperature. The amount of hygroscopic water is a variable quantity. Upon incomplete saturation of the atmosphere with water vapour (relative humidity lower than 100%), the surface of the soil particles becomes coated with a thin layer of molecules of hygroscopic water. When the atmosphere is completely saturated with water vapour (relative humidity equal to 100%), the surface of the soil particles gets coated with the thickest possible layer of water molecules (maximum hygroscopicity). Hygroscopic water is held with a force reaching a value of 17-37 thousand atmospheres. The moisture of soil corresponding to maximum hygroscopicity is, according to Fageler, held with a force of up to 50 atm. and, upon decrease of moisture down to 0.1 of maximum hygroscopicity, it is held with a force of up to 50 thousand atm. In consequence, hygroscopic water is not capable of moving within the soil layer without passing first into free gaseous moisture. The layer of ad-

---

* The relative humidity of air is expressed in fractions of unity or in percentages, corresponding to the ratio of the amount of water contained in air at the given moment to the maximum amount of water which may be contained in air at the same temperature when it is completely saturated with water vapour.

sorbed water on the surface of the particles consists of rows of molecules clothing the soil particles. Its thickness is all the smaller as the particle is smaller. This water possesses greater density and viscosity than free dropping-liquid water.

Hygroscopic water freezes considerably below O° and the last portions of it freeze only at —78°C. This water does not dissolve salts. Its density goes up to 2.4, its dielectric constant equals 2.2, its thermal capacity equals 0.7. Hygroscopic water does not play any role in the swelling of soil. Under the influence of adsorption it becomes orientated, reminding a solid body with a crystalline structure. The quantity of hygroscopic water in the soil, given one and the same air moisture, depends on the mechanical and colloidal composition of the soil, humus content, absorbed bases and hygroscopic salts. Hygroscopic water is all the more abundant in soil as its mechanical composition is heavier and as it is richer in organic matter, i.e., the larger the total surface of the particles (Table 18). This water is unavailable to plants and forms part of the dead reserve of moisture in soil.

Fluctuations in maximum hygroscopic moisture depend on the amount and composition of absorbed bases and soluble salts. After the leaching of salined soil, the value of its maximum hygroscopic moisture goes sharply down (by 2-3 times).

Table 18

Hygroscopic Water Content in Percentages

Mechanical composition of soil	Hygroscopic water as % of weight of absolutely dry soil	Maximum hygroscopic water, %
Sand	0.5-1.5	1.5 and more
Light loam	1.5-3.0	3.0-5.0 (10)
Medium loam	2.5-4.0	5.0-6.0 (12)
Dust-like clay	6.0-8.0	8.0-10.0 (20)
Peat	18.0-22.0	—

For example, one salined clayey soil was capable of holding 27% of maximum hygroscopic moisture, and after leaching, only 14.4%; another salined, loamy soil before leaching held 16.5%, after leaching only 6.7% (S. V. Astapov). Maximum hygroscopic moisture is determined in practice after prolonged (several weeks) saturation with water of a soil lot placed in an exsiccator over 3% (10%) H_2SO_4 (after Mitsherlich) or over a saturated solution of K_2SO_4 (after Nikolayev), the relative humidity being approximately 94%. Maximum hygroscopic moisture is somewhat below the maximum quantity of adsorbed (molecular) water.

The moisture held by molecular forces around solid soil particles is called *molecular (film) water*. When soil particles saturated with moisture up to maximum hygroscopic moisture come into contact with liquid water they draw upon it until the molecular forces of attraction reach complete equilibrium (molecular condensation). This water on the surface of soil particles forms a sort of film, which term, however, does not correspond to any geometric notion and has therefore but a conventional meaning. It is held with less force than hygroscopic water. That is why it belongs to the category of loosely held water. The limit which divides maximum hygroscopic from film water is not sufficiently clearly determined. The attraction of molecular water and its fixation on the surface of soil particles is due to the field of forces formed by the electric charges of the soil cations.

The effect of the molecular forces of attraction on the surface of soil particles is felt outwardly all around the particles to a distance of approximately 0.06-0.006 μ. This range of action of the forces embraces up to 100 molecules of water and the molecular forces of the soil particles are fully saturated. The inner layer of water molecules corresponding to the hygroscopic moisture is tightly held by the soil particles. It is immovable. The loosely held outside layers of film water may move downwards under the influence of gravity or be displaced in any direction under the effect of molecular forces upon drying out of the soil from above, due to the effect of suction, etc. The less molecular water remains in soil the more firmly it is held by the soil particles. And as the water film grows thicker, moisture is held by the molecular attraction of the soil particles with decreasing force until, finally, upon further thickening of the film of water, there comes a moment when these forces cease to exert any effect and the water particles are freely separated from a layer exceeding a certain limit of thickness.

Loosely held (molecular, film) water possesses a number of distinguishing properties. Its density and viscosity are considerably higher than those of free water. The density of its layers close to the surface of the soil particles goes up to 1.7, which corresponds to a pressure of several tens of atmospheres. In that case it ceases to obey gravity. In some of its properties it is akin to solid bodies. Its heat conductivity is close to that of ice (0.5), its dielectric constant close to 2.2. This part of the molecular water loses its solving power. Due to the fact that it does not dissolve electrolytes, this water is devoid of electroconductivity. It does not yield to centrifugal force when gravity is accelerated up to 18,000-70,000 dynes.

The amount of film water reaches 1.5% of the weight of soil in sandy soils, up to 15-17% in loamy and up to 30% in clayey soils. This amount depends upon the mechanical composition, the origin of the soil, the osmotic pressure of the soil solution and other factors. The content of loosely bound water changes appreciably even within one and the same soil, depending on the moisture content, the concentration and composition of the soil solution and other factors. In soils poor in water-soluble salts, water so held may exceed the amount of maximum hygroscopic water by 2-4 times. In salined soils it may be quantitatively less than maximum hygroscopic water. Maximum molecular moisture goes down upon packing of soils, even though its amount in relation to pore-space goes up. A measure of the amount of molecular water is obtained from the volume remaining undissolved, using highly concentrated solutions, as well as from the negative adsorption of sugars, the heat of moistening, by biological methods, etc.

Maximum molecular moisture corresponds to the moisture at which plants wilt, known as the wilting coefficient, i.e., the "dead" (unavailable) reserve of water in soil. This water is held by soil with forces appreciably higher than the suction force of roots [14-16 (25) atm.]. In order to be able to compare the reserves of unavailable moisture of soils in a more satisfactory way, use is made of what is known as the moisture coefficient (*Eh*), which is the ratio of the weight of water (*a*) which soil is capable of holding when subjected to a centrifugal force 1,000 times greater than gravity to the weight of dry soil (*b*), expressed as a percentage, $Eh = \frac{a}{b} 100$.

Knowing the moisture equivalent (*Eh*) one can determine the wilting coefficient of plants $= \frac{Eh}{1.84}$, which is close to the maximum molecular moisture capacity. Wilting of plants is spread over a long period during which moisture appreciably changes and it is therefore difficult to determine just what is the amount of water present in soil when the plant dies. One may distinguish: a) moisture of stunted growth of plants (MW_1), i.e., moisture which corresponds to the first symptoms of wilting; b) moisture of delayed growth, of serious (dangerous) wilting (MW_2); c) moisture of total (irreversible, steady) wilting (MW_3). Plants utilise that moisture which exceeds by approximately 1.5-2.5 times maximum hygroscopic moisture. This amount corresponds approximately to maximum molecular water. When the water present in soil falls below that amount, the plants die (Fig. 21).

According to S. I. Dolgov, maximum molecular moisture cannot be identified with the maximum amount of sorbed water. Neither can it be taken as a measure of the maximum content in soil of film water unavailable to plants, owing to the fact that it may include, in addition, a certain amount of capillary-condensed or meniscus water. Another source of water unavailable for plants is the intracellular water of dying plants in peaty and soddy-podzolic half-boggy soils.

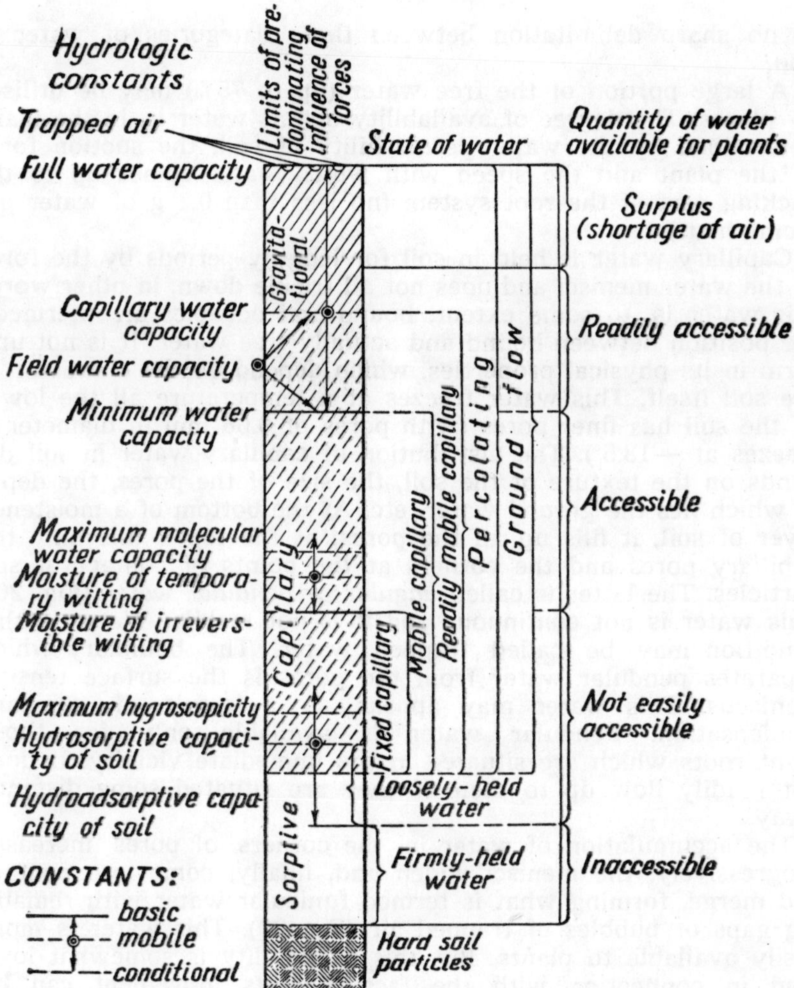

Fig. 21. Diagram illustrating the various forms and states of soil moisture (after S. I. Dolgov)

Free water. Water contained in soil besides that which is physically bound is called free water. It accumulates in soil once the adsorptive (molecular) forces of attraction have been completely used up (balanced) and it is outside their sphere of influence. Loosely held (film) water passes imperceptibly into *capillary*, retained in soil as a result of the differences in surface pressures set up by the convex surfaces separating water from air. There is no sharp delimitation between these categories of water in soil.

A large portion of the free water (up to 75%) may be utilised by plants. The degree of availability of this water is, in the main, determined by the water permeability of soil, the suction force of the plant and the speed with which water comes up to the sucking part of the root system (no less than 0.1 g of water per 1 cm^2/hour).

Capillary water is held in soil for lengthy periods by the force of the water menisci and does not all trickle down, in other words this water is, to some extent, bound and occupies an intermediate position between bound and actually free water. It is not uniform in its physical properties, which depend on the character of the soil itself. This water freezes at a temperature all the lower as the soil has finer pores (with pores of 0.06 mm in diameter it freezes at —18.5°). The distribution of capillary water in soil depends on the texture of the soil, the size of the pores, the depth at which lies the ground water, etc. At the bottom of a moistened layer of soil, it fills up all the pores; at the top it fills only the capillary pores and the corners at the points of contact of soil particles. The latter is called *angular* or *pendular water* (Fig. 20). This water is not continuous and it is not mobile. Water in this condition may be called trapped water. The boundary which separates pendular water from the pores is the surface tension meniscus. This water may appear as a result of capillary condensation. Pendular water is available only for those plant roots which are situated in the immediate vicinity. It does not readily flow up to them if they are situated some distance away.

The accumulation of water in the corners of pores increases progressively, the menisci widen and, finally, come into contact and merge, forming what is termed funicular water with remaining gaps or bubbles of trapped air (Fig. 20). This water is more easily available to plants. But this availability is somewhat lowered in connection with the fact that its movement can be achieved only once the appreciable pressure offered by the imprisoned air has been overcome.

With the ousting of part of the trapped air, we get sections of capillary water, with menisci forming at the ends. Continuous capillary water in fine-grained soils may be in the shape of isolated microaccumulations, separated by bulkheads of bound water. This water is referred to as *capillary sorptively locked water*. And finally, capillary mobile water fills up the capillary pores completely. Capillary water is subjected to the (capillary) meniscus forces and, depending on the actual conditions, is, to some extent, subject to gravity. The term "capillary forces" or "capillary potential" is used by a number of research workers to designate a certain resultant of forces: of surface tension, osmotic and sorptive forces.

When all the capillary pores are filled with water and the capillary potential is reduced to zero, water in soil begins to move down along the large pores under the influence of gravity. This water is referred to as gravitational, or phreatic water. Gravitational water in soil appears after rain, watering, thawing of snow and defreezing of the soil and may appear at the expense of a certain portion of capillary water freed from the forces which were holding it. The latter occurrence can be observed for example upon a sharp change of the atmospheric pressure, upon increases in the size of pores, as a result of a drop of the soil's temperature or other causes.

Gravitational water may be seepage water or supported water, it may percolate downwards or be stagnant. This water is available to plants. But its utilisation may be limited due to its rapid movement. Furthermore, it is to some extent all the less utilised as it infiltrates more intensely. Penetrating through soil, gravitational water reaches the water-bearing horizon, where it fills up all the pores and collects in the form of what is known as the *ground water*. This water saturates the water-bearing horizon and can, under the influence of hydrostatic pressure, flow along subterranean slopes. Ground water can present a head or not, be immobile or mobile.

Gaseous water is always present in all the free pores or those that are partially filled with water. It is a constituent part of soil air. In all soils, when the moisture content exceeds maximum hygroscopic moisture, as has been established by A. F. Lebedev, the relative humidity of air is always equal to 100%.

When the temperature falls below zero, liquid water in soil passes into the solid state. Turning to ice, it increases in volume by $^1/_{11}$. The density of water at 0° is 0.999968 g/cm^3 and the density of ice at the same temperature is 0.9168 g/cm^3. The thermal capacity of ice is two and a half times less than that of water, i.e., it is equal to 0.487 cal/g degree (at 0°). The melting heat of ice at 0° is equal to 79.4 cal/g. Ice possesses plasticity. As it crystallises in the soil pores, it draws apart the soil particles but, at the same time, it cements them, forming frozen soil and subterranean ice. The breaking up effect of freezing water in soil voids, especially of clayey soils, is attended by a redistribution in the density of the soil mass.

As a result of quantitative changes in the moisture content of soil, some forms of water pass into other forms, as if following a certain sequence. Gaseous water turns to hygroscopic, the latter increases and reaches the value of maximum hygroscopic. The latter passes into molecular, then into maximum molecular and, finally, into capillary, pendular to begin with and then funicular

9—891

and locked capillary, reaching maximum capillary readily mobile water. The latter, increasing quantitatively, passes into gravitational water which trickles down to the waterproof bed on which rests the water-bearing horizon of ground water. Ground water, like all other forms of liquid water, passes into the gaseous form, which may, upon adsorption and condensation, pass into the liquid form and so on. Each separate form is sufficiently clearly distinguishable from the contiguous form, thus exhibiting a certain discontinuity. At the same time, they form a certain sequence and pass imperceptibly one into the other, thus showing a certain continuity.

Soil Moisture

The amount of water contained in soil at a given moment is its moisture content. This varies from an insignificant amount to full water capacity. Extreme shortage of water is harmful for the development of plants, halting their growth and causing their death; as for excess of water, it has an adverse effect upon soil, promoting reduction processes, hampering nitrification, etc. The moisture content and the overall water content in soil at any moment is governed by the amount of water coming in and going out of soil. The moisture content of soil varies with the physical properties of soil, its water permeability, water capacity, capillarity, evaporation, etc. Soil moisture is also tied with the local relief. The soils of the raised elements of the relief are usually characterised by a lower moisture content and, conversely, due to lateral ("translational") inflow of water, more humid soils—which may go as far as the accumulation of stagnant water and the formation of swampy soils—are found at the foot of slopes, at the bottom of the lowermost parts of gorges and ravines, in depressions on watersheds and plains. Soil moisture is a dynamic feature. In the course of the vegetative period, the dynamics of soil moisture, apart from the properties of the soil itself, are linked with agrotechnical measures and with the phases of plant development. Through proper treatment, agro- and hydromelioration, it is possible to bring about the utmost favourable water regime.

Every soil possesses its own dynamics of moisture throughout its genetic horizons. Moisture can be absolute (natural), represented by the gross (absolute) amount of moisture in soil at a certain point, at a certain time, expressed as percentages of the weight or volume of soil, and relative moisture, measured as percentages of the pore-space (full water capacity).

Volumetric, gravimetric and relative moistures are calculated from the formulas:

$$W_{vol} = 100 \frac{V}{p} q; \quad W_g = \frac{p}{100\,a} q_1; \quad W^r = \frac{W_a 100}{p},$$

where W_{vol}—volumetric moisture, in percentages;
W_g—gravimetric moisture, in percentages;
W_r—relative moisture, in percentages;
W_a—absolute moisture, in percentages;
p—pore-space, in percentages;
V—apparent density;
q—moisture, in percentages of the weight of dry soil;
q_1—moisture, in percentages of pore-space.

Relative moisture is often expressed in relation to the capillary porosity or to the limit field moisture capacity according to the formula:

$$W_r = \frac{W_a 100}{C_p},$$

where C_i—capillary porosity (limit field water capacity), in percentages.

Soil moisture is often expressed as the amount of the water reserve in tons or cubic metres down to the depth of the soil layer, according to the formula:

$$W = W_1 hd \times 100,$$

where W—weight of the water in tons, or its volume, in m³;
W_1—soil moisture in percentages by weight of absolute dry soil;
h—depth (thickness) of soil layer, in m;
d—apparent density of the soil;
100—a constant factor, obtained by dividing 10,000 m² by 100 from the ratio of the weight of water to the weight of absolutely dry soil $\frac{W_1}{100}$.

To express the water reserve in millimetres of a water column, the amount of water obtained in tons or cubic metres should be divided by 10, due to the fact that a layer of water of 1 mm on an area of 1 hectare amounts to 10 m³ or 10 tons. Soil moisture in millimetres of a water column is calculated from the formula:

$$A = W_1 hd \times 10,$$

where A—water reserve in mm; the rest as above. For example, a 50 cm soil layer with 20% of moisture and an apparent density of 1.3 will contain A— $20 \times 0.5 \times 1.3 \times 100 = 1{,}300$ t/m³, or 130 mm of water. Soil moisture is determined by various methods.*

Water Capacity of Soils

The water capacity of soils is their capacity to contain and retain the maximum amount of moisture upon the given conditions. It is dependent on the forces which exert their effect in soil and on the factors of soil formation (structure, capillarity, condition of moisture, regime of ground water, evaporation of water from soil, temperature of soil, concentration and composition of soil

* They are the drying, carbide, alcohol, pycnometric, electrometric methods; moisture content can also be determined from changes in the electroconductivity of special blocks (of gypsum and other materials) placed in soil, from changes in the electromotive force, from changes in the suction capacity of soil (tensiometric method), from the absorption of gamma-rays, by the neutrons method, etc. At the present time, a micromethod for the determination of soil moisture is being worked out using heavy water as indicator, etc. The methods are described in the practical course of meliorative soil science.

solution, etc.) and it changes under their influence. For example, the water capacity is all the lower as the temperature of the soil and air is higher, but this does not apply to soils rich in humus in which, in that case, the water capacity goes up. Nitrates, chlorides and lime raise the water capacity, whereas potassium and sodium carbonates, on the contrary, lower it, etc. Water capacity changes through the soil profile according to its depth, the genetic horizon and the height of the water column.

This can easily be ascertained from the following simple experiment. Let us filter water or send it under pressure from underneath through a two-metre column of homogeneous soil, of sand for instance, uniformly distributed in a wide glass tube, until the whole of the free air from the interstices has been completely eliminated. After the supply of water has been stopped, during a certain time, water will be trickling down from the column. Once a state of equilibrium has been reached, water will no longer be seen to trickle down. From the data collected upon the determination of the water throughout the layer, its distribution in the column can be represented by the diagram shown in Fig. 22*a*.

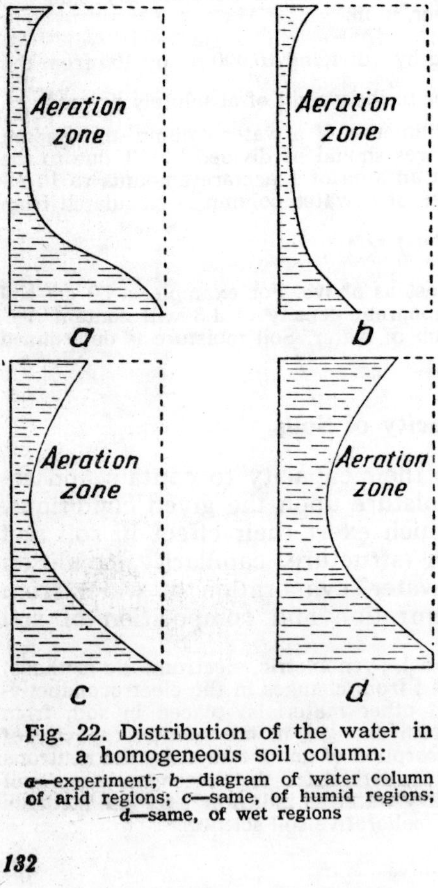

Fig. 22. Distribution of the water in a homogeneous soil column:

a—experiment; *b*—diagram of water column of arid regions; *c*—same, of humid regions; *d*—same, of wet regions

Three zones can be distinguished in a homogeneous ground: the first is the upper zone of uniform low moisture content; the second is the intermediate zone of upwards decreasing moisture; in the upper part of this zone, only the fine capillaries are filled with water, in the lower part, water fills the small and medium capillaries; the third is the lower zone of uniform high moisture, the small, medium and large capillaries of which are filled with water. At the bottom of the column, all the pores are filled with water.

It can be seen from the diagram (Fig. 22) that, as if enclosed within the soil or ground column, there is a water column which changes considerably depending on the height of homogeneous ground or soil situated above the water table (the depth at which lies the level of the ground water), as well as on the conditions of moisture supply from the surface. A certain average expression of the forms of this column will correspond to the complex of the natural factors of the given spot and native zone. Under native conditions, these columns will vary according to the season of the year and in connection with the variations in the weather conditions and changes in soil moisture. Under native conditions, after certain disturbances, their form is always restored in one way or another. The water column undergoes radical changes when the soil is under cultivation and all the more so when it is subjected to hydromelioration.

The behaviour of the water differs in the various sections (zones) down the soil column and the forms of the water in it differ too. In the two lower zones, soil becomes overmoistened under the marked effect of gravity, whereas in the upper zone, water hardly submits to gravity at all. If the lower part of the column were lengthened, both lower water zones would also be found to move downwards, remaining parallel to themselves as it were, in view of the fact that water will move downwards under the influence of gravity. But if we were to lengthen or shorten the column within the limits of the upper zone, the degree of moistening would remain practically unchanged; in other words, upon a change in the height of the water column, water does not trickle down. Down below, in the zone of capillary rise, water is held by capillary forces; in the upper zone, water bound in molecular fashion is held by the molecular attraction forces of the particles.

Experiments conducted by the Soviet scientist A. F. Lebedev have confirmed that in a column of ground remains only that maximum amount of water which it is capable of retaining, and any additional amount trickles down completely. This maximum amount corresponds to the maximum field water capacity of the soil in the given conditions. In columns of this kind, we get the formation of zones of varying moisture. At the bottom of the column, when it is supported, the water fills all the pores, which corresponds to full water capacity. At the top of the column, the pores are free from water, moisture is held by molecular forces on the surfaces of the soil particles. This moisture condition corresponds to maximum molecular water capacity. Any other, intermediate, section of a column of this kind exhibits what is known as field (minimum) water capacity, depending on the conditions of deposition, the texture and properties of the ground. In a stratified column, the water column acquires a more complex character.

The value of the minimum field water capacity along the column in homogeneous grounds is determined by the height of the column above the water table. In heterogeneous stratified grounds and soils, each respective layer of a certain mechanical and microaggregatory composition possesses its own definite minimum field water capacity. In columns of homogeneous ground, the overall water reserve is all the greater as it is higher, but increases of the reserve are not proportional to increases in height, in view of the fact that the water capacity of separate sections of the column goes down as they are situated further up from the water table. With an increase in the height of the column, the relative water reserve per unit of length goes down, nearing maximum molecular water capacity but never reaching it, because a certain amount of capillary perched or capillary supported water may, in addition, be present in soil. For that part of the column which is situated in the middle of the zone, between the maximum height of capillary rise from underneath and the depth reached by the capillary perched water from above, the moisture will be almost constant and there the reserve of moisture is directly proportional to an increase in length of the column (minimum water capacity). In heterogeneous, stratified columns, occurs an increase of the water capacity of separate strata at the planes of separation, and for that reason, the water reserve in a stratified column is higher compared with a homogeneous one.

The following types of water capacity of soil are therefore distinguished: maximum hygroscopic, maximum molecular, maximum capillary, minimum field, limit field and full field water capacities.

Maximum hygroscopic water capacity corresponds to maximum saturation of soil with hygroscopic water.

When liquid water, which is characterised by a loose arrangement of its molecules, becomes subjected to the influence of the field of forces of other molecules (adsorption forces), it is capable of becoming denser and of being firmly held by the soil particles. The maximum amount of moisture which soil can retain on account of these forces, i.e., the maximum amount of bound water, is called the maximum molecular (adsorption) water capacity. Quantitatively it corresponds to the amount of maximum molecular moisture upon which the soil particles are coated with a film of water of maximum thickness. This water is under the influence of molecular forces of attraction between the soil particles and the water molecules.

Maximum capillary water capacity of soil is its capacity to imbibe water to capillary saturation and to retain it in the capillary interstices. Quantitatively it changes within the interval from above maximum molecular to just below limit field water capacity. A distinction is made between capillary water capacity in the upper and in the lower parts of the capillary fringe above the water table and in the zone of capillary perched water in soil. Their amount will depend on the thickness of the layer of soil and its properties, on the depth at which lies the ground water, on the time which elapsed since the surface was moistened and other conditions.

The full water capacity of a soil is the content of moisture equal to the volume of the pores. This corresponds to a condition whereupon the soil is entirely saturated with water and water withdraws with difficulty or not at all. Under field conditions, this can come about when the ground water presents a head or upon excess of moisture on the surface. The full water capacity of a soil corresponds to that set of conditions when all the pores are filled with water and the capillary potential is reduced to zero. In such a case, with the elimination of the head, all the water, except film water, obeying gravity, is set into motion.

At full water capacity, in practice, the volume of water is usually somewhat lower than the full pore-space (in volumetric units). Full water capacity of soil creates conditions of minimum aeration. When the soil is oversaturated with water, the particles of peaty and peat soils may be somewhat moved apart by water.

Thus, full water capacity is realised when soil contains its largest possible amount of water.

When the outflow of water is unrestricted, the capacity of soil to retain water is governed by the existing native (field) conditions and the manifestation of the corresponding forces. The conditions and forces change and consequently also the value of field water capacity, ranging from minimum to maximum (highest potential) field water capacity. Minimum field water capacity is a variable quantity and fluctuates from maximum hygroscopic to capillary water capacity.

Limit (overall according to N. A. Kachinsky) *field water capacity* is the maximum degree of moistening of the soil upon free outflow of the excess of water. Such a condition becomes possible when the water table is near the surface and upon excess of moistening of the soil from above. Limit field water capacity is the capacity of soil, once it has been moistened and the excess of water has trickled down, to retain that maximum amount of water which, under the concrete conditions of deposition of the soil, remains for a prolonged period in a relatively or practically motionless state. This corresponds to the maximum amount of perched moisture which the soil is capable of holding in its capillary and partly in its noncapillary pores in a state of relative quiescence. According to S. I. Tyuremnov, the product of limit field water capacity in percentages multiplied by the apparent density is a relatively constant quantity. For loamy and clayey soils it approximates 30-32.

In capillary pores, water is held by the meniscus forces and, in certain noncapillary ducts, it may be partly retained mechanically, as if "scooped in" upon moistening from above. This is, in fact, what distinguishes field water capacity from capillary water capacity. Determination of the limit field water capacity from separate soil samples, particularly those with a disturbed structure, is useless. The energy with which water is retained in soil in a quantity corresponding to limit field water capacity is less than one atmosphere. Limit field water capacity determines the boundary differentiating water which is relatively motionless in a given soil horizon from water which is noticeably moving under the influence of gravity, i.e, it corresponds to the amount of water above which relatively motionless water passes into a different qualitative state, viz., gravitational water. The value of the limit field water capacity changes through the soil profile depending on: a) character of stratification of soil-forming rocks and soil genetic horizons, irregularities in mechanical and aggregatory composition; b) height above water table; c) state of cultivation of the soil; d) rate and technique of watering.

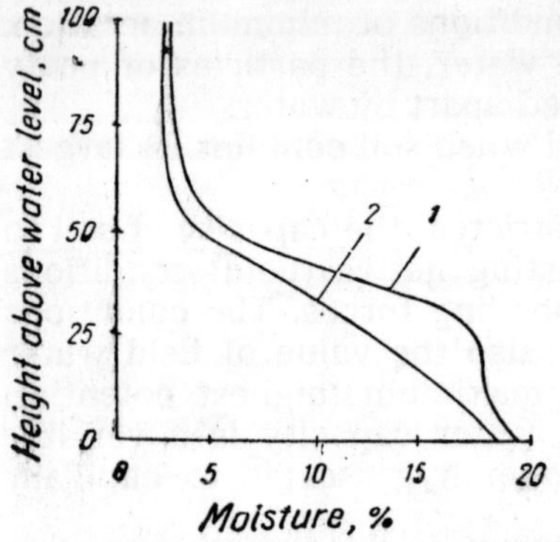

Fig. 23. Distribution of water in a column of sand upon trickling down and upon capillary rise (after A. F. Lebedev):

1—trickling down; 2—rise

Investigations point to a noticeable increase of the limit field water capacity in the upper soil horizons. When moisture penetrates into soil from above, the layer of capillary perched water formed is thicker than upon capillary rise from underneath. This is confirmed by a simple experiment: moistening of sand alternately from underneath and from above (Fig. 23). The presence of perched capillary water of this kind conditions a more productive utilisation of the water reaching the surface of the soil. The value of the field water capacity varies depending upon the agrotechnical treatment and methods of watering. In a deeply worked tame soil it is higher than in a similar but unworked or badly worked soil. Limit field water capacity may serve as the initial quantity for determining the state of moistening and the reserve of water in soil during any given period. Furthermore this quantity is relatively easily determined experimentally and changes from one set of cultivation conditions to another are not so marked and are easy to determine. The value of the limit field water capacity corresponds to the maximum rate of watering. But the actual rate of watering is lower than the limit field water capacity due to the fact that soil usually has, in addition, a certain reserve of moisture. This rate is expressed by the difference between the limit field water capacity and the water reserve present in soil at the given depth prior to watering, a difference which is referred to as the moisture deficit:

$$D = P - m,$$

where D—moisture deficit;

P—field limit water capacity;

m—reserve of moisture present in soil at the given time.

Furthermore, taking into account the fact that part of the water is lost through direct evaporation from soil and through transpiration, these losses are added to the amount of the watering rate. But it is difficult to get an accurate estimation of these losses. In summer, losses of this kind sometimes reach 50-100 m^3/ha. The rate of watering is calculated so as not to allow water to percolate below a set, convenient depth. A rational rate of watering should moisten that layer of soil which is reckoned to contain the roots (80-100 cm).

The value of the rate of watering may be expressed in a simplified form by the following equation:

$$M = P - m,$$

where M—rate of watering, in m^3;

P—value of field limit water capacity, in m^3;

m—water reserve prior to watering in the given layer, in m^3.

The rate of watering in its general form is, according to A. N. Kostyakov, expressed by the formula:

$$m = HA(\beta_{max} - \beta_0),$$

where m—rate of watering, in m^3/ha;

H—estimated depth of moistening, in m;

A—pore-space, in percentages;

β_{max} —maximum moistening of the H layer in percentages of pore-space;

β_0 —moisture content of H layer in percentages of pore-space, prior to watering.

A. N. Kostyakov rightly gives to field limit water capacity the name of potential water capacity, which any given soil may possess. In order to free soil from salts, rates of watering are used which are referred to as leaching rates. This is made up of field limit water capacity less the reserve of water present in soil plus that additional amount of water which is necessitated by the actual circumstances and conditions of leaching. From the difference between the initial reserve of water in soil and the limit field water capacity, one can determine the amount of surplus moisture. From the difference between the overall pore-space and the limit field water capacity, expressed in percentages of the soil volume, one determines the aeration upon irrigation. The aeration of a soil at a given moment is equal to the volume of the pores which remain free from water. It changes in connection with the fluctuation of soil moisture. Optimum aeration of soil is an indispensable condition for plant development. It constitutes 20-40% of the soil's pore-space and upon an aeration inferior to 6%, crops do not grow.

According to L. P. Rozov, conditions are favourable when the limit field water capacity for an average mechanical composition constitutes 65-70% of the overall pore-space; at 80-85%, conditions are satisfactory and at 90%, they are unsatisfactory.

The limit field water capacity is determined directly in the field in the vicinity of a typical soil section by flooding with water a well determined area of soil bounded by a special frame or by a bank of earth. The water capacity of soils with surplus moisture may be determined from monolithic samples, once gravitational water has trickled through. The water capacity of a peaty soil can be determined, after the water has been pumped out from the ditches bounding the area, from the moisture of the drained monolith or lysimeter. A description of the methods for determining water capacity will be found in the practical course of meliorative soil science.

Chapter VIII

MOVEMENT OF WATER IN SOIL

The movement of water in soil has been studied by many Russian scientists: N. M. Sibirtsev, P. A. Kostychev, A. A. Izmailsky, G. N. Vysotsky, A. F. Lebedev, V. R. Williams, L. P. Rozov, S. V. Astapov, F. E. Kolyasev and others. At the present time, the problems relating to water regime are being studied by N. A. Kachinsky, A. A. Rode, S. I. Dolgov and others.

The movement of water in soil depends upon the degree of its moistening and the manifestation of the heterogeneous forces which condition the following mechanisms of the movement of water: a) gaseous, b) sorptional, c) capillary, d) hydrostatic, e) gravitational, f) diffusion and g) osmotic. A sine qua non condition for the movement of water in soil is a difference between the forces, or gradient, to which it is subjected (gravity, meniscus forces, sorptional, osmotic). All the forces act simultaneously but depending upon the moisture content of the soil, one or another category of forces becomes predominant.

A distinction is made between readily mobile moisture, when the moisture of soil lies between full and limit field water capacity, medium mobile moisture, i.e., from limit field water capacity to film-meniscus (angular), trapped, and hardly mobile (angular and lower). When it is not above maximum hygroscopic, moisture moves only in a gaseous condition.

A part is played in the movement of water by the osmotic pressure of solutions, or otherwise the capillary-osmotic filtration in salined soil, which takes place under the influence of the concentration gradients of dissolved substances. Also of great importance are the thermoosmotic and thermocapillary movements of water into which water of low mobility is drawn. In winter, thermocapillary flows may move to the soil's surface large amounts of dropping-liquid water, resulting even in the formation within soil of ice lenses.

The movement of dropping-liquid water in soil is slowed down by friction, which increases as the moisture goes down.

As for the penetration of water into a dry soil, it is hampered by hysteresis, a phenomenon which has been only insufficiently studied so far. Apparently it consists in that the moistening of the surface of the soil particles is, at first, hampered by a layer of firmly adsorbed air or of repellent (hydrophobic) substance covering the particles.

An important part in the retention of moisture by soil is played by sorptional forces, which create pellicular plugs of bound water in the narrowest parts of the pores and interstices between the soil particles.

Continuous dropping-liquid water present in soil is capable of transmitting hydrostatic pressure, which stops as soon as the continuity of the liquid body is interrupted and the pores freed from water become filled with air. But the manifestation of hydrostatic pressure, as well as its interruption, may bear a local character, i.e., may be met with at some isolated part of the territory and even in limited areas of it (potuskuls) and in separate horizons of the soil. In soils of fairly heavy mechanical composition, hydrostatic pressure is not manifested, in spite of the continuity of the body of water. The speeds, direction and character of the distribution of water in soil may be determined using radiometric apparatus, by following the movement of the front of water marked with some isotope.

Movement of Gaseous Moisture

Gaseous moisture is present in soil no matter what amount of moisture it may contain and occupies all the pores free from dropping-liquid water. The movement of this moisture (distillation of water vapour) is the most widespread. It can be active and passive. Active movement of gaseous moisture is connected with diffusion, passive movement with the movement of air. In spite of the fact that the amount of gaseous moisture at any given moment does not exceed 0.001% of the soil's weight, this moisture plays a fairly important role in the distribution of water in the soil layer. All the forms of water in soil including physically bound ones take part in the process of the movement of gaseous moisture. When water vapour escapes from the soil into the atmosphere, newer and newer portions of the soil's moisture pass into a gaseous condition. The temperature remaining unchanged, gaseous moisture moves from places which are more saturated with water vapour of higher elasticity to places which are less saturated with water vapour of lesser elasticity. Changes in the pressure of the vapour are directly proportional to temperature changes. Consequently, moisture moves from soil areas with a high temperature to areas with a lower temperature and not necessarily from a moist layer to a drier one. In all soils with a moisture higher than maximum hygroscopic, the relative moisture of soil air equals 100% and, consequently, the elasticity of the water vapour is governed only by temperature. When the soil's moisture is below maximum hygroscopic, the relative moisture of the air is always below 100% and is controlled by the soil's initial moisture and temperature. In that case, the relative moisture of air, the temperature remaining unchanged, decreases with a decrease of the soil's moisture, and the initial moisture of the soil remaining unchanged, it goes up with a rise in temperature. Moist soil air moves as a result of contraction or dilation upon temperature fluctuations. It is blown away from the upper layers of soil as a result of the mechanical action of wind, or is displaced by water upon changes in atmospheric pressure, etc.

Gaseous water in soil may condense, i.e., pass into the form of dropping-liquid water. Condensation may be brought about by molecular forces of attraction on the surface of the soil particles. This type of condensation is referred to as molecular condensation. Its limit is maximum hygroscopic moisture and practically it does not go beyond that. In quite fine pores (<0.0002 mm) formed by silty particles, we get what is called capillary condensation, which occurs under the effect of meniscus forces and is connected with changes in the elasticity of the vapour. This condensation is limited by the volume of the fine capillary pores of the soil. The elasticity of the vapour in locked soil pores depends on the temperature of the medium and on the curvature of the surface of the water at the separation boundary with the soil air. Above a meniscus with a concave surface, the elasticity of the vapour is less than above a meniscus with a convex surface. The absolute elasticity of the vapour is all the lower as the concavity of the meniscus is more pronounced. The elasticity gradient of the vapour conditions the distillation of water from concave, flattened out, as well as less convex surfaces, to more convex surfaces, until complete exhaustion of the moisture which is being distilled or until the curvatures of the menisci are the same.

Gaseous water which moves in soil from warmer to cooler zones is subjected to thermal condensation. Given the proper conditions, thermal condensation goes on without limitation but the capacity of soil to retain dropping-liquid water is not unbounded, being limited by the amount of moisture equal to the limit field water capacity. If the soil's moisture reaches that limit and thermal condensation continues, then the water thus formed inevitably trickles through or, upon low temperatures, is transformed into ice. Flowing out of the water may bring about leaching out (podzolisation, solothisation) or leaching of salined soils. Conditions arise in nature when the elasticity of the vapour of the atmosphere is higher than that of the vapour in the upper layers of the soil, whereupon the soil is enriched with water forming upon the condensation of water vapour from atmospheric air.

The daily and seasonal pressures of the vapour present in soil are constantly changing. At night, in summer, a distillation of vapour from the lower, warmer soil horizons to the upper ones, which are cooler, may occur. Here, on the surface of the structural aggregates of soil particles, it condenses, passing into the condition of dropping-liquid water. This phenomenon can be termed underground dew. In the daytime, on the contrary, moving downwards from above, the water vapour may condense on the surface of the soil particles deeper down in the soil. In wintertime, water in the shape of vapour moves from the lower horizons into the upper ones and this brings about a substantial increase of the amount of moisture at the beginning of the vegetative period in the roots horizon. Translocation of water vapour in soil also takes place as a result of thermodiffusion, which sets up temperature gradients. Thermal gradients cause the light humid soil air to move, bringing it nearer to the walls of the pores, where the water vapour condenses.

The overall annual amount of water obtained through condensation may be considerable, reaching 60-100 mm and more (Odessa). The process of condensation of atmospheric water vapour on the surface and within the soil (usually down to a depth of 0.8-2.0 m) has a practical significance. It may substantially replenish the reserves of useful water and influence the soil's water balance, particularly in arid regions, where condensation at night may reach 4-8 mm, i.e., an amount of water sufficient to allow the growth of plants of dry habitats (xerophytes). The condensation of water vapour plays an important role in the local and general moisture cycles on the earth.

As a result of the convection and diffusion of the vapour of the soil air and the inflow of dropping-liquid water, evaporation takes place, which affects deeper and deeper layers of the soil. The annual loss of water from the soil through evaporation into the atmosphere may, for southern districts, be several times higher than the amount of atmospheric precipitations. In northern districts, this loss is considerably smaller. Evaporation is a zonal phenomenon since it is functionally dependent on the temperature of the given place and the amount of atmospheric precipitations. The annual amount of moisture evaporating from the surface of the earth in dry and hot zones depends on the amount of annual precipitations; in cold and humid zones it depends on the value of the annual radiation balance of the soil's surface, and in temperate and moderately humid zones, on the annual precipitations as well as the radiation balance of the soil's surface. The actual evaporation of water from the soil may sharply differ from the potential evaporation upon an unlimited inflow of water. In a waterless desert, for example, the actual evaporation is quite insignificant, due to the absence of water, whereas potential evaporation may reach quite large proportions. In forest zones, there is hardly any difference between actual and potential evaporation.

The approximate correlation between precipitations and potential evaporation in the various climatic zones may be expressed by the values given in Table 19.

Table 19

Approximate Correlation Between Precipitations and Potential Evaporation in Various Climatic Zones (after P. S. Makeyev)

Soil	Precipitations, mm	Potential evaporation, mm	Ratio of precipitations to potential evaporation
1. Tundra			
Maliye Karmanuly (Novaya Zemlya)	261	153	1.72
Salekhard (estuary of river Ob) . .	287	229	1.25
2. Taiga			
Surgut-on-Ob (Western Siberia) . .	500	351	1.42
Kirov (European U.S.S.R.)	587	459	1.28
3. Broad-leaved forests			
Vladivostok (Eastern Asia)	537	459	1.17
4. Steppes of temperate latitudes			
Melitopol (European U.S.S.R.) . .	398	856	0.46
Kokchetav (Kazakhstan)	258	670	0.39
5. Deserts of temperate latitudes			
Nukus (Central Asia)	79	1,407	0.06
6. Subtropical deserts			
In-Sala (Sahara)	28	3,464	0.01
7. Savannahs			
Bombay (India)	1,880	1,233	1.54
8. Humid equatorial forests			
Singapore (Malay peninsula) . . .	2,356	1,014	2.32
Para (Belem, estuary of the Amazon)	2,277	630	3.60

These correlations of annual precipitations are very approximate, in view of the fact that the most important factor is not the annual amount of precipitations, but the seasonal correlation between temperature and moisture during the vegetative period. Of great importance are the seasonal correlations between these values and the conditions for maintaining moisture in soil. Various areas of soils of one and the same zone from the physico-geophysical standpoint (deposition and evaporation of water) may range from excessively moist to extremely dry.

Soils in which the ground water lies close to the surface lose far more water through evaporation than those whose ground water lies deep down. But gaseous water circulates throughout the whole of the soil layer irrespective of its thickness and the depth of the water table.

The heavier the mechanical composition of soil, the more pronounced the evaporation. Evaporation from finer or more compact (structureless) soils is also more pronounced than from structural soils. This is due to the fact that in a heavy and compact structureless soil, as the water evaporates, there is an uninterrupted flow of more water up the capillaries from down below.

The evaporation from different soils saturated with water is the same, but upon drying out, the rate of evaporation differs. Sandy (warm) soils dry out quicker. Swampy, humous, clayey (cold) soils dry out slower. The evaporation from humid dark-chestnut soils is greater than from light coloured ones. As the concentration of salts in soil goes up, the evaporation goes down. The evaporation from a soil depends upon the shape and area of the surface subjected to evaporation. Evaporation from a soil with a wavy, irregular surface is all the greater as the surface of evaporation is larger. The evaporation from the irregular surface of a soil saturated with water may be greater than that from an open surface of water. Evaporation is affected by weather conditions. Evaporation is all the greater as the temperature and wind velocity are higher. Furthermore, evaporation from soils of southern slopes, all other conditions being equal, is greater than evaporation from northern slopes. The soils of eastern and western slopes occupy an intermediate position. The more pronounced the incline of a slope, the greater the evaporation.

An important role in the reduction of relative evaporation from soil is played by the live and dead plant covers (fallen leaves, forest litter) which protect the soil from losses of water. A live plant cover lowers the rate of evaporation of water from the immediate surface of the soil in comparison with a surface devoid of vegetation. But the evaporation from a soil possessing a plant cover, including evaporation through the plants (transpiration), is substantially higher than the loss of water through evaporation of a soil ploughed up and left fallow. Covering the surface with mulch paper (special paper or cardboard) or loose material (muck, straw, grass, peat) reduces evaporation and leads to moistening of the soil and improvement of its thermal regime. A similar effect is obtained through surface hoeing of the soil. The loosened layer of soil, which itself dries up, protects the lower lying horizons from evaporation, due to the breaking up by hoeing of the capillary interstices of the soil and the formation of noncapillary pores along which no capillary rise of water can take place. The same thing occurs upon structurisation of the soil. But upon strong predominance of noncapillary porosity as well as upon excess capillarity, water evaporates in an equally unproductive fashion. Minimum evaporation occurs when the ratio of capillary interstices to noncapillary ones is about one and a half to one. In the absence of irrigation, this ratio is somewhat higher than 1.5 and where irrigation is practised, it is equal to one.

In arid regions, plants suffer not so much from a shortage of atmospheric precipitations as from the unproductive loss of moisture from the soil's surface through evaporation instead of this moisture passing into their water-conducting system. A soil possessing a plant cover loses water through transpiration (evaporation from leaf surface). The loss of water through transpiration depends on the transpiration coefficient, i.e., the ratio of the weight of the water absorbed by the plant, passing through it and evaporating into the atmosphere, to the weight of the dry matter of the plant.

The weight of water passing through a plant is 200-1,000 and more times higher than the unit of weight of dry matter, depending on the particularities of the plant itself as well as on the weather and soil conditions. The amount of water lost through transpiration is enormous. Lack of transpiration during the nonvegetative period or during a longer period (felling or forest fires, bare fallow) accounts for the marked accumulation of moisture in soil, which may bring about a rise of the water table and swamping.

Movement of Molecular Water

Soils of average loamy mechanical composition contain up to 15-20% by weight of molecular (film) water, held by molecular forces. The molecules and ions which form a surface layer on the soil particles are energetically unsaturated. They are endowed with certain residual forces of attraction. These forces attract dipole water molecules towards the surface of the particles, forming a kind of water film. The molecular attraction forces are far from saturated when maximum hygroscopic moisture has been reached and continue to attract molecules of moisture when gaseous water and dropping-liquid water come into contact with the soil particles. In the latter case, the thickness of the layer of water molecules goes up considerably. It is obvious that due to physical conditions, the molecular forces of different particles are not equally saturated, that is why a gradient of these forces arises in soil, which causes film water to move. The force which causes the movement corresponds to the difference of potential of the molecular forces. A certain role in the movement of molecular water is played by osmotic pressure and indirectly by meniscus forces. Film water may move upon the existence of a moisture gradient due to pressure in the films, in connection with a certain degree of unsaturation of the molecular forces. Maximum saturation of these forces occurs when the soil particles are in contact with capillary (free) water, in the first place with angular (abutment) water. In that case, a certain additional amount of film water joins the molecular water, increasing the thickness of the films until they are in equilibrium with the abutting water "cuffs". The movement of film water begins from the formation of a continuous multimolecular layer of water around the soil particles and continues until maximum molecular water capacity is reached.

The mechanism of the transfer of film moisture in a simplified form may be explained as follows: in view of the existence of a molecular gradient and different pressures, the soil particles, which are coated with a film of water, cannot fail to exchange moisture. One part of the film slides, so to speak, over another until a state of equilibrium is reached, i.e., the film water is transferred towards the predominating force, until a state of equilibrium is reached between the attractions. Incidentally the difference in thickness of the "films" is not the decisive factor in determining the direction followed by the molecular moisture in motion.

To be strict, the movement of film water does not depend upon gravity. The velocity with which this water moves is expressed in so many millimetres per 24 hours. This movement is easier to observe in stratified soil-forming rocks. It goes on until the maximum molecular water capacity values in strata of different mechanical compositions become equivalent.

Molecular (film) water in the adsorbed condition possesses a number of particular properties which free water does not possess. It is of low mobility, most of it is unavailable to plants, it is a poor solvent, etc. However, cases of salinisation of soils where the water table is situated too deep for possible capillary rise point to the existence of a marked mobility of film water and the translocation with it of soluble salts. The latter may, however, take place through diffusion and osmosis across the "membranes" separating the soil pores, as well as through the crystallisation of salts in the form of a progressive growth of tiny needle-shaped crystals, etc.

The peripheral part of maximum film water at the limit of exhaustion of molecular forces is endowed with normal mobility but does not flow below the menisci of the cuffs of abutting (angular) water, due to the fact that the manifestation of the molecular forces still prevails over the meniscus forces. The properties of peripheral film water differ from those of intrafilm water. It possesses normal fluidity and according to Pascal's law, the pressure exerted on its surface by external forces is transmitted uniformly in all directions.

The source of the replenishment of film-molecular water is capillary water.

Capillary Movement of Water

The movement of water within the soil pores, which is due to the pressure of meniscus forces is usually referred to as capillary movement. But this movement has its particularities and differs from the movement of water along capillary tubes. The main laws governing capillary movement of water can, in practice, only roughly be applied to the understanding of the mechanism of the movement of water in the soil pores, without equating them with capillary tubes of a different diameter. This movement is a complex process which depends upon the moisture content of the soil, the character of the porosity, the moistening, the temperature, the temperature gradient, hysteresis, trapped air, the character of the water supply, sorption, etc. The leading role belongs to the specific surface of the soil particles and the surface tension of the water.

Due to surface tension in pores and in corners, at the points of junction of soil particles and aggregates, menisci are formed of various sizes and curvatures. These depend, in the main, on the degree of moistening and the physico-chemical conditions. The curvature of the menisci is dependent upon the direction and range of action of the meniscus forces. In isolated cases, the menisci may be due to the capillary movement of water. In the case of an overall capillary movement in a closed front, menisci may be lacking.

In general outline, the manifestation of meniscus forces comes to this: the uppermost layer of water of a meniscus at the limit of separation with soil air differs from the rest of the water in that here, as a result of reciprocal attraction between the water molecules, arises what is called a surface tension which exerts its influence on the whole of the rest of the water, tied in some or other way with the limiting layer. The water molecules on the surface, as distinct from the rest of the molecules within the mass of water, are subjected to unilateral attraction (pull) inwards of the liquid, on the part of the neighbouring molecules (Fig. 24). This explains the tendency that is exhibited by water to take the shape of drops with the smallest possible spherical surface. With an increase in size of the pores and of the area of the surface of separation between air and water, part of the molecules from the inner layers come to the surface, producing a certain amount of

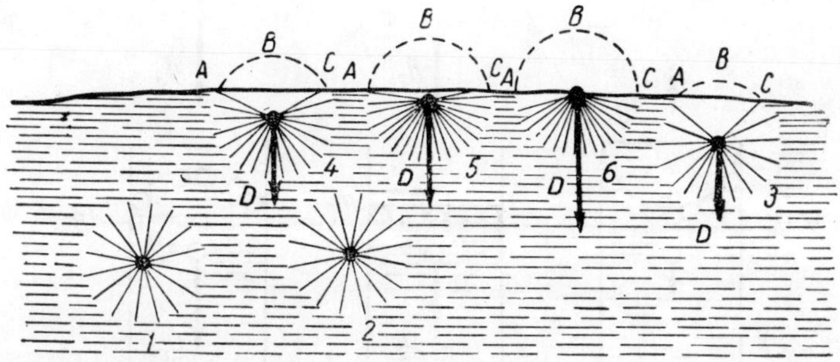

Fig. 24. The effect of molecular forces on a water molecule in the surface and inner layers:

1,2—molecules situated at a distance from the surface greater than the radius of molecular forces; *3, 4, 5, 6*—molecules situated at a distance from the surface smaller than the radius of molecular forces; *D*—value and direction of resultant force

work. This work is transformed into an additional potential energy of the surface molecules, called surface energy. Upon a further decrease of the surface, the additional (surplus) energy may execute work equal to that spent towards the increase of the surface, etc. Surface energy (F) is obtained by multiplying surface tension (σ) by the value of the surface (S):

$$F = \sigma S.$$

As a result of reciprocal attraction, the molecules of the surface layer exert a surface pressure on the mass of the water, equal to the surface tension. Surface tension is expressed in dynes per centimetre or in ergs per square centimetre. At 0° it is equal to 75.64 e/cm^2. With a rise in temperature, the interaction between the molecules decreases and with it also the surface tension, which falls down to 69.56 e/cm^2 at 40°. Over a flat surface of water, the surface tension (pressure) reaches 10,700 atmospheres. This pressure depends upon the form of the surface of the meniscus, which can be seen from the diagrams showing the correlation between the values of inner pressure and surface tension (Fig. 25). The molecules lying above plane AB attract the molecule upwards, whereas the molecules which lie below this plane draw this molecule inwards. In view of the fact that a greater number of molecules act from within (from below on the diagram) than from above, the resultant corresponding to the value of the inner pressure will everywhere be directed inwards (downwards on the diagram). Furthermore, it will be smallest (P_1) under a concave surface and greatest (P_3) under a convex surface. Under a flat surface, the surface pressure will be normal or greater than under a concave surface and smaller than under a convex surface.

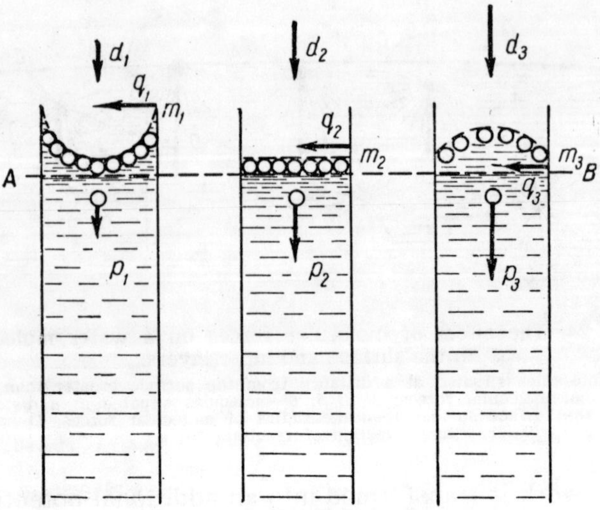

Fig. 25. Correlation between the values of inner pressure and surface tension:

$P_1<P_2<P_3$—inner pressure; $q_1>q_2>q_3$—surface tension; $d_1<d_2<d_3$—vapour pressure; $m_1>m_2>m_3$—number of surface molecules

In contradistinction to inner pressure, surface tension is two-dimensional—flat and directed in a plane tangent to the meniscus. It acts invariably like a spring or a constantly elastic film, not subject to linear shortening. Surface tension is conditioned by the existence in the surface molecules of surplus free energy. More water molecules find place on a concave surface than on a flat one. The smallest number of molecules is found on a convex surface. Hence, the surface tension will be highest over a concave surface and lowest over a convex one, and the inner pressure becomes correspondingly smaller under a concave meniscus surface and larger under a convex surface than in the case of a flat surface. The surface tension force is directed towards concavity, i.e., when the meniscus is concave, towards the space outside the liquid and when the surface is convex, inwards. As a result, there occurs, in the main, a capillary movement of water. This movement is tied with the capillary pressure equal in value to the difference of pressures at both sides of the curved surface. With a decrease in diameter of the menisci, the surface tension force goes up and, up to a certain limit, there is also an increase of the capillary movement. The pressure of the vapour over the concave surface of a water meniscus is smaller than over a flat surface and all the smaller over a convex one. That is why condensation of vapour begins earlier over a concave meniscus.

Any curved surface separating water from air being regarded as a film, exerts upon the water an additional pressure in comparison with that experienced by water under a flat surface. In the case of a convex surface, this additional pressure is positive, in the case of a concave surface, it is negative. As it tends to become flat, a convex film presses on the lower lying layers of water, whereas a concave film distends them. In the first case, the pressure tends to displace the mass of water from the limiting surface inwards, whereas in the second case, it forces it to move away from the concave surface on account of the air that is driven away. Both phenomena proceed until the free energy is exhausted and the surface of the meniscus gets completely flattened out, provided no other opposing forces intervene to counterbalance that energy.

Capillary movement is limited by the length of the body of water in the soil interstices which is equal to the distending force of the meniscus. Upon the formation of a convex meniscus, a movement arises which proceeds until it turns to a concave meniscus. Upon widening of pore interstices or upon merging with wider pores, the adjacent narrower portion is being intensely fed until the moisture from the widened portion becomes exhausted. The condition for an intensive capillary movement is the presence of gaps between the films in the pores and a local focus of water in soil.

Capillary movement of water in soil is to some or other degree influenced by the walls of the soil interstices and pores. Depending on the properties of the soil and on the manifestation of forces of mutual attraction between the water molecules and the substances on the walls of the pores, in most cases the water in the pores moistens the walls but not always.

The following dependency is set up: the smaller the diameter of the tube, the more pronounced the curvature of the menisci, the more its surface pressure differs from the normal and the higher the water will rise if the meniscus has a concave shape, and the lower it will fall if the meniscus has a convex shape. Water thoroughly moistens the soil particles, forming concave menisci whose pressure is lower than unity.

Moisture retained due to the difference of pressure between bottom and top menisci corresponds to the capillary perched moisture. The length of the little column of perched water cannot exceed the corresponding difference between the surface pressures of the menisci.

In soils of light mechanical composition, the menisci formed in the large pores retain the water arriving from above only in the places of contact of the soil particles and in corners (abutment water). The body of water surrounding the point of contact of the particles acquires the shape of a doubly concave lens or cuff. The abutment water may also be localised at the points of contact of micro- and macroaggregates.

Usually, such lens-shaped accumulations of water are not linked with one another, in contradistinction to capillary water proper, which is in the shape of numerous ramified, unbroken water threads filling in the fine pores or ducts.

Capillary water proper which fills up the soil pores may be found in soil in a relatively stable state of equilibrium. The difference of surface pressures corresponding to the difference of curvatures of antipodal menisci counterbalances gravity and ensures the suspended or relatively stable state of the water.

A change of temperature of the soil may cause capillary water to move towards the current of heat. This movement arises due to a change of the capillary potential, in connection with the fact that upon a temperature rise, the coefficient of surface tension falls and moisture which, prior to that, was in a relatively immobile or slightly mobile condition, begins to move appreciably in the soil interstices. Similar movements occur upon daily changes in the temperature and moisture content of the soil.

According to investigations conducted by S. I. Dolgov, in capillary water, the cations of the salts of the soil solution condition an appreciable manifestation of osmotic pressure, which correspondingly affects the mobility of capillary water.

The formation of capillary water is, in some or other way, linked with a source of dropping-liquid water. Usually it rises from the water table or infiltrates from the surface. The movement of capillary water is governed by the following laws:

1) The speed of the capillary movement is directly proportional to the diameter of the pores.

2) The path of the capillary movement or height of rise is inversely proportional to the diameter of the pores.

Capillary movement corresponds to three conditions:

1. Pendular, angular—when water occupies only the corners of the pores, forming water rings (cuffs) around the points of contact of the particles. This is capillary water interspaced by air, composed of separate drops bounded by menisci whose negative pressure may cause water to move within the boundaries of the drops (trapped water condition). The air in the pores communicates with the atmosphere. This water has little mobility and is of relatively low availability for plants, but it may move, turning into film water and vapour.

2. Funicular, when not all the pore-space is occupied by water, and part of it becomes filled with air in the form of bubbles isolated one from the other and from the atmosphere (trapped air condition). This water moves relatively easily within a short distance. It is fully available to plants.

3. Strictly capillary frontal movement, when water completely fills all the capillary pores. This is capillary readily mobile water. It is not devoid of the capacity to transmit hydrostatical pressure. This water lies immediately above the ground water or is formed in the surface layer upon abundant infiltration. It is readily available to plants.

Capillary supported and perched water. Of great practical significance are capillary supported water and capillary perched water. Capillary supported moisture is formed either above the surface of free water in soil or immediately above the waterproof layer of the soil-ground. In the first case, capillary supported water accumulates as a result of infiltration from above or the raising above the surface of free water upon the formation of what is termed capillary fringe. Moisture in the lower part of the capillary fringe comes near to full water capacity and capillary pressure approaches zero. On an inclined (irregular) surface of free water, in accordance with the formation of inclines, occurs a lateral displacement of supported capillary water due to the pressure gradients which are set up. Capillary supported water above a waterproof layer may arise as a result of infiltration insufficient for the formation of full capillary fringe above the surface of accumulated water or upon an intense loss from the reserve of free water through evaporation, outflow, etc. This supported capillary water represents a sort of undeveloped capillary fringe. Subsequently, upon the intensification of the infiltration or decrease of the outflow, it may develop into a capillary fringe or occupy its upper part over the local (soil) accumulation of water.

Capillary perched water is localised in separate portions of the soil and stays there, overcoming gravity with capillary forces. Capillary perched water is also upheld inside structural aggregates and in groups of them. This water is not linked in any way with the ground water. In sandy soils, perched capillary water consists of abutment water. It is formed only as a result of the moistening of initially dry sandy soil. The body of water arising thereupon is bounded by complex surfaces from above and from underneath. When the thickness of the soaked layer increases substantially and the pressure of the perched water exceeds the maximum possible value of the difference of surface pressures, a large part of the water trickles down rapidly in separate tongues. At the places of formation of percolation tongues, the pressure of the perched water quickly increases and intensifies the downward flow of water, bringing about a sort of wavy movement of this water under the influence of gravity. After the water has trickled through from a sandy soil, there remains in it the retained abutment water, which is held in an appreciably smaller amount than before trickling through took place. In a moist sandy soil, it is impossible to raise once again the moisture up to the value of the maximum amount of abutment water retained previously. Apparently, the significant factor for the lag in the progress and formation of perched capillary water in a dry sandy soil is the hysteresis* of wetting.

---

* Hysteresis—from Greek *hysterein,* to be behind, to lag.

The phenomenon of hysteresis of wetting can be observed when it begins to rain and perched water drops are formed on a window pane. After the pane has been wetted by rain, the new drops spread as a fine film. Hysteresis of wetting plays an important role in the formation of a layer of capillary perched water in soil. Hysteresis is manifested only when a dry soil is moistened and, apparently, does not intervene when a moist soil is moistened further.

It has been experimentally established (by A. F. Lebedev) that when moisture penetrates from above, a thicker layer of capillary perched water is formed than in the case of capillary supply from underneath. Capillary perched water is formed: a) at the contact of soil horizons or grounds of different mechanical composition; it depends on the difference of capillary potentials; b) at the soil's surface or the inner bounding planes of the strata upon weak infiltration; it depends on the difference of capillary potentials in the dry and moist conditions; c) above capillary supported water upon resistance of compressed air.

Being continuous, perched moisture is capable of moving to the places of evaporation from soil of dropping-liquid water, within the limits of the whole of the wetted layer. When the continuity is interrupted, the movement of the water slows sharply down and almost ceases. In that case, the water breaks up into separate portions and remains in the shape of cuffs around the points of contact of the soil particles and partly occupies separate pores or isolated systems of pores. Capillary perched water also includes maximum film water.

According to A. A. Rode and S. I. Dolgov, sorption forces also take a part in the retention of perched water in soils of heavy mechanical composition, by creating "plugs" of bound water in the most narrow sections of the soil pores, which retain the free water in the widened parts. The lower surface of the body of water may, in that case, be made up not of menisci but of plugs of film (molecular) water. Capillary perched water migrating in soil in all directions, takes part in the runoff within soil, which usually exceeds the surface runoff, especially in soils of lighter mechanical composition.

The capacity to retain perched water increases with the taming of soils.

Suction capacity of soil. Soil unsaturated with water possesses the capacity to suck up through capillarity, moisture with which it comes into contact. This can easily be ascertained by placing in a relatively dry soil a porous vessel filled with water hermetically connected with a mercurial manometer. The instrument will reveal a negative pressure, caused by the suction by soil of water from the vessel. This phenomenon is due to the fact that the soil mass together with water form, as was established earlier, a coherent system. Water is absorbed by soil and is held by it with a certain force, representing an aggregate of capillary, adsorption, sorption, osmotic and other forces. The suction of water by soil is also influenced in varying degree by the temperature fluctuations of soil and water, changes in the composition and concentration of dissolved and suspended substances, hysteresis, the force of gravity, etc. All these forces are interdependent and interact with one another. The sum of these forces which act in soil, drawing and retaining moisture, is referred to as the soil-water tension or capillary potential. This force is all the greater as the soil is drier and it is almost inexistent in a soil saturated with water.

The soil-water tension (capillary potential), which corresponds to a certain value of the soil's free energy, is expressed in units of pressure in centimetres of a water column and is designated by the symbol F. Schofield proposed to introduce a measure termed pF, which is the decimal logarithm of this quantity (1 atm or 1033 cm corresponds to $pF = 3$; upon lowering of the moisture down to zero, the pF value tends towards its upper limit of 7, which corresponds to 10^4 atm). From the pF values one finds the equivalent quantitative values of the moisture of various soils and of their various degrees of wetness. Equivalent values of moisture correspond in energy to the soil-water tension.

To determine the moisture content of a soil which is not saturated with water, use is made of moisture-measuring instruments called tensimeters. The range of tension of soil moisture as measured by a tensimeter lies between 0 and 850 cm of a water column or approximately 0.85 atm.

The suction capacity of soil serves as the basis for subsoil irrigation, using porous earthenware pipes or artificial mole drains dug by special ploughs. The principle of the suction by soil of water from porous vessels was first worked out and applied in practice in 1924 by the Soviet scientist V. G. Kornevoi.

The water-raising capacity of soil. Soil possesses a water-raising capacity, i.e., the property to raise water up the capillaries. The speed and height of the capillary rise of water and solutions of salts depends upon the character of the soil and the regime of the ground water as well as upon the thermal condition of the atmosphere, the relative moisture of the air and other factors. The larger the soil particles, the faster the rise but the height of the rise is then smaller and, conversely, the finer the pores and the smaller the soil particles, the slower the upward movement of the water, but then, it rises higher. Packing of the soil gives rise to an increase of the height to which moisture rises, whereas loosening and structurisation of the soil halts the rise of capillary water.

When there is evaporation of water from the surface, a homogeneous finely porous soil may dry out to a considerable depth. An increase in the soil's moisture content causes the speed with which water rises to go up, due to the fact that, in that case, the soil is capable of absorbing more water. Capillary water rise is less pronounced in cool soils, due to the fact that at low temperatures, water becomes more viscous and consequently, the surface tension at the boundary separating water from air increases. This explains why capillary rise of water is depressed in autumn and winter and immediately above permanently frozen ground. Capillary rise is less pronounced in a dry soil than in a wet one. In a very dry soil, in a period of drought, the supply of water ceases practically completely.

Soluble salts diminish the rise of water, increasing the density of the rising water and lowering the surface tension. A special role is played by Na_2CO_3, which promotes a higher rise of the water, by saponifying fats.

The height of the capillary rise of water increases from sandy to loamy soils. But as the mechanical composition of clay becomes heavier, the height of the capillary rise is depressed. In the heavy clays of the illuvial horizons of soil, capillary rise practically stops. The structurisation of a soil may depress the water-raising capacity, in connection with the formation of large interaggregatory interstices. Grass sod, forest litter, dry peat are characterised by an insignificant water-raising capacity. The height of the capillary rise for different rocks may be expressed by the following approximate values (in centimetres)

sand	50-100	clay	300-400
sandy loam	100-150	loess	250-350
loam	150-300	peat	120-150

In sandy soils, the height of maximum capillary rise is reached in a matter of a few hours, whereas in loams and clays, water rises for several months and even years. The height of the rise for loess-like loams, in laboratory conditions, reached 3.5 m at the end of 5 years and 6 m in field conditions.

Gravitational Movement of Water

Gravitational movement of water takes place under the force of gravity. This movement involves water which is not retained in soil and grounds by molecular, meniscus, osmotic and other forces. Gravitational water moves in pores whose diameter exceeds 30 μ. In pores ranging in size from 3 to 30 μ, water is drawn off under a pressure of 0.1-1.0 atm and possesses normal mobility. In pores smaller than 3 μ water possesses low mobility and is not drawn off by plants.

Firmly bound water takes no part in the gravitational movement. A. A. Rode distinguishes between true and potential gravitational water. It is truly gravitational when there is no impermeable layer or other obstacles on the path of the downward percolating moisture and the direction of its movement is determined by the force of gravity. Capillary forces and friction force exert an influence on the rate of movement of this type of water. It is also called percolating water.

Impermeable strata situated on the path of moving gravitational water cause it to become supported. That part of the layer of supported water which is situated above the water table is called the capillary fringe or down flowing moisture, because it is capable of flowing along an incline, should the limit field water capacity be exceeded. Part of this layer may be represented by stagnant water. Gravitational water flows down along a column of soil when the ground water lies deep down, or moves when there is a difference in the levels of the water table. In particular, gravitational movement may take place along large noncapillary pores, cracks and voids in soil. Not infrequently, in the course of the movement, these passages are disturbed, due to the fact that large interstices collapse or become filled with suspensions translocated from above (phenomenon known as colmatage).

In the process of gravitational movement in dry soil, a substantial amount of water is lost in the wetting of the enormous surface area of the soil particles and in the formation in their vicinity of films as well as in the feeding of capillary perched moisture. When the water table lies deep down, gravitational movement is possible if, under the given conditions, it exceeds minimum field water capacity. In that case, additional water must travel a long distance in a vertical direction through the column of soil before the appearance of runoff.

In soil saturated to limit field water capacity, runoff occurs as a result of the transmission of hydrostatic pressure. In that case, it is not the water newly arrived which runs off, but that which was already present in soil and is now dislodged by the water coming from above.

In heavy clayey structureless soil even upon full saturation, nearly all of the water will be in the bound condition. The moisture in a soil of this kind, in spite of its continuity, is incapable of transmitting hydraulic pressure simultaneously. In such horizons of the soil column, we get the formation of supported gravitational moisture, which is of two kinds: a) water of the zone of saturation and b) water of the zone of capillary fringe (*CW*).

Two subzones may be distinguished in the zone of saturation: of periodic saturation and of constant saturation. Trapped air is present in soil in the zone of saturation, which may be gradually eliminated from the soil layer due to its solution in water and slow diffusion into the atmosphere. Supported water is usually directly connected with ground water.

The water associated in soil with the less water permeable rocks and the permanently frozen strata or the B soil genetical horizons, which flows out in section, is known as soil or top water.

Soil-ground saturated with water is capable of giving out varying amounts of water through translocation. The maximum yield of water is equal to the difference between full water capacity and minimum field water capacity. The value of the maximum amount of water given out will be smaller than the estimated value by the value of the volume of the pores occupied by trapped air.

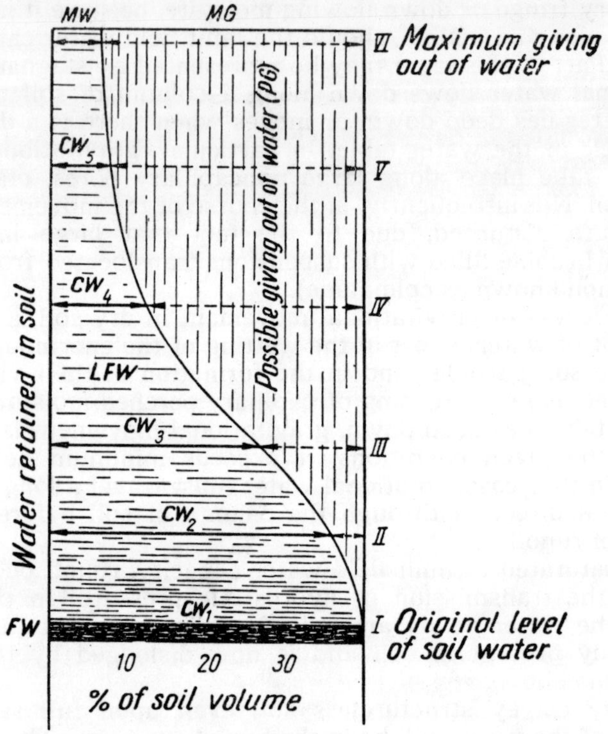

Fig. 26. Correlation between full water capacity, field water capacity and giving out of water (after A. A. Rode)

The water thus given out will be different in different zones of the capillary fringe, gradually increasing as we rise higher from the level of the ground water. This can be seen in the diagram (Fig. 26) where the amount of water given out (*MG*) is represented on the right side as the difference between full water capacity (*FW*), limit field water capacity (*LFW*) on the left bottom side and minimum water capacity (*MW*) at the top of the soil column.

Water permeability. Water permeability is the soil's capacity to let water pass through it. Water permeability depends upon the porosity of the soil, its structure, the mechanical composition, the temperature of the water and soil, the manifestation of the force of gravity, the presence of salts, the surface of wetting, capillary forces, the initial moisture content, etc. Water permeability depends not only upon the overall quantity and volume of the pores (pore-space) but on the way the pores are combined, on their sizes, as well as on the stability of the separate pores. Water permeability is all the greater as the pores are larger. Hence, sandy soils are more permeable than loamy and clayey soils, in spite of the fact that the overall pore-space of the former is substantially lower; this is because their pores are larger (the pore-space of sandy soils may approximate 30% of the soil's volume, whereas it is more than 50% in clayey soils). The water permeability of sand decreases substantially upon the admixture of a small quantity of clay or loam. The tiniest pores of fairly fine heavy clays are well nigh impermeable to dropping-liquid water. Soils rich in silty particles ($d < 0.001$ mm) are of low permeability; if the particles cannot be moved apart by water, the soil is almost impermeable. Large particles admixed in small amount to small particles do not increase permeability. Thus, the pebbly layer in the silty deposits of flood-plain sediments in the widened part of the valley of the river Vakhsh is impermeable to water and even acts as an impervious horizon. The overall water permeability of stratified soils of heterogeneous mechanical composition is determined by the strata least permeable to water.

The speed of the arrival and movement of water in soil depends on the number of large through soil pores, on the rapidity with which the soil particles become wet, and on the resistance offered by the trapped soil air. The influence of the latter on water permeability is quite significant. But this property of soil has not yet been properly studied. Water permeability is a complex phenomenon, it follows a definite sequence, passing through several stages: seepage, brought about by sorption and capillary forces; thorough wetting, under the influence of capillary forces; filtration, under the influence of gravitational forces. These stages are somewhat conventional and there is no sharp delimitation between them. The water permeability of soil undergoes substantial changes in connection with the progress of soil formation under native conditions and conditions of cultivation.

A distinction is made between water permeability and water conductivity. Water permeability refers to a certain physically measurable speed: length divided by time. Water conductivity is defined as the volume of water passing per unit of time through a column of soil with a cross section equal to one unit of surface. It is proportional to water permeability but is not equal to it, due to the fact that water conductivity is measured in terms of time.

Water does not penetrate into soil in a uniform fashion but in the form of tongues, in connection with the existence of granulometric and other differences. Water permeability is substantially lower in stratified soils. An increase in the number of the layers accompanied by a decrease of their thickness in peaty-sandy multilayered screens greatly lowers water permeability. It has been established that upon one and the same thickness of a layered screen, its water permeability is all the lower as the number of its strata is larger. N. A. Kachinsky explains the efficiency of screens of this kind by the formation of streaks containing trapped air, an increase of water friction and water viscosity on account of a greater amount of air disseminated within it and the formation in the screen of surfaces of separation water-air at the places of contact of layers of different composition. In the water permeability of a soil we find expressed the sum-total of its physical, chemical and biological properties. The presence in soil of swelling substances (colloids) worsens and lowers water permeability. Salts which coagulate colloids ($CaCl_2$, $CaSO_4$) improve water permeability. Salts which do not coagulate colloids adversely affect water permeability.

The speed with which water percolates and the amount which does so are directly proportional to the head and inversely proportional to the thickness of the layer traversed. The amount of that water is directly proportional to the area of the section and the square of the radius of the soil particles. When grains of fine sand are lodged in the interstices between the large grains, the amount of percolating water is the same as for fine sand alone. The amount of percolating water is intermediate between those for coarse and fine sand when the fine grains cannot lodge themselves in the interstices between the large grains of sand.

Present-day agricultural and meliorative practice is in great need of a differential study of the phenomena relating to water permeability in soils in its various phases: seepage, thorough wetting and filtration.

Seepage and thorough wetting. The term seepage of water into soil designates the filling up in a definite sequence of the free soil pores with moving water. In seepage, there occurs a vertical and lateral flow of free water in soil under the influence of the head gradient, the pull of gravity and meniscus forces. Menisci are formed at the boundary of wetting. The curvature of the menisci changes under the influence of the weight of the water. Meniscus forces as well as the head gradient are, in the main, directed downwards. The process of seepage of water into soil begins with the absorption of water, followed by saturation and movement in the form of film water, then of capillary and, finally, of gravitational flow. During absorption of water, the properties of the soil change appreciably. Sometimes there is a certain disruption of the structure, which is due to the fact that water which penetrates in the submicroscopic pores as a consequence of a series of conditions and of a change of the dielectric constant becomes nonpolar and in conjunction with the compressed air blows up the clods at the fine pores. This results in a change (disruption) of the structure and structural condition of the soil.

As soil absorbs water and becomes wet, it partly swells and its volume and properties change. A clayey soil may absorb up to 100% of its own volume of water and more and the figure is even higher for peaty soils. When dust-like sands get saturated with water, they turn to quicksand and become mobile like water due to the fact that the fine grains of sand become surrounded with a film of water. But in that case, sand does not actually swell, it is only diluted, turning to a flowing mass. When the rock grains are fine, the force of attraction of the water particles which surround the points of contact of such grains can overcome the weight of the grains and disunite neighbouring grains. The speed of seepage into a moist soil decreases in connection with the fact that the water which occupies part of the soil cavities hampers the penetration of new amounts of moisture. Water permeability is characterised by a certain speed of seepage. According to the speed of seepage all soils are divided into the following groups: a) highly permeable, b) medium permeable (characterised, according to L. P. Rozov, by a speed of seepage of approximately 0.05 to 0.15 m during the first hour of seepage), c) of low permeability.

The speed of seepage in time *t*, when watering by flooding, refers to the layer of water which seeps through in a unit of time during the period of seepage. The average speed of seepage in time *t* for watering in furrows refers to the layer of water which seeps through in the furrow of a given section in a unit of time of the period of seepage. When sprinkling, the average speed of seepage represents the layer of water which seeps through in a unit of time of sprinkling.

Seepage is more pronounced on virgin, deeply ploughed and on structural old arable soils and less pronounced in soils of low structure and ploughed shallow. On salined soils the speed of seepage is higher than on nonsalined ones but in solonetzic soils and solonetzes, it is fairly low. The speed of seepage goes down considerably in the packed horizons of the soil. The seepage of water into soil is a dynamic phenomenon. It changes appreciably in the process of soil formation, particularly under conditions of taming and melioration of soils.

The speed of seepage (v) of water into soil can be determined from the formula:

$$v = KI^{\alpha},$$

where

$$I = \frac{h + a}{a},$$

where h—layer of water on the soil's surface;

a—thickness of the layer of soil absorbing the water;

α — exponent, ranging from 1 for fine-grained to 0.5 for very coarse-grained soils;

I — head gradient, gradually falling from a greater value in the beginning to near zero in the end. At the same time the speed of absorption of the water also goes down, approaching a certain constant corresponding to the filtration coefficient of soils.

The speed of seepage gradually decreases with depth and time, approaching a certain constant quantity K which characterises the coefficient of filtration.

In the process of seepage, the soil becomes soaked (saturated) to some or other depth with water obtained from capillary water. Hence is also recognised the speed of saturation of the soil with water, which is determined at the same time as the speed of seepage.

The speed of saturation of the soil with water is determined under a constant head. The speed of saturation relative to the unit of area of the section is expressed by the formula:

$$v = \frac{Q}{St},$$

where Q—water used up in cm³;

S—area of the monolith's cross section, in cm²;

t—time, in seconds.

Academician A. N. Kostyakov has proposed to express the seepage into soil by the mean speed of seepage (v_m) from the beginning of the watering to a given time, to begin with, every 10 m, for example, then 20, 30, etc., to the end of the experiment:

$$v_m = \frac{v_1}{(1 - \alpha)\, t^{\alpha}} = \frac{v_0}{t^{\alpha}},$$

where v_1—speed of seepage at the end of the first unit of time, in sec, min or hours;

t—time of seepage, in sec, min, hours;

α—exponent characterising the dynamism of the seepage of water into the soil and the properties of the soil;

v_0—average speed of seepage of water into the soil during the first unit of time.

From the data obtained relative to seepage, one determines the overall amount of the water of seepage and the average speed of the seepage (v_m) for several moments of time (10-15). The results are plotted on a graph where on the axis of ordinates the plots v_m and on the axis of abscissas the time in minutes. On the same graph one also plots successively the growing overall quantity of water of seepage (cm³) (Fig. 27).

Filtration. The seepage of water into soil is accompanied by moistening or infiltration, which then passes into filtration. Infiltration or moistening designates the not yet fully established downward flow of water in soil. When the speed of arrival of the water on the surface (watering, rain, melting snow) exceeds the speed of infiltration, there occurs a surface runoff of the water which has not the time to percolate into the lower horizons of the soil.

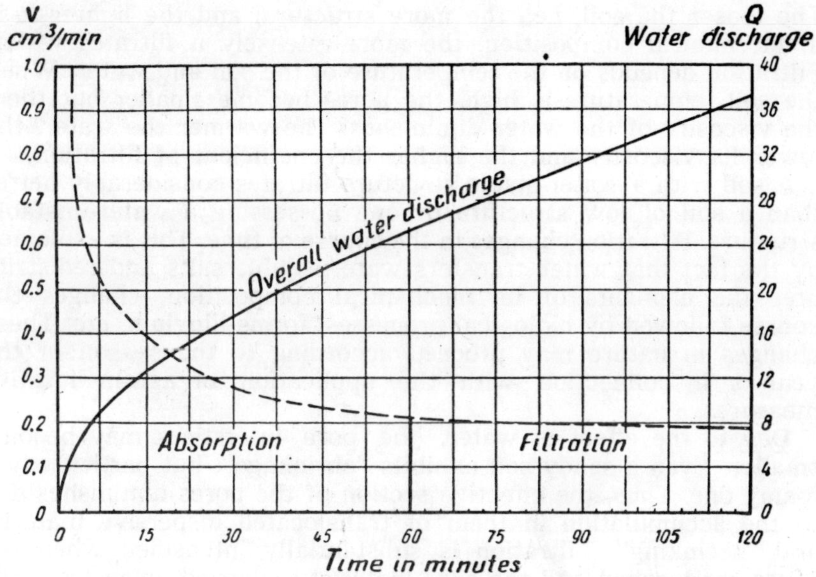

Fig. 27. Curve illustrating seepage and filtration

As the water permeability of the soil decreases with depth or when arises an obstacle in the shape of an impervious layer on the path of the downward movement of the water, the percolating water may form supported soil water. Soon after the wetting of the top layer of soil (2-3 cm), the surface tension may fall to zero, whereas at a substantial depth from the surface it may remain fairly high. This creates a substantial force, directed downwards and conditioning infiltration. After a certain time, this force may become insignificant and the infiltration will be maintained by the prevalent gravitational forces.

Filtration refers to the uninterrupted movement of water in an overmoistened soil under the influence of the head gradient and the pull of gravity. Filtration begins from the moment all the soil pores become filled with water and gravitational flow begins. If infiltration persists, the discharge of water may continually go up. Having reached its maximum, it becomes relatively constant. From that moment on, infiltration passes into filtration. The latter is not altogether stable either. Its dynamism depends on the change of the length and effective section of the path followed by the percolating water as well as on the properties of the latter. The filtration of water in soil is governed by factors such as the soil's mechanical, aggregatory and chemical composition, and physico-chemical and biological processes taking place in it, which are determined by the physico-geographical native conditions, the soil's cultivation condition, the degree and character of its moisture, etc. The looser the soil, i.e., the more structural and the lighter it is in mechanical composition, the more intensely it filtrates water. Filtration depends on the temperature of the soil and water. When the soil temperature is high, the pores become smaller but then, the viscosity of the water diminishes. The warmer the water, the lower its viscosity and the higher the coefficient of filtration.

A soil with a water-stable structure filtrates considerably better than a soil of low structure or one possessing a water-unstable structure. Filtration changes in the course of time. This is explained by the fact that water transfers water-soluble salts and redistributes the elements of its mechanical composition, changes the course followed by biological processes, forms illuvium, etc. These changes in nature may proceed according to the season of the year or in connection with the application of agromeliorative measures.

Due to the effect of water, the pore interstices may become smaller. Even a sandy soil exhibits "shrinkage", but particularly a loamy one. Thus, the effective section of the pores diminishes due to the accumulation in them of translocated dispersive material and "shrinking". Filtration is substantially intensified when the pores are leached and the silty material is carried away from the filtrating column of soil and ground. Soil acting as a filter does not allow the passage of particles whose diameter exceeds that of the interstices. With changes in the speed of filtration, even particles of a diameter as would freely pass through the interstices between the grains become deposited, in places, along the curvilinear movement of the water.

The process of filtration in stratified grounds proceeds in a more complex fashion than in homogeneous ones, but the amount of filtrating water is governed by the thickness of the layer or the sum of the strata with minimum water permeability. Filtration of soil may change in conformity with the course followed by the soil-forming process, especially in connection with the formation of an illuvial horizon. Packed soil horizons (illuvium) are characterised by lower water permeability. In case of flooding, these horizons may condition the stagnation of water and the development of reduction processes and gleisation in the upper part of the soil.

In view of the fact that filtration is fairly dynamic, the characteristic of the filtration capacity applies to the given moment only, reflecting the process of the change of the water permeability of soil. This characteristic of the filtration of soils has, therefore, a relative meaning.

The flow of water referred to as filtration obeys Darcy's law. It is conditioned by the difference of heads. The speed of the waterflow will be all the higher as is shorter the path of filtration which determines the resistance met by water upon flowing round the soil particles. The law applies integrally to sandy grounds and for laminar movements. The expenditure of filtrating water in soil is proportional to the head and inversely proportional to the length of the path $Q=KF\frac{H}{L}$; designating $\frac{H}{L}$ by I, we get:

$$Q = KFI,$$

where Q—discharge of water in cm^3 per unit time, through 1 cm^2 of soil cross section in 1 sec;

K—coefficient of filtration in cm/sec, which depends on the physical properties of soil and water [on porosity, effective (averaged) diameter, molecular water capacity, temperature of the water];

H—head, difference of levels at inflow end and outflow end;

L—flow length, thickness of the layer, in cm;

I—hydraulic or head gradient, incline;

F—cross-sectional area, in cm^2.

$$Q = vF = KFI \text{ cm}^3/\text{sec},$$

$$v = \frac{Q}{F},$$

where v—flow velocity.

$$\text{When } I = 1 \text{ and } F = 1\ Q = K \text{ cm}^3/\text{sec};\ v = K\frac{H}{L};\ \text{or}$$

$$v = KI \text{ cm/sec};\ K = \frac{v}{I};\ K = \frac{Q}{F};$$

$$\text{when } I = 1,\quad v = K.$$

Tho coefficient of filtration represents the flow velocity of water in active pores and cracks of the soil when the hydraulic gradient equals 1. The coefficient of the speed of filtration K_v is equal to the ratio of the coefficient of filtration K to the active pore-space (P_{act} in fractions of unity):

$$K_v = \frac{K}{P_{act}}.$$

The coefficient of filtration varies within a wide range. With time, the value of K usually goes down, less frequently goes up, for example upon dissolving of water soluble salts in the soil pores. The values of the coefficient of filtration (K) for various grounds are given in Table 20.

Table 20

Value of the Coefficient of Filtration for Various Soil-Grounds

Soil-ground	K, cm/sec
Pure sand	1.0-0.01
Clayey sand	0.01-0.005
Sandy loam	0.005-0.003
Loam	0.001-0.00005
Clay	0.0005-0.000005
Carbonate loess	0.0005-0.0001
Noncarbonate loess	0.00005-0.00001
Younger peat: a) sphagnum peat	0.002-0.0008
b) hypnum peat	0.006-0.002
Medium old peat: a) sphagnum peat	0.0002-0.0001
b) hypnum peat	0.0008-0.0002

The process of filtration is of great practical importance with regard to soil melioration, in particular for the determination of the leaching rates for salined soils. Methods for determining the coefficient of filtration are described in the instructions on field and laboratory investigations.

Soil-Ground Water

The water regime of soils and the regime of ground water develop in interdependence. The vertical soil column of water found in soils of homogeneous mechanical composition rests on the table of the main water-bearing horizon.

In stratified soil-forming rocks, the water column is correspondingly more complex and there may even be a formation of secondary (local) water-bearing horizons of a temporary or permanent character. These local water-bearing horizons arise either in connection with changes in the physical properties of the ground (formation of deep illuvium under solonetzic, solodised or podzolised soils), or in connection with disturbances in the everyday water regime of soils upon irrigation and the formation of a more compressed water column in the soil.

On the depth at which lies the ground water depend the moistening and distribution of moisture in the various soil horizons as well as the content and distribution of water-soluble salts and plant nutrient elements.

For agricultural purposes, the ground water must lie at a depth from the surface of the soil not exceeding a critical (tolerable) limit.

The critical depth at which the ground water lies is, as a rule, somewhat less than the height of maximum capillary rise of water (0.5-3.5 m and more, depending on the ground).

In steppe regions, in case of low mineralisation of the ground water, its optimum level is 1.5-2 m. With a level of this order, the soil does not dry out and the plants are constantly supplied with water without resorting to artificial irrigation. The depth at which lies the ground water also exerts an influence on the evaporation of moisture from the soil. The evaporation is all the more intense as the ground water lies closer to the surface (Table 21).

The influence of the ground water on evaporation is lessened if the depth at which it lies is substantially greater than the height of the capillary rise of water. In addition to raising the evaporation of water from the soil, a rise of the level of the ground water exerts a marked influence on the accumulation of water-soluble salts in soil. The depth at which lies the ground water in the south-eastern regions (2.5-3.5 m) is a critical depth, at which mineralisation proceeds intensively. At a lesser depth, the evaporation of water and the migration of water-soluble salts, especially from a depth of 2 m, are several times greater. With a lowering of the

Table 21

Changes in the Evaporating Capacity of Soil-Grounds in Connection with the Depth of the Water Table (after B. S. Kravkov)

Soil and its state of cultivation	Depth of the water table	Amount of evaporated water	
		in m³/ha	coefficient
Wasteland, strongly salined, devoid of vegetation	0.5 1.0 1.5	2,538 1,020 385	6.5 2.6 1.0
Wasteland, medium salined, devoid of vegetation	0.5 1.0 1.5	3,542 1,330 495	7.2 2.7 1.0
Wasteland, slightly salined, with native vegetation	0.5 1.0	20,265 12,942	1.59 1.0
Arable land, under lucerne	1.0 1.5	20,450 16,990	1.2 1.0
Same, under cotton	1.0 1.5	6,155 3,665	1.68 1.0

ground water below the critical depth (> 2.5-3.5 m) mineralisation is usually slowed down and becomes stabilised.

The movement and regime of the ground water depends on the water properties (water permeability, water-raising capacity, water capacity), on the composition and texture of the deposits in which the ground water is contained. The height of the water table depends on the underground relief, the impervious rocks, on the general and local influx of water. The relief of the water table changes constantly in connection with the condition of the water balance of soil and ground water.

From the contours of water table one determines the height at which lies the ground water at any given time. This height depends, together with the other sources of supply of the ground water, on the influx of the gravitational soil water trickling down along the column of soil. At the same time, the ground water, in turn, influences the water regime of soils, conditioning the inflow and rise of capillary water and with it of salts. If the water reaching the roots is sweet capillary water, it is wholesome for plants but if the water contains salts in amounts sufficient to depress the growth of plants, then it is harmful.

The composition and amount of the salts present in the soil solution differ markedly from their composition and amount in the ground water or from salty exudations on the soil surface. The level of the ground water fluctuates in accordance with changes in the weather, seasonal and cultivation conditions and corresponds to a certain average annual everyday level. In accordance with changes in the level of the ground water, the capillary fringe zone is displaced along the absolute and relative height of the column of water, periodically increasing or decreasing in thickness. The displacement of these levels depends first of all on the climatic and hydrogeological conditions. The drier the climate, the deeper the ground water and the lower the level of the water in rivers under native conditions. In native zones where evaporation from the surface of the soil is less than infiltration, there occur a secular progressive accumulation and rise of the level of ground water. In zones where the evaporation from the soil's surface is greater than the amount of annual atmospheric precipitations in a proportion of 5-6 to 1 and more, and with a secular widening of this proportion, the level of the ground water progressively falls.

The level of the ground water depends to a great extent on the geological conditions (structure and physico-chemical properties of the rocks) and the influx of water from the surface of intake, the places of infiltration, which are not infrequently situated away from the main contour of the ground water. In that case there is a movement of the ground water along the incline of the underground relief of the impervious layer. The direction and relative speed of the movement of the ground water is determined from the contours of water table map and directly in the field using special methods (chemical method, pumping method, etc.).

The speed with which the ground water moves is not high (0.5-1 m in 24 hours) but depending on temperature, sizes and number of the pores, underground incline, it may vary within a wide range.

The level of the ground water goes down upon lowering of the basis of erosion and, conversely, goes up upon its rise. The position of the basis of erosion determines, therefore, the water regime of soils and consequently the development of soil formation (salinisation and desalinisation, swamping or drainage of soils). The formation of carbonate and sulphate horizons is linked with the water regime. When soils are being moistened by the ground water, there is a formation of hydromorphous half boggy, peaty, peaty-gley and solonchakous soils. The level of the ground water may be raised or lowered in varying degrees as a consequence of the application of meliorative measures, such as irrigation or drainage of soils. The water and with it the air and nutrient regimes of soil are amenable to radical changes. The regime and level of the ground water are fairly often tied with the vegetation growing on the soil's surface, which transpires moisture obtained from the soil. Forests may promote the lowering of the level of the ground water but it is a well-known fact that afforestation in steppe and forest-steppe zones promotes the overall moistening and improvement of the local water regime. Among the plants which depress the level of the ground water we should mention lucerne.

Afforestation, the creation of large reservoirs, irrigation and drainage of large tracts of land, river control, the lessening of the surface runoff will radically change the water regime of soils and with it the water regime of the ground water. It is thus possible to regulate the regime of the ground water, which can go as far as the creation of underground water reservoirs. The ground water is a huge reservoir which can be drawn upon freely not only for water supply but also for irrigation purposes. A detailed examination of the problems relative to the ground water regime is the subject of a special branch of study, viz., hydrogeology.

Chapter IX

WATER REGIME AND WATER BALANCE OF SOILS

Water regime is the term used to designate the sequence and sum-total of all the phenomena connected with the entrance of water into soil, its condition in soil and its withdrawal from soil. The water regime is characterised by the seasonal water balance. The water balance is the quantitative expression of the dynamic condition of the reserve of water in soil in a definite period of time. The amount of water flowing into soil and its expenditure from soil during the period considered characterise the moisture cycle of soil. The sum-total of the quantitative changes undergone by soil moisture during that period is referred to as the moisture regime.

The water regime governs the water-air and thermal properties of the soil. Through the water regime one may influence the raising of the effective and potential fertility of soils. In order to ensure this, one should be able to determine at any moment the amount of water present in soil and available for crops. The amount of water present in the various genetical horizons of the soil and at the depth considered is determined from the formula:

$$Q = M - y$$

or

$$Q = \frac{W - 2H}{2},$$

where Q—available (useful) water, in m^3/ha;
M—reserve of water in soil, in m^3/ha;
y—wilting coefficient, in m^3/ha;
W—full water capacity, in m^3/ha;
H—maximum hygroscopic moisture, m^3/ha.

The maximum amount of water which a soil can contain does not exceed the overall pore-space (A m³/ha). The maximum possible amount of water available for plants in soil (Q m³/ha) can be expressed by the formula $Q = A - y$. Upon this amount of water, the content of air in soil will be at its minimum. The optimum available moisture in soil can be approximately expressed according to the formula:

$$Q_{opt} = \frac{60}{100} A - y$$

or

$$Q_{opt} = \frac{80}{100} P - y,$$

where A—pore-space;

P—limit field water capacity.

But $\frac{60}{100} A$ and $\frac{80}{100} P$ are only the approximate optimum quantities, varying within more than 10%.

These rates can only be considered approximate because they vary considerably in relation to one another in accordance with the changing properties of soil and plants as well as with the weather conditions. That is why we cannot limit ourselves to a mere determination of the initial properties of soil but we must predict and take into account the changes which will take place in it in the process of agromeliorative measures.

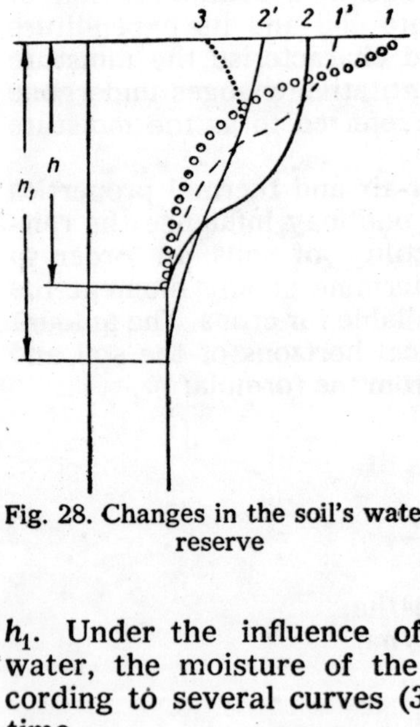

Fig. 28. Changes in the soil's water reserve

The amount and distribution of water in soil are closely tied with its arrival and translocation along the soil profile as a whole and in the separate genetical horizons.

The change with time of the reserve of water in soil following watering with different rates and amounts of entry of water are illustrated in a diagram by A. N. Kostyakov (Fig. 28). Upon a rapid supply of water to soil, the distribution of soil moisture passes from layer h on curve 1 immediately after watering to curve 2 after a certain time has elapsed. Curve 2^1 corresponds to limit field water capacity of the given soil in layer h_1. Under the influence of evaporation and spreading of the water, the moisture of the soil decreases distributing itself according to several curves (3 for example), depending on soil and time.

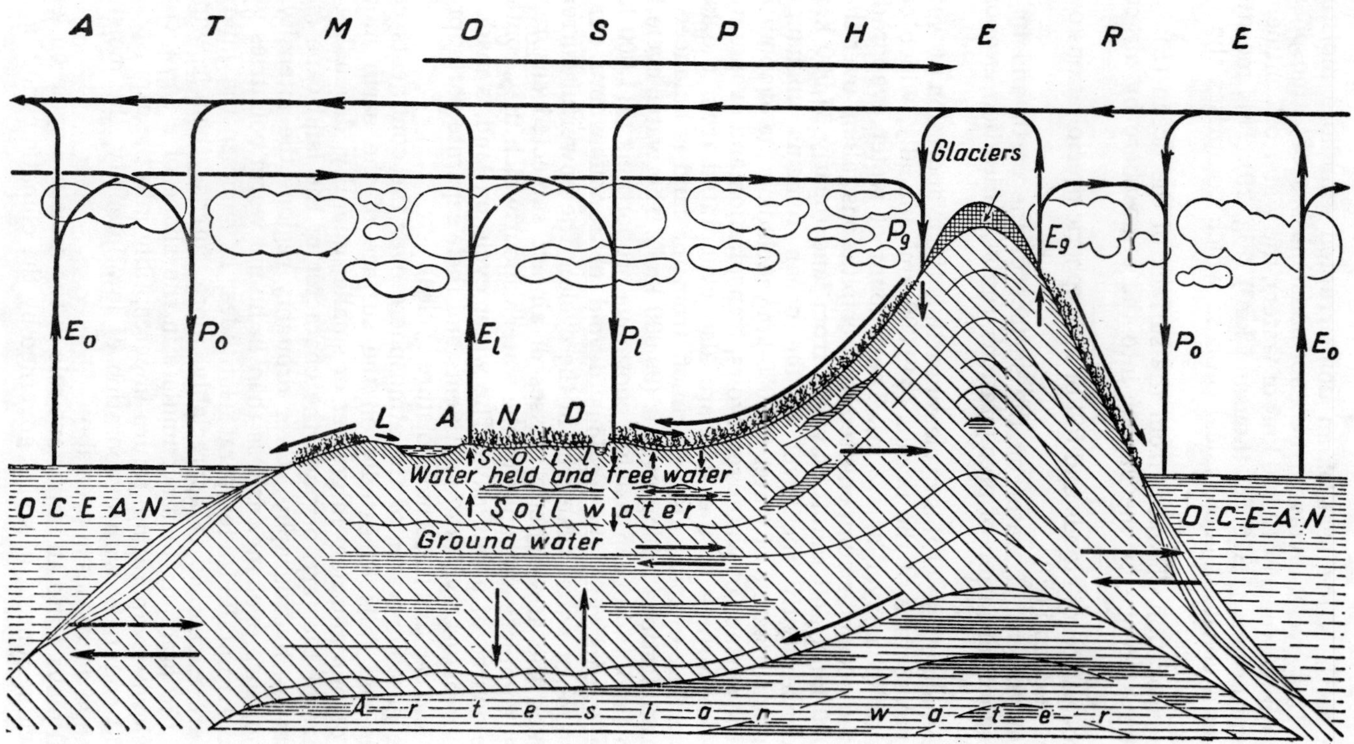

Fig. 29. Diagram illustrating the water cycle on earth

The earth is the seat of an uninterrupted grandiose moisture cycle. This cycle involves the enormous masses of atmospheric water vapour (nearly 12,300 km^3 of water), the water of continents, seas and oceans (1,370,323 thousand km^3) (Fig. 29). This moisture cycle is a constant geophysical process which comprises the following phases:

a) evaporation of water from the surface of the ocean (E);

b) transference of the vapour into the atmosphere by air currents;

c) formation of clouds and fall of atmospheric precipitations over oceans and land (P);

d) movement of water on the surface of the earth and in its depths (accumulation of precipitations, runoff, infiltration, evaporation).

The main phase of this cycle on earth is the evaporation of water from the surface of the oceans (449,000 km^3 annually, which corresponds to a layer of 1.2 m). The amount of water evaporating from the oceans is compensated by precipitations falling over them (412,000 km^3) and runoff of water from land (nearly 37,000 km^3), and consequently the level of the ocean remains unchanged. Enormous amounts of water (62,000 km^3, including the evaporation from the surface of water reservoirs, transpiration and evaporation from the surface of the soil) are also evaporated from the land surface. The evaporation and runoff from the land is balanced by the precipitations falling over it (99,000 km^3). The water cycle between land and ocean reaches enormous proportions (511,000 km^3 annually excluding the regions devoid of runoff where nearly 8,000 km^3 of water take part in the cycle). The powerful currents to which the atmospheric masses of air are subjected ensure an almost complete exchange of the water vapour which they contain, in the course of a few days. The water cycle on land is made up of the moisture cycles of continents and their separate parts: provinces and regions (internal moisture cycles).

The gaseous water of the atmosphere over the continents condenses and falls repeatedly on the surface of the earth in the form of rain and dew in greater or smaller quantity, depending on the size of the territory and the character of the landscape (soil, vegetation, hydrography). Over droughty regions the intensity of the moisture exchange is less than in humid warm countries with a heavy rainfall. In temperate latitudes, in summer the internal moisture exchange is appreciably more intense than in winter when the precipitations accumulate in the shape of a snow cover and, slowly evaporating, return imperceptibly into the atmosphere. In cold countries with evaporation of low intensity, the moisture exchange decreases appreciably.

Soil plays quite an important role in the moisture cycle on land. The water content of soil is governed by the climatic conditions of the zone (Fig. 30). The average value of the active moisture (available for plants) in a soil layer one metre thick during the spring-summer period increases as we move from the south-east to the north-west.

In north-western regions, soil is characterised by maximum moisture, fluctuating during the warm period of the year from 180 to 240 mm. Fluctuations in the average moisture diminish in a southward direction in connection with pronounced dryness and in a northward direction in connection with the steady moistening of the soil.

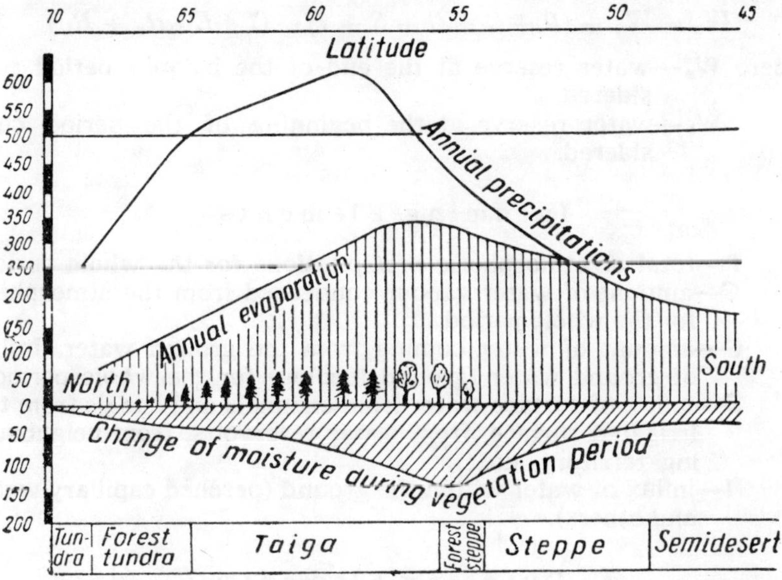

Fig. 30. Diagrammatic profile showing changes of climatic elements and soil moisture

The role of the soil in the external moisture cycle and internal moisture exchange is considerably enhanced by its cultivation, whereupon there is a marked increase of the moisture, water permeability and water capacity but a decrease of the surface runoff and useless evaporation. An improvement of fertility and yields is attended by a corresponding increase of the transpiration.

The various native and cultural changes in the condition of the surface of the earth on large areas (thawing and formation of ice, change of the coastline, formation or disappearance of water reservoirs, large-scale melioration, etc.) lead to corresponding hydro-climatic changes in the distribution of precipitations, evaporation and runoff. These changes in the moisture cycle proceed at a fairly slow pace in nature, as a consequence of the gradual changes of the native factors. More rapid and directed changes of the same kind in the moisture cycle are brought about by the productive activity of man.

Elements of Water Balance of Soils

The water balance of soils of separate districts and regions is made up of many variable quantities. It can be approximately expressed by the following equation (after A. A. Rode):

$$W_e = W_b + (P + C + G + S + I) - (T + E + R_s + R_i)$$

where W_e—water reserve at the end of the balance period considered;

W_b—water reserve at the beginning of the period considered.

Incoming Elements:

P—total of atmospheric precipitations for the whole period;

C—amount of water vapour condensed from the atmosphere for the whole period;

G—amount of water coming from the ground water. Influx of ground water (ground feeding) for the whole period;

S—influx of surface water from channels and canals from the irrigated surface, from water reservoirs, from neighbouring territory, etc.;

I—influx of water within the ground (perched capillary water and others).

Outgoing Elements:

T—transpiration for the whole period;

E—amount of water evaporated for the whole period;

R_s—value of surface runoff;

R_i—runoff within the soil (filtration, etc.).

All the items are best expressed in millimetres of a water column. In certain cases some items of the water balance may be of insignificant value and it is almost impossible to take them into account. Frequently, the amount of condensation is insignificant. The water balance may be determined for any period or for a year. Not infrequently the annual balance is in equilibrium, i.e., $W_b = W_e$, if the moistening or drying out of the soil does not progressively increase. In the opposite case, the water balance is positive or negative (exhibiting a deficiency of water).

After eliminating the small values, the equation will read:

$$P + G = T + E + R.$$

(inflow) (outflow)

In specific cases, in connection with weather conditions, the reserve of water in soil may be somewhat raised or lowered at the end of the yearly period in comparison with the previous reserve. For sod-podzolic soils, for example, a water balance in equilibrium is possible in the following form:

$$P_{450mm} + G_{50mm} = T_{200mm} + E_{80mm} + R_{190mm} + v_{30mm}$$

where v is a certain variable quantity which can be positive or negative, depending on the specific local physico-geographical conditions.

The water balance of soils and its quantitative expression are determined from the analysis and precise computation of all its elements. The influx of atmospheric water is calculated from the data provided by meteorological stations in millimetres of the water column of precipitations. Local conditions may exhibit marked deviations from the average amounts of atmospheric precipitations in one or the other direction. For a more accurate estimation it is therefore necessary to be in possession of the data of the direct recording of the amount of atmospheric precipitations and other meteorological data.

An attempt is also made to determine the amount of the water supplied upon the condensation of vapour, from direct observations in the field.

However, the methods for effecting these observations have not yet been perfected. The inflow of surface water and its runoff from the given area is determined either from a coefficient or from the runoff modulus.*

The inflow of water from the channels, canals, from the whole irrigated area, from water reservoirs and from neighbouring territories is arrived at from direct observations. The internal inflow in soil of water from the ground water and the lateral inflow from temporary local accumulations of water can be found from the moisture dynamics of the given district.

In order to determine the quantity of water getting into soil, all the named sources of the period considered are added up together with the initial quantity of water in the soil at the beginning of the balance period. The latter is also determined directly in the

*Runoff coefficient: $K = \frac{S}{P}$, where K—the runoff coefficient; S—the amount of water running off, in mm; P—the amount of precipitations, in mm. Runoff modulus: $r = \frac{Q}{F}$, where r—runoff modulus, quantity of water in l/sec running off from 1 km^2 of the basin surface; Q—runoff in l/sec from the whole surface; F—surface of the basin, in km^2.

field, before the experiment, on the basis of the water regime of the region and the given area. Together with the determination of the inflow it is necessary to find all the losses of water from the soil and to express them quantitatively. The amount of water lost through evaporation from the surface of the soil is calculated from the data supplied by the meteorological observations of the research stations or directly on the spot.

Soil is the factor which limits and regulates the utilisation of water by plants. Not infrequently there is a gap between the needs of the plant in water and the actual possibilities for satisfying them existing in soil (speed of water influx to the roots, water availability, etc.). If during the vegetative period, the necessary requirements of plants in water are met, then all the positive sides of the soil are fully realised and the yield goes up. If the consumption of water during the vegetative period falls below a certain critical minimum, it is an indication that the soil is drying out and this leads to a decrease in yield. The critical minimum of water supply fluctuates considerably from zone to zone. Optimum moisture is, on the whole, determined by the soil properties. Its fluctuation from culture to culture is insignificant. Wilting of plants may, however, be caused by an exceptionally low moisture content of the air, the soil's moisture content being relatively adequate, when the suction force of the roots is unable to cope with the increased requirements of plants in water due to excessive transpiration. In the latter case, the speed with which the water is consumed sharply exceeds its supply. The plants are "seized" as it were and they die. This is brought about by arid winds in periods of aerial drought.

According to their degree of availability to plants, the following types of soil moisture are distinguished:

a) unavailable, which has a value of one and a half that of maximum hygroscopic or molecular water capacity or less—this is firmly bound moisture, not mobile in the liquid condition (dead water reserve in soil);

b) moisture of low availability, intermediate between moisture of total wilting WM_3 and moisture which brings about symptoms of wilting WM_1. This is moisture of low mobility. It corresponds to maximum content in soil of abutment (angular) water;

c) medium available, intermediate between WM_1 and CW (capillary water capacity). The plant is constantly supplied with water. High productivity;

d) readily available, intermediate between capillary and full water capacity. With moisture higher than limit field water capacity we get free gravitational readily mobile water, passing into surplus water;

e) surplus water, exceeds limit (potential) field water capacity, approximating full water capacity and over. This water becomes insufficiently available due to the fact that it is flowing. In the first case, the plants have the time to absorb an insignificant amount of it and in the second case, the anaerobic conditions which are set up do not favour the development of plants in general and the productive utilisation of water in particular.

It can thus be seen that the range of available active water lies between moisture of wilting and maximum water-holding capacity (Table 22).

Table 22

Range of Active Moisture in the Soil Horizons of Certain Soils (after A. A. Rode) in Percentages of Soil Weight

Soil	Horizon	Field water capacity	Moisture of wilting	Range of active moisture
Soddy medium podzolical sandy loam	A_1	31	11	20
	A_2	23	6	17
Deep rich loamy chernozem	0-10	48	14	34
	10-20	35	13	22
	50-60	29	12	17
Southern clayey-loamy chernozem	0-10	33	16	17
	10-20	31	16	15
	30-40	26	15	11
Typical loamy sierozem	0-10	19	7	12
	10-20	18	8	10

In poorly drained soils, when the level of the ground water is high, soil moisture may come to exceed limit field water capacity, with attending extremely poor aeration. In such conditions, the plants react at first by slower growth which is followed by death. Of great importance is the depth of distribution of the roots. A shallow penetration of the root system on humid poorly drained soils may prove disadvantageous in such places where, in the second half of the vegetative period, the top layers of the soil dry out. Plants which, in the beginning, developed a shallow root system and adapted themselves to very moist conditions will suffer from drought in the second half of their vegetative period due to the fact that the roots have not the time to follow the lowering of the level of capillary fringe and may find themselves in the dry horizons of the soil at a time when the moisture content in the lower part of the soil profile is relatively adequate. This applies also to sprinkling irrigation with frequent light waterings at the beginning of the vegetative period. Subsequently the plants may suffer from shortage of water, should the waterings become less frequent. When the ground water remains at a high and stable level, it can serve as the main source of moisture. Plants whose roots remain within the zone of capillary fringe, cease to be affected by changes of the moisture content in the top soil horizons.

For meliorative purposes, we cannot limit ourselves to the water balance of the rated horizon (0.8 m, 1 m, 2 m, etc.).

The water reserve in soils of different native zones and local conditions varies substantially. The maximum possible reserve of water in soil is determined by the limit field water capacity. For a two-metre layer of loamy and clayey soils, it can be expressed approximately as follows: in the forest-steppe zone from 7,000 to 7,400 m^3/ha, in the steppe zone from 6,200 to 7,000 m^3/ha. Approximately 2,800-3,400 m^3/ha of it are unavailable to plants (dead reserve) and only the remaining 3,400-4,000 m^3/ha are useful. A fairly large number of cultures which possess a deep lying root system are not limited to a reserve of water of 2 m and draw water from lower lying soil horizons (lucerne and others).

One cannot judge of the properties and behaviour of the soil moisture from the absolute moisture. It is indispensable to know the relative moisture, from which one may establish the force with which soil moisture is retained and determine its availability to plants, confronting it with the corresponding water constants, i.e., establishing the margin of moisture. In comparison with another, a soil with lower absolute moisture may contain a greater amount of available moisture, and vice versa.

From the analysis of the water balance of a soil one may determine the reserve of water useful to plants at any given moment of the vegetative period. The water balance is subjected to constant, regular changes in time and from one area to another. These changes can be represented graphically by curves, diagrams, cartograms, etc. (Fig. 31). Knowing the laws which govern them, one is in a position to foretell changes in the water balance, which makes it possible to regulate the water balance of irrigated lands. Every soil undergoes an annual water balance cycle. The water balance of the soil of any portion of the earth should be considered not in isolation but against the background of the water balance of the surrounding territory.

Types of Water Regimes of Soils

The mean annual water balance determines the type of water regime of the soil. As a result of the manifestation of one or other water regime, dissolved and dispersed substances become distributed in the soil profile, genetical horizons arise and the general aspect (type) is formed of soils of one zone or another. The average type of water regime of any soil is made up of the season-

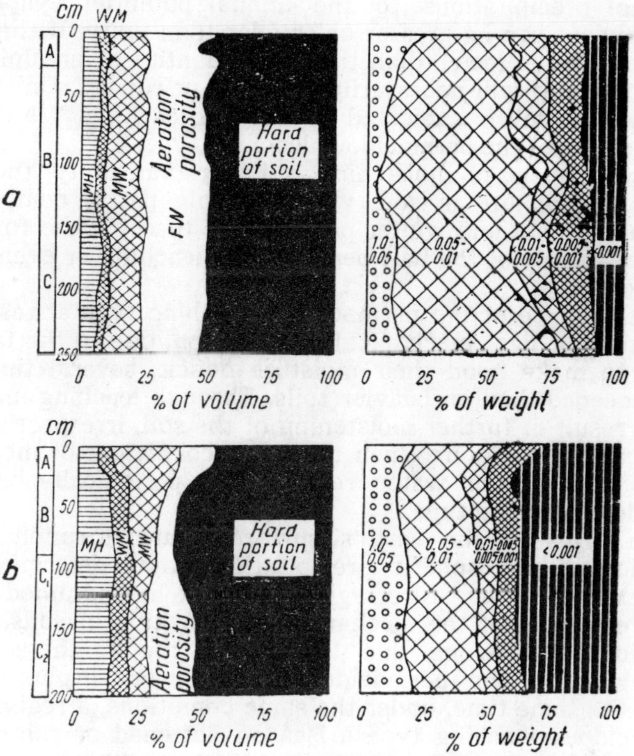

Fig. 31. Water properties and mechanical composition of soils:

a—of a typical sierozem (after Bolshakov); *b*—of a deep-columnar solonetz (after Malyanov); *MH*—maximum hygroscopic; *WM*—wilting moisture; *MW*—minimum water capacity; *FW*—full water capacity

al water regimes or seasonal distribution of water in soil. The type of water regime of soil and its elements are characterised by a certain dynamism.

The general and seasonal water regimes of different soils show marked differences. These differences are determined in the first place by the following conditions in which the soils develop: a) climatic, b) geomorphological, c) hydrogeological, d) biological (chiefly phytological), e) soil (chiefly the physical and water properties), f) permafrost, g) the production activity of man and so on.

The climate determines the amount of atmospheric precipitations and the amount of possible evaporation from the soil surface. The influence of this factor is approximately determined by the moisture coefficient of the soil or the ratio of the annual total quantity of precipitations to the annual potential evaporation. This coefficient can be greater or smaller than unity. If the potential evaporation is lower than the total quantity of precipitations, part of the water will necessarily infiltrate or run off. Under these conditions, the soil is subjected to through leaching. We then get a water regime of the permacidous leaching type.

If the coefficient of moistening is lower than unity, there is a shortage of moisture in soil, which is only partly compensated during the autumn and winter periods. We then get the formation of a water regime of the half permacidous leaching or even impermacidous type.

A water regime of the permacidous leaching type arises principally in soils of light mechanical composition, due to the fact that in order to make good their moisture deficit, several times less water is needed than on heavier soils. Through leaching may arise also as a result of further moistening of the soil, irrespective of its mechanical composition. When the mean coefficient of moistening is equal to one, the water regime may periodically be of the permacidous leaching type.

As the forms of the earth's surface influence runoff and the distribution of atmospheric precipitations, they determine, to a certain extent, the moisture regime of soils. In undrained depressions, irrespective of the mechanical composition of soils, in case of additional moistening due to surface runoff, there arises a water regime of the permacidous leaching or surplus moisture type. At the same time, under the same conditions, a regime of the impermacidous leaching type arises on flattened or raised areas.

Hydrogeological conditions in combination with geomorphological and climatic ones may condition the appearance of water regimes of the permacidous leaching, impermacidous leaching and ascending (exudational) type. Upon lowering of the coefficient of moistening, the significance of the ground water goes up. When the coefficient of moistening is higher than unity, the ground water maintains a high moisture content in soil. When the coefficient of moistening is lower than unity, there arises an ascending current of water from the ground water, which is all the more marked as the capillary fringe rises closer to the surface. By constantly wetting the lower soil horizons, the ground water may contribute to the formation of a water regime of the half-boggy type. Moistening of the soil by the ground water up to the surface gives rise to a water regime of the boggy type. A water regime of the boggy type arises also under a humid climate and when the ground water presents a head.

The water regime of soil is also influenced by the local soil water, which sometimes determines a stagnant water regime of soil. Such water-bearing horizons may disappear in periods of high desuction* and in the presence of runoff within the soil, and are reconstituted upon favourable weather conditions.

An important place in the outgoing side of the water balance is assumed by plant transpiration. By absorbing water (desuction), the vegetation may bring about the formation of a dessicated soil horizon at a certain depth from the surface. Replacement of the vegetative cover, fluctuations in its composition or its destruction bring about radical changes in the water regime of the soil.

The water regime of soil is also determined by the granulometric composition and its changes.

All the above mentioned conditions are interrelated. But in some cases, it is the soil-physical conditions that come to the fore, whereas in others, it is the phytological ones that do so, etc. However, it is the climatic conditions that prevail. According to the value of the coefficient of moistening, the Soviet scientist G. N. Vysotsky has established the following types of water regimes:

upon a coefficient of moistening of	1.33	for forest zones—permacidous leaching type
,, ,, ,,	1.00	for forest-steppes—transitional type
,, ,, ,,	0.67	for chernozemic steppes—impermacidous leaching type
,, ,, ,,	0.36	for dry steppes—impermacidous leaching type

On a territory with a prevalent distribution of the transitional and impermacidous leaching types, when the ground water lies close to the surface, a water regime of the exudation and stagnant type may arise. In dry steppes and semi-deserts, the water regime is of the ascending type with a tendency towards accumulation of salts. A similar type of water regime arises also in places where the ground water, with or without head, lies in an up-and-down fashion.

In time, the character of the water regime of soils changes, in connection with changes in the physico-geographical conditions. The water regime also changes appreciably in the course of a period of successive dry or humid years and according to the season of the year.

The water régime obtaining under native or cultivation conditions is not the same. But upon abundant atmospheric precipitations or watering, in all types of water regime, with the exception of the exudation type, a downward translocation of moisture prevails in the root zone.

The existing distribution of land over the elements of the relief may exert different influences on the formation of the water regime

* Desuction—from the Latin "desuctio"=suction, dessication of the soil by the vegetation. Term introduced by G. N. Vysotsky.

of the area. Forests and swamps regulate the surface runoff. The downward movement of water under a forest, which prevails in places, increases the overall reserve of ground water. Forests, particularly coniferous ones, lengthen the period of spring melting of snow. The melting of snow may be delayed by 10 days in a deciduous forest or brush, by 18-20 days in a pine forest and 25 days in a spruce forest, in comparison with open land.

The water regime of soils is influenced by the character of the forms of the earth's surface and the elevation of the relief. The more rapid melting of the snow on southern slopes and their drying out, in contradistinction to northern slopes, creates a deficient water regime. Some soils are uniformly wetted throughout their depth, in others water penetrates extremely irregularly in connection with differences in the microrelief, the ground cover, structure, composition, etc.

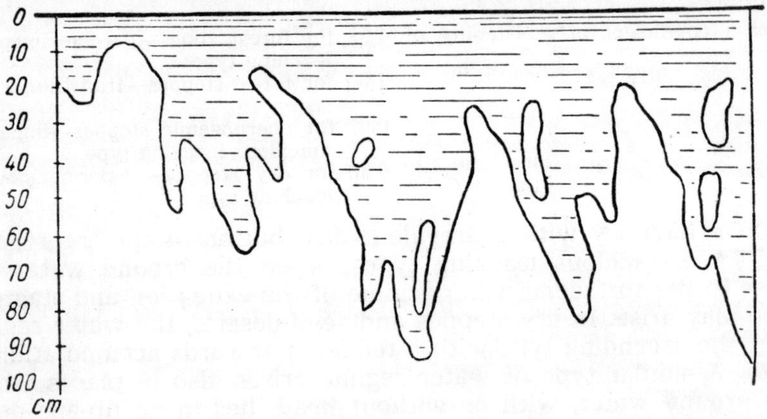

Fig. 32. Irregular wetting of soil (upon rainfall and irrigation)

Irregularities in the penetration of rainfall water in a dry soil are, according to N. A. Kachinsky, characterised by the diagram in Fig. 32. The same thing occurs but on a larger scale upon the melting of snow. In places of accumulation of snow, more intense infiltration is observed and there is a formation of "potuskuls"—places of through soakage. G. N. Vysotsky illustrates the water regime of soils from zone to zone and from one element of the relief to another by the diagram represented on Fig. 33.

Apart from infiltration, evaporation of water from soil and transpiration, A. A. Rode includes in the factors which determine the water regime, the source and degree of moistening of the soil. The sources which supply soil with moisture comprise atmospheric precipitations, the ground water, flood-water and irrigation

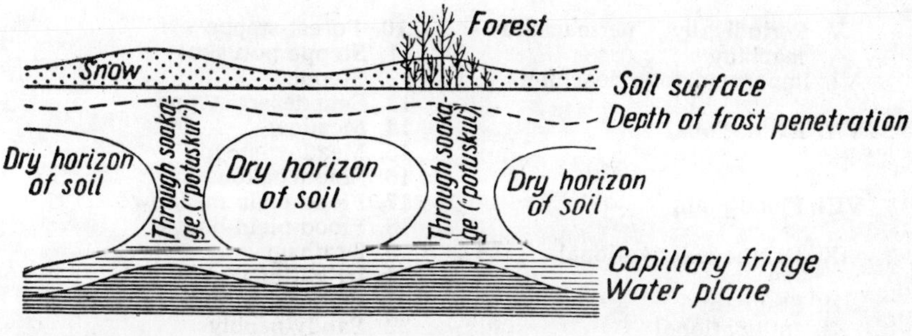

Fig. 33. Diagram illustrating the recharge of the ground water by "potuskuls" in the steppe zone (after G. N. Vysotsky)

water. In polar regions and in the permafrost zone, the factor which determines the water regime of a soil is the frozen ground, in the form of an impervious layer lying close to the surface. As the frozen ground melts and this impervious layer is lowered, the thickness of the root zone increases. Here obtains a water regime characteristic of soils lying above frozen ground, which depends mainly on the soil temperature and the depth of thawing.

The water regime of soils is strongly influenced by the production activity of man. The drainage of swamped land and swamps lowers the level of the ground water and eliminates the surplus moisture from the soil. A high agrotechnical standard directed towards the conservation of moisture promotes greater and more uniformly distributed moisture of soils. Agro-, forest and hydromelioration radically improve the water regime of soils even to the extent of transforming one type into another. Irrigation raises the soil's moisture content, which, not infrequently, is attended by a rise of the water table and creates an irrigational type of water regime.

In accordance with the above exposed facts, we may distinguish the following types and subtypes of soil water regimes (based mostly on A. A. Rode's diagram and other sources):

Type	Subtype
I. Above frozen ground	1. Tundra-boggy 2. Forest-tundra
II. Permacidous leaching	3. Taiga 4. Half-boggy
III. Stagnant weakly permacidous leaching	5. Boggy 6. Ground-boggy
IV. Ground-permacidous leaching	7. Taiga deep permacidous leaching 8. Ground-taiga 9. Ground half-boggy

V. Periodically permacidous leaching	10. Forest-steppe
	11. Steppe potuskul
VI. Impermacidous leaching	12. Steppe
	13. Semidesert-steppe
VII. Exudational	14. Meadow
	15. Meadow-steppe
	16. Solonchakous
VIII. Flood-plain	17. Flood-plain-meadow
	18. Flood-plain-boggy
IX. Drainage-irrigational	19. Drainage
	20. Irrigational
X. Stagnant	21. Solonchakous-boggy
XI. Infiltrational	22. Sandy-pebbly

Types and Subtypes of Water Regime

Type I. Above frozen ground regimes. Coefficient of moistening $\lessgtr 1$, varies from 1.3 to 0.3, being 0.9-0.2 during the summer months. Return of moisture to atmosphere higher or lower than infiltration. In the cold period, permanent frost merges with seasonal. As a result of freezing of soils from the surface and tightening pressure on lower lying overmoistened horizons, chasms and mud-like effusions are formed. From spring onward, an overmoistened horizon is formed above the frozen ground, whose thickness increases as the layer of frozen ground retreats lower down.

Subtypes: 1. Tundra-boggy. 2. Forest-tundra. Are characterised by the thickness of the thawed out horizon and the duration of the period of above-zero temperature of the soil.

Type II. Permacidous leaching regimes. Coefficient of moistening > 1. Return of moisture to atmosphere less than infiltration. A characteristic feature of such regimes is the annual moistening with water of the whole of the soil-ground layer down to the ground water. The amount of water leaving the soil through subsurface drainage exceeds the amount of the water reaching the soil from the ground layer. Outflow greater than inflow, which conditions leaching of the soils.

Subtype 3. Taiga (forest). Recharge by atmospheric precipitations with, in places, additional surface recharge. As a rule, the soil is moistened with water in spring and autumn, sometimes in summer. The water leaves the soil through subsurface drainage in spring, sometimes in autumn and summer. The lower half of the soil profile is periodically moistened along the capillaries. The ground water sometimes gets within the boundaries of the soil profile with capture by the capillary fringe of the whole of the soil profile. Moisture fluctuates from limit field to molecular water capacity and lower. Typical for podzolic soils.

Subtype 4. Half-boggy. Same recharge regime. Thorough moistening with water is frequently repeated during the yearly cycle. There is an uninterrupted slow subsurface flow. Constant soil-capillary moistening. The ground water penetrates periodically into the soil profile, sometimes right up to the surface. The capillary fringe usually reaches the soil's surface. Moisture fluctuates from full to limit field water capacity. Typical for podzolic-boggy taiga soils and reclaimed peaty bogs.

Type III. Stagnant weakly permacidous leaching. Is characterised by extremely low flowage or stagnancy, which conditions the formation of bogs.

Subtype 5. Boggy. Atmospheric recharge plus additional surface recharge. Soil water stagnant. The ground water constantly wets the soil, rising not infrequently to the surface. The upper limit of the capillary fringe is always close to the surface. The moisture in the upper part of the soil profile fluctuates from full to limit field water capacity; in the lower part, it is seldom lower than full water capacity. Typical for peaty-boggy soils of high bogs.

Subtype 6. Ground-boggy. Recharge as in previous subtype. Thorough moistening with water is repeated in the course of the year. Soil runoff insignificant, uninterrupted. The soil is constantly moistened. The level of the ground water frequently rises to the surface. The capillary fringe is always close to the surface. Moisture fluctuates from full to limit field water capacity. Characteristic of peaty-boggy soils of depressions.

Type IV. Ground-permacidous leaching regimes. Coefficient of moistening>1. Yearly return of moisture to atmosphere less than infiltration recharge from ground and atmosphere, frequently with additional surface recharge. The level of the ground water is sometimes raised.

Subtype 7. Taiga deep permacidous leaching (forest) regime. Recharge from the atmosphere with additional surface recharge. The soil is moistened in spring, sometimes in autumn, more seldom in summer. There is an insignificant periodical subsurface drainage. The level of the ground water and the capillary fringe are always below the C horizon of the soil profile. Moisture fluctuates from limit field water capacity to moisture of wilting. At the bottom of the profile, there is somewhat more moisture. Characteristic of sod-podzolic soils of south of taiga zone.

Subtype 8. Ground-taiga. Recharge from atmosphere and ground, often with additional surface recharge. Filtration occurs during the whole of the warm period of the year. There occurs an outflow of soil water. Ground-capillary moistening. The level of the ground water and the capillary fringe rise periodically into the soil profile. The moisture content fluctuates from field limit to maximum molecular water capacity. Characteristic of dark coloured podzolic and half-boggy soils.

Subtype 9. Ground half-boggy. Recharge similar to previous one. Thorough moistening is frequently repeated in the course of the year. Subsurface drainage is to some or other extent constant. Soil is constantly moist. The level of the ground water rises periodically into the soil profile. The capillary fringe in the soil profile is constant, frequently reaching the surface. The moisture content fluctuates from full to limit field water capacity and lower. Characteristic of dark coloured gleyey-podzolic soils and low lying peaty-gleyey soils.

Type V. Periodically permacidous leaching regimes. Coefficient of moistening $\leqq 1$. Return of moisture in isolated years is less than infiltration.

Subtype 10. Forest-steppe. Atmospheric recharge with and without additional surface recharge. Thorough moistening occurs in spring, not every year. Subsurface drainage of short duration in spring. The level of the ground water is much below the soil profile. Moisture content fluctuates between limit field water capacity to moisture of wilting. Characteristic of grey forest soils.

Subtype 11. Steppe potuskul subtype. Differs from the previous one by the absence of soil runoff. The level of ground and soil water and the upper limit of the capillary fringe are always below the boundary of the soil profile. Characteristic of soils at the edge of forest, forest belts and depressions of steppe and semisteppe zones.

Type VI. Impermacidous leaching regimes. Coefficient of moistening <1. Return of moisture to atmosphere comes near to amount of infiltration. Atmospheric precipitations moisten the soil thoroughly to a depth not exceeding 3-4 m. The ground water lies deep (12-15 m and more). Between lower boundary of the zone of thorough moistening and upper boundary of capillary fringe lies the so-called dead horizon of drying out of varying thickness (up to several metres). Low moisture content of soil is maintained by desuction. No connection between soil and ground water. This type of water regime is characteristic of soils of the steppe type (chernozems, chestnut soils, sierozems). The water regime of grey forest soils and degraded chernozems in the forest-steppe region follows this pattern.

Subtype 12. Steppe. Atmospheric recharge with hardly appreciable additional surface recharge. The soil does not become thoroughly moistened. There is no subsurface drainage. Sometimes capillary perched moisture is present. At the base of the soil profile lies a thick (several metres) dry horizon, with a moisture content less than maximum molecular water capacity. There is no exchange of water between soil and soil-forming rock. The ground water and the capillary fringe lie deep, below the soil profile. The moisture content fluctuates between field water capacity and moisture of wilting. Characteristic of chernozems, chestnut and brown soils.

Subtype 13. Semidesert-steppe. Atmospheric recharge. No thorough moistening. No runoff. There is a thick drying out horizon. The soil loses considerable amounts of moisture through evaporation. The ground water and the capillary fringe always iie substantially below the base of the soil profile. The moisture content fluctuates from capillary water capacity to moisture of wilting. Characteristic of brown soils and sierozems with deep lying ground water.

Type VII. Exudational regimes. Water-ground regimes where moistening can occur in the presence or the absence of head. Coefficient of moistening$<$1. Return of moisture to atmosphere substantially exceeds infiltration; the difference is made good by inflow from the ground water. The soil profile is constantly in the zone of capillary fringe. The moisture in the soil is usually capillary supported. Ground and soil runoffs are merged.

Subtype 14. Meadow. Is characterised by ground-atmospheric incoming of water, sometimes with additional surface inflow. The soil gets thoroughly moistened in winter and spring. There is a constant subsurface drainage. The soil is moistened throughout the profile. In the summer period, as a result of transpiration and evaporation of moisture from the soil, there occurs a certain lowering of the level of soil-ground water, which is restored in autumn.

The level of the ground water usually comes close to the soil profile. The capillary fringe is constantly in the soil profile and sometimes rises right up to the surface. The moisture content in the upper part of the profile fluctuates from full water capacity to moisture of wilting and in the lower part, from full to limit field water capacity; in a period of drought it goes lower down. Typical of chernozem meadow soils and meadow-solonetzes.

Subtype 15. Meadow-steppe. Recharge as in the previous subtype. The soil is thoroughly moistened in spring, in relatively wet years. Subsurface drainage occasionally occurs only in the lower part of the soil profile, where the ground is constantly wet. There is an insignificant desuctive lowering of the level of the ground water in summer, which in autumn is replaced by a rise. The ground water lies deeper than the soil profile; only the capillary fringe reaches its lower part and higher. The moisture content in the upper part of the soil profile fluctuates from limit field water capacity to moisture of wilting and in the lower part, from full to maximum molecular water capacity and lower. Characteristic of meadow-chernozemic, meadow-chestnut, meadow-sierozemic soils and meadow-steppe solonetzes.

Subtype 16. Solonchakous. Recharge from the atmosphere and ground. Shallow moistening in winter and spring or spring. Subsurface drainage is almost inexistent. The soil is constantly wet. In summer, the level of the ground water goes down somewhat and rises again in autumn. The level of the ground water and capillary fringe do not leave the soil profile. The moisture content fluctuates from full to limit field water capacity and lower. Characteristic of meadow soils of the chernozemic zone and meadow salined soils of other zones.

Type VIII. Flood-plain regimes. Coefficient of moistening $\lesseqgtr$ 1. Return of moisture to the atmosphere usually lower than infiltration, with the exception of flood plains of deserts. The water regime of flood-plain soils is, in the main, determined by the hydrologic regime of rivers. In southern regions, it represents an alternation of movement of capillary perched water with capillary supported water. Prior to spring flooding, the rise of the ground water is accompanied by a rise of the capillary supported water, which merges with the capillary perched water. During the flood period, the soil reaches its maximum saturation with water. In the absence of flooding, in the drier years or on high ground which escapes flooding, the soil is moistened by capillary supported water.

Subtype 17. Flood-plain-meadow. Recharge from the ground and atmosphere with additional flood-plain recharge. Thorough moistening is repeated during the year in the taiga zone and occurs during flood tide in other zones. Soil runoff and capillary-ground moistening go on without interruption. The level of the ground water usually reaches the soil profile. The capillary fringe sometimes rises close to the surface. The moisture content fluctuates from full to maximum molecular water capacity. Characteristic of flood-plain-meadow and flood-plain forest soils.

Subtype 18. Flood-plain-boggy. Recharge from the ground and atmosphere with additional flood-plain recharge. Thorough moistening of the soil is repeated during the year. Soil runoff is continuous. The soil is moistened throughout the entire profile. The level of the ground water lies in the soil profile. The capillary fringe reaches the surface of the soil. The moisture content fluctuates from full to limit field water capacity. Characteristic of flood-plain-boggy soils.

Type IX. Drainage-irrigational regimes. Coefficient of moistening $\lesseqgtr 1$. Appreciable return of moisture to atmosphere and lateral outflow.

Subtype 19. Drainage. Recharge from the ground and atmosphere. The soil gets periodically thoroughly moistened. There is a constant subsurface drainage. Constant ground-capillary moistening. The level of the ground water and the capillary fringe lie in the soil profile. The moisture content fluctuates from full to maximum molecular water capacity. Characteristic of soils of reclaimed swamps.

Subtype 20. Irrigational. Recharge from the atmosphere and mostly from additional artificial sources as a result of frequent waterings. The soil is often thoroughly moistened deeper than the root zone. Due to watering, the level of the ground water rises and partly returns to its usual level between the waterings and more so in winter. The moisture content fluctuates from full water capacity to moisture of wilting.

Type X. Stagnant water-ground regimes, not infrequently with additional moistening. Coefficient of moistening $\lesseqgtr 1$. Return of moisture to atmosphere often exceeds infiltration. The soil profile is constantly in the zone of supported capillary fringe. The level of the ground water usually comes close to the surface. The amount of water penetrating into soil exceeds outflow.

Subtype 21. Solonchakous-boggy. Soil water of a stagnant character with high concentration of the solution. Typical of mineral bogs and solonchakous soils. Akin to this subtype is the boggy subtype with a regime of soil and ground waters of a stagnant character.

Type XI. Infiltrational. Clearly marked downward flow of the water. Most characteristic of sands and sandy soils, which are noted for their high water permeability, low water capacity and slight water raising capacity.

Subtype 22. Sandy-pebbly. Atmospheric recharge, more seldom with additional surface recharge. The soil is easily thoroughly wetted even by small precipitations. The whole of the water is absorbed and penetrates into the soil with great speed forming ground water on impermeable rocks. Akin to this subtype are the karst-suffosive regime and the water regime of fissured rocks.

Chapter X

THERMAL AND AIR REGIMES OF SOIL

Thermal Properties and Thermal Regime

Soil possesses certain inherent thermal properties and a thermal regime, which is the aggregate of the thermal phenomena taking place in soil. The thermal regime of soil depends, in the main, on the heat it receives from the sun or, to be more precise, on its capacity to absorb radiant energy, which becomes transformed into thermal energy. Every normally sunlit square centimetre of the earth's surface receives up to 1.946 calories of solar energy per minute (solar constant). However, the solar energy which actually reaches the earth's surface is usually from 2 to 4 times less, due to it being scattered by the atmosphere, nebulosity, inclination of the rays relative to the earth's surface, etc.

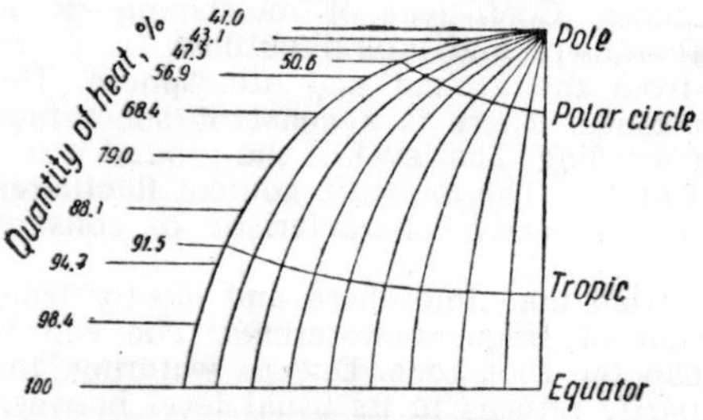

Fig. 34. Distribution of the radiations and thermal energy reaching the surface of the earth

The amount of heat received by the earth's surface decreases from the equator to the pole. If to the amount of heat received at the equator per surface unit we give a value of 100%, the value at the latitude of the tropics will be approximately 91.5%, approximately 50% at the latitude of the Polar Circles and will fall down to 41% beyond that (Fig. 34).

The soil absorbs an enormous quantity of solar heat, reflecting from 0.1 to 0.3 of radiant energy. The amount of reflected radiant energy, i.e., the ratio of the amount of radiant energy (A) reflected by the earth's surface to the amount received (ε) expressed in percentages is called reflectance or surface albedo. Albedo $\left(\frac{A}{\varepsilon}100\right)$ is measured with special apparatus, albedometers. Albedo is determined from the character of the surface in conformity with the data given in Table 23.

It can be seen from the data given in the table that albedo can be regulated through agromelioration. An average of less than 50 cal/cm^2 is available for warming up the soil and air during the warm period, north of 60° of northern latitude. In the central regions of the U.S.S.R., the corresponding amount is 100-150 cal/cm^2, in the south-east it is 200 cal/cm^2 and more. In the latter case, this is 60% and more of the total radiation and in the north, 25-30%.

Table 23

Surface Reflectance (Albedo)

Surface	Albedo, %
Moist chernozem	8-9
Dry chernozem	14
Recently ploughed sierozem	17
Sierozem cotton field	20-22
Sierozem rice field	12
Dry surface of clayey desert (sierozem)	29-31
White sand	40
Snow	70-80
Grassy vegetation	20-26
Crowns of arboreal vegetation	14-18
Water surface	10

In spite of their shallow penetration, the light rays exert a certain photochemical influence on the surface horizon of the soil: they oxidise ammonia, alter the composition of humus, exert a lethal influence on the microorganisms, etc.

Apart from the main source of radiant energy, soil receives heat from exothermic, physico-chemical and biochemical reactions. However, the amount of heat received as a result of biological and photochemical processes exerts almost no influence at all on soil temperature changes. In summer, the temperature of a dry warmed soil may rise as a result of wetting. This heat is known as the heat of wetting. It is manifested upon a very slight moistening of soils rich in organic and mineral (clayey) colloids.

Heat of wetting sometimes raises the soil's temperature by several degrees (up to 10°) which may scorch the delicate root hairs and cause them to wilt. The heat of wetting given off upon swelling of colloids is conditioned by physico-chemical reactions between the mineral particles of the soil and water, characterised by the transformation of the water aggregates with destruction of the hydrogen bonds. The heat given off is the residual heat of hydration of ions on the surface of the soil's hard phase.

Very slight warming up of the soil may be due to the internal heat of the earth, arising as a result of radioactive and other processes going on in its depths. This source of heat is small, it amounts to 54 cal per 1 cm^2 of soil surface per year—the amount of heat which is necessary to melt a layer of ice 6-8 mm thick (A. V. Klossovsky).

Among the other sources of heat, we must mention the "latent heat" liberated in the process of crystallisation, condensation and freezing of water, etc.

Depending upon the mechanical composition, the humus content, the colour and the moisture content, soils are considered as warm or cold. Dark soils rich in organic matter absorb substantially more heat than light ones. It has been proved experimentally that soils coloured with some dark substance (soot, gumbrine) may, in summer, possess a temperature exceeding by several (up to 8°) degrees that of soils of the same composition coloured in white (magnesium, chalk). This may be of great practical significance with regard to the application of thermal meliorative measures.

In view of the fact that water is a better conductor of heat than air, swamped and boggy soils exhibit lower temperatures in the course of the year than dry soils but are characterised by a narrower amplitude of daily and annual fluctuation. The temperature of the soil depends on the presence and character of the vegetative cover. The vegetation acts as a screen and retains part of the sun-rays, reducing warming up of the soil in daytime, protecting it from heat losses at night and thereby reduces the amplitude of daily temperature fluctuations of the soil.

Soils covered by forests or a thick stand of grass are warmer in winter than soils devoid of vegetation. In summer, the vegetation protects the soil from the drying and cooling effect of winds. The difference of temperature reaches 3° and more. During the cold part of the year, the vegetation, acting as a cover of low heat conductivity, decreases cooling down of the soil. Under a vegetative cover, the amplitude of temperature fluctuations of the soil is somewhat narrower than in the absence of such a cover.

The maximum temperature reached by soils covered with vegetation is lower than that of bare soils. A snow cover has a similar effect, preventing excessive cooling and deep freezing of the soil.

The soil's thermal properties are governed by their mineralogical and mechanical composition. In agricultural practice, clayey and loamy soils are known as cold, not easily warmed, and sandy soils as warm, easily warmed. This distinction is particularly marked in spring and summer. Spring cultivations on sandy soils begin almost ten days earlier than on clayey ones.

The thermal differences exhibited by soils are also conditioned by the orientation and angle of incline of the surface. In temperate latitudes, the soils of southern slopes are warmer than those of northern ones. It should be noted that an increase of the angle of incline towards the south is equivalent to a transfer to the south as it were (approximately 100 km for every 1°).

It is also obvious that the temperature of the soil goes down as the absolute height of the place increases, a fact which accounts for the formation of snow covers on high mountains. The thermal properties are influenced by the elements of the macrorelief and even microrelief, giving rise to a correspondingly variegated soil climate.

The thermal regime of a soil depends on its heat capacity, heat conductivity and temperature conductivity.

Heat capacity. The heat capacity is determined by the amount of heat in calories necessary to raise the temperature of the unit of weight (1 g) or volume (1 cm^3) of soil by 1°C. The heat capacity of the unit of mass (1 cm^3) is referred to as the specific heat capacity. A distinction is made between gravimetric heat capacity (0.2-0.4) and volumetric heat capacity (0.5-0.6). The volumetric heat capacity of a soil is equal to the gravimetric heat capacity multiplied by density. The volumetric and gravimetric heat capacities of the solid phase of soil are smaller than those of water and greater than those of air. The specific heat capacity (in cal/g deg) of heated soil is 0.179 at a temperature of 25°C, of organic matter 0.3 at a temperature of 25-50°C; at a temperature of 0°C, the specific heat capacity of quartz is 0.174, of water 1.00, of air 0.0003 cal/cm^3. The volumetric heat capacity of soil depends on its content of water (after D. G. Vilensky) (Table 24).

Table 24

Volumetric Heat Capacity, in cal

Mechanical composition of the soil	Water content, in % of pore-space							
	0	20	30	40	60	70	80	100
Sand	0.30	0.39	0.43	0.47	0.55	0.59	0.63	0.72
Clay	0.24	0.36	0.42	0.47	0.59	0.65	0.71	0.83
Peat	0.15	0.30	0.37	0.45	0.60	0.68	0.75	0.91

It can be seen from the data given in Table 24 that upon an increase of moisture content, the heat capacity rises less in sands, more in clays and still more in peat. For this reason, peaty and clayey soils are not easily warmed and are therefore cold whereas sandy soils are warm.

The heat capacity of a soil in a dry and relatively compact condition corresponds to the heat capacity of its solid phase, in other cases it depends on the heat capacity of the water and air contained in its pores. Hence the heat capacity of a moist soil is higher than that of a dry one. However, the higher specific heat capacity of a moist soil prevents it from ever getting very warm. Furthermore, the temperature of a moist soil goes substantially down on account of the water evaporating from it (the latent heat of evaporation of soil at ordinary temperatures is about 580 cal per gram of evaporated water). In spring, moist soils remain cold on the surface longer than dry ones, although the former are more uniformly warmed. During the cold hours (at night) or during the relatively cold periods of the year (in spring and autumn), the deep horizons remain warmer in moist soils than in dry ones.

Due to their high heat capacity and low heat conductivity, in summer, the top layers of peaty soils are subjected to extreme overheating in the daytime and cooling at night.

The vegetation is adversely affected by the sharp temperature fluctuations exhibited by soils of this type. In order to reduce the amplitude of temperature fluctuations on the surface, one resorts to shadowing of the inter-row space with a cover of plants.

Heat conductivity and temperature conductivity. Heat conductivity is the capacity to conduct heat or, to be more precise, the speed with which the temperature rises within the soil. Heat conductivity is determined by the amount of heat in calories passing through a unit of surface, through a unit of thickness in a unit of time, upon a temperature gradient equal to unity. Heat conductivity is usually expressed in calories per second per 1 cm^2 through a layer 1 cm thick, the temperature gradient between the two surfaces being 1°C. The heat conductivity of the mineral part of the soil exceeds the heat conductivity of water by 3-4 times and that of air by 100 times and more. The heat conductivity of water is 0.00135, of ice 0.00573, of air 0.0000557, of dry soil 0.0003-0.0005. The heat conductivity of soil goes down with an increase of its dispersiveness and goes up with moistening (Table 25). The smaller the heat conductivity of soil, the higher the amplitude of temperature fluctuations and the deeper down the fluctuations are felt.

Table 25

Changes in Heat Conductivity of Soil (after D. G. Vilensky)

Mechanical composition of soil	Heat conductivity of soil	
	air-dried	saturated with water
Coarse sand	0.00047	0.0041
Fine sand	0.00046	0.0039
Sandy loam	0.00045	0.0032
Loam	0.00033	0.0021
Peat	0.00027	0.0011

Air-dried soil possesses lower heat conductivity than moist soil. The heat conductivity of a moist soil is all the higher as it contains more water. The heat conductivity is tied with the forms and categories of soil moisture, increasing accordingly from hygroscopic on to film, capillary, limit field and full water capacity, and this, in turn, is tied with the dispersiveness, chemical composition and properties of soil. The higher heat capacity of moist soil is due to the close thermal contact between the separate soil particles which are united by water pellicles. Water being a body endowed with a relatively high heat conductivity, communicates this property to soil. It cools soils, if it is stationary. Moving water transfers warmth from warmer horizons to less warm ones and warms up the soil. A moist soil loses more heat than a dry one.

The soils which best conduct heat are the sandy soils. In warm periods, they are more quickly warmed and in cold periods, they are, on the contrary, more intensely and deeply cooled. The soils which conduct heat worst are the clayey and peaty soils. Compared with other soils, peaty soils possess a number of particularitics. Pcaty soils absorb more and give up less heat. On dry and friable soils, the amplitude of temperature fluctuations is greater than on moist and compact soils.

The turf-peats of reclaimed swamps, possessing low heat conductivity, are characterised by a delayed exchange of heat with the lower layers. Dry peat is a good heat insulator. It hampers deep cooling and warming of the soil. In medium latitudes under peat, the soil remains frozen until the beginning of summer and in the north, to the middle or the end of summer.

Due to low heat conductivity and weak exchange of heat with the lower layers, frosts are more frequent on the surface of peat-turfs and reclaimed swamps.

The heat conductivity of a soil is greatly influenced by its looseness or compactness. The heat conductivity of soil goes down when it is loose and goes up when the soil is compact. This is due to the fact that loose soil contains more air, which is a worse conductor of heat than the mineral, solid part of soil.

With compactness, the heat conductivity of a snow cover may be increased 80 times, thus sharply lowering its heat insulating capacity.

In addition to heat conductivity, we have temperature conductivity, which is the temperature change of 1 cm^3 of soil, brought about by the arrival of a certain amount of heat, viz., that which passes in 1 second through a cross section of 1 cm^2 upon a difference in temperature equal to 1° at a distance of 1 cm. The temperature conductivity of a soil is equal to the heat conductivity divided by the specific heat capacity and density. The relationship between heat conductivity and temperature conductivity is expressed by the equation:

$$K = \frac{\lambda}{cV},$$

where K is the temperature conductivity;

λ—the heat conductivity;

c—the gravimetric heat capacity;

V—the soil's apparent density;

cV—the volumetric heat capacity.

The heat conductivity, temperature conductivity and heat capacity of a soil determine its temperature and the heat exchange between the horizons.

The soil's temperature exhibits rhythmical changes through its genetical horizons, according to daily and annual cycles. This is due to the alternation of day and night and of the seasons of the year, or the periodicity in the arrival of radiant energy on the earth's surface. The difference in temperature between the upper and lower sections of the soil is governed chiefly by the daily alternation of heating of the surface by day and its cooling by night, as well as the lag in the transmission of the thermal effect down and inwards.

The soil's daily temperature fluctuations in the course of the year are recorded by routine measurements down to a depth of 50-100 cm. In summer, when the sky is clear, the soil's temperature is considerably higher than in cloudy weather. The most pronounced temperature fluctuations are observed on the soil's immediate surface; they are attenuated with depth and, at a depth of approximately 1m, they become relatively constant. With depth, the daily penetration of soil temperature is delayed on the average by 2.5 to 3.5 hours to every 10 cm of depth.

In summer, the daily temperature fluctuations in soil are considerably greater than in winter when, under a snow cover, the soil exhibits a relatively constant temperature, somewhat below zero.

The seasonal temperature fluctuations exhibited by soil affect a depth of several metres. The annual penetration of temperature is delayed by 20-30 days to every metre of depth. The constant temperature of a soil, which approaches the mean annual temperature of the air at that spot, is located, in a temperate climate, at a depth of 20-30 m, whereas in the tropics, it is found at a depth of about 8 m. In that layer of the earth's crust (8-30 m), there occurs an annual heat exchange. The value of the heat flow is proportional to the temperature gradient and depends on the heat conductivity of the soil-ground.

In the temperate belt, in connection with the seasonal freezing, soil formation is interrupted for several months in winter (t lower than 0°) (Fig. 35). The further we go in a northward direction, as can be seen from the diagram, the longer the interruption. In the zone of permanent snow and ice, there is no soil formation. Thus, the speed of soil formation increases from the pole to the equator. In the desert zone, it is slowed down in connection with the intense drought and heat.

The value of the heat flow is proportional to the vertical temperature gradient and depends upon the soil's heat conductivity. In temperate latitudes, during the warm period of the year, there is an average accumulation of 1,000 to 3,000 cal in a column of soil layer harbouring roots with a cross section of 1 cm^2. In the

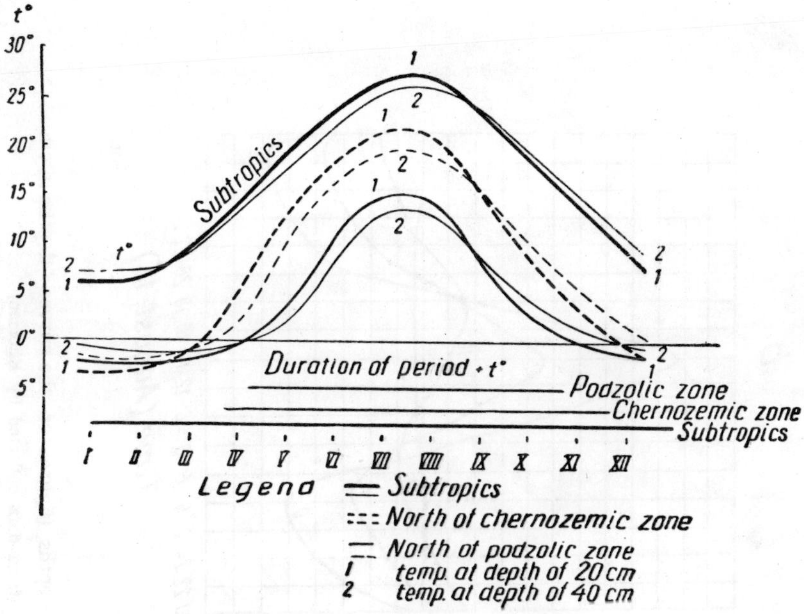

Fig. 35. Annual soil temperature fluctuations (from data collected by A. M. Shulgin)

European part of the U.S.S.R., the accumulation of heat in soil begins on the average around March 15th in the south and around May 25th in the north. The heat accumulated through summer is lost during the cold half of the year.

The seasonal temperature fluctuations of the soil right next to its surface can be represented diagrammatically (Fig. 35). A similar curve would express the daily temperature fluctuations on the soil's surface between day and night, which are more pronounced in summer, affecting the soil to a lesser depth (80-100 cm) (Fig. 36).

The soil's temperature regime is influenced by the snow cover, which is interdependent with and produced by the climate. Loose snow possesses a fairly high heat reflectance and a low heat conductivity hardly exceeding that of air. Snow of this kind reflects 80-90% of the radiant energy falling on it. The snow cover brings about cooling of the layer of air lying close to the ground. In the absence of a snow cover, in depressions, over the moist surfaces of soils, the fog which forms close to the earth prevents this cooling down. In that case, the temperature of the air lying close to the soil's surface and of the soil is 1-2° above that of higher areas.

Freezing and thawing of soil. With the onset of constant below-

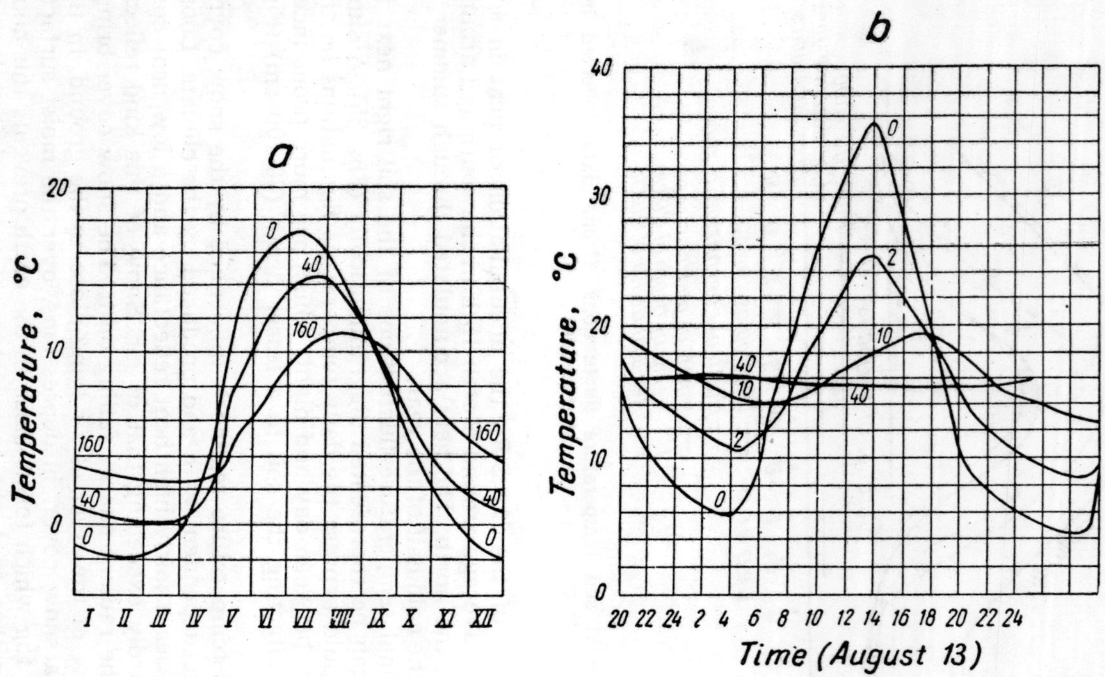

Fig. 36. Temperature fluctuations of the soil at different depths (figures on the curves—depths in cm): *a*—annual (after G. N. Lyuboslavsky); *b*—diurnal (from data collected by Khomen)

zero temperatures in the layer of air close to the earth, the soil freezes. To begin with, frost affects the surface and then gradually penetrates deeper, reaching a depth of 1.5 m and more. In the southern regions of the U.S.S.R., frost penetrates to a lesser depth and, in places, it does not occur every year or is even lacking altogether. The depth to which the soil freezes fluctuates considerably from year to year in connection with changes in the winter weather conditions.

Frost affects soils containing liquid moisture in smaller or larger amount, which freezes and cements the soil particles and aggregates into a solid mass, more difficult to break up than pure ice. A physically dry soil, despite the low temperatures which penetrate into its depth, does not give rise to such a continuous dense mass, although the life processes in it, just as in a frozen ground, come to a temporary standstill.

The water contained in soil freezes at a temperature somewhat below zero: at —4 or —4.5° in sands and clays, at —5° in peat. This is due to the fact that soil moisture, as a rule, contains dissolved substances which depress the freezing temperature of water and the more so as the concentration of the solution is higher. Upon freezing, the concentration goes still higher and depresses still further the freezing point. Thawing, on the contrary, in view of the lowering of the concentration, is accelerated. The soil freezes all the less readily as it is less moist, due to the fact that, in such a case, some of the water is in the bound form, freezing at lower temperatures. Furthermore, the more dispersed the soil, the more intensely are manifested the surface phenomena, which retain moisture with great vigour, and the lower the temperature at which it freezes. But in connection with the high heat capacity of water, an overmoistened soil freezes later than a less moist one. A compact soil freezes earlier and deeper than a loose one. Cold penetrates deeper in a dry soil and peat than in a moist soil, furthermore, the larger the amount of water which has to be frozen, the more heat is used up. The larger the quantity of water turning to ice, the more cold is accumulated, which prevents as it were any deeper penetration of it. But on the other hand, heat conductivity, whose increase is directly proportional to the amount of water, promotes the penetration of cold.

The water present in soil is supercooled to —2.6 or —4.5°. The first to freeze is the free water, which is contained in the large noncapillary pores. To begin with, there is a formation of small ice crystals (0.005-0.01 mm) of various shapes,* the amount and sizes of which rapidly increase, causing further pulling up of water towards the colder part of the soil. The linear speed of crystallisation increases considerably with an increase of the degree

* At low temperatures, there is a predominance of flaky ice crystals.

of supercooling (230 mm/min at —0.9°, whereas it reaches 2,800 mm/min at —7°). With the crystallisation of ice in the soil pores, a crystallisation force is set up and, as a consequence, the soil pores become stopped up and forced apart giving rise to what is known as the "heaving" effect of frost. The growth of ice crystals in large pores involves, as a consequence, a flow of water from the small capillaries where, in connection with their decreasing size, the freezing of water is delayed. This process spreads correspondingly still further, involving yet finer and finer pores if there is still some water left in them. The crystals of ice in the soil pores also grow on account of the moisture of condensation, which comes up as vapour from the lower horizons of the soil, free from frost. The distillation of water vapour from the surface of supercooled water and condensation on the surface of ice crystals is connected with the fact that water vapour possesses less elasticity over ice than over water.

The growing ice crystals in pores and cracks, which move apart and pack the soil particles and units, condition the formation of a water-unstable "frost" structure, which usually disappears soon after the thawing of the soil. As a result of the freezing of soil layer after layer and stratification due to crystallisation forces, freezing of the water leads to the formation, clearly shown in places, in the A_2 horizon, of a stratified and even laminated structure. In places where there is an energetic thermal condensation of water vapour and crystallisation of ice, the soil may become waterlogged upon thawing. This is often seen on roads, threshing-floors and other soil platforms which get packed on the surface.

More often than not, the formation of frozen ground precedes the appearance of a stable snow cover and snow comes to lie on frozen soil. More seldom the snow cover comes to lie on a cold but not yet frozen soil. Snow falling after the soil has been frozen and covering it with a thick loose layer warms it up as it were and prevents it from growing much colder, although it continues to freeze.

Soil freezes to a greater or smaller depth, depending on the number of days during which the temperature remains below zero, the actual temperature reached and the amount of water in soil, in other words, this depth is governed by the overall expenditure of cold.

The process of the freezing of soil is attenuated by forest litter, standing timber and grass cover. The soil of a forest freezes less intensely and to a lesser depth and thaws earlier than in a field. In a taiga forest it usually thaws before the snow begins to melt. That is why in a forest, melted snow is mostly absorbed by the soil and runs off to a lesser extent. Land under a grass cover or under tall stubble and covered with a thicker snow cover freezes to a lesser depth than ploughed up land. On slopes and high ground, the soil gets frozen down to a greater depth. The depth to which soil freezes is inversely proportional to the thickness of the snow cover.

The soil usually thaws in an irregular fashion, in connection with its properties, weather, relief, cultivation and other conditions. When the temperature of the air rises, even in winter, the soil may partly thaw at the lower boundary of freezing. Sometimes the soil is freed from frost before all the snow has melted. This occurs when the snow cover reaches a great thickness without becoming packed and the soil does not freeze deep. When the thickness of the snow cover is relatively smaller and the soil is frozen to a greater depth, thawing is somewhat delayed. When the effect of cold from above on the surface comes to an end, the thawing which began from underneath has not the time to reach the upper horizon than thawing sets in from above. When frost in soil retreats from underneath and from above, sometimes, at a certain distance from the earth's surface, a frozen layer persists for some time. In that case, the melted water, being arrested by the frozen layer, overmoistens the unfrozen layer but most of it simply runs off from the surface and is lost.

The peat horizon of soil decreases the depth of freezing and delays the thawing of the soil. The frozen layer under peat sometimes persists into the middle of summer and beyond that. The slow thawing of reclaimed peat bogs is tied with the decrease of heat conductivity of the ploughed up peat horizon, which dries out to a considerable depth in summer. Deep winter freezing and late thawing of reclaimed peat bogs cause a worsening of the air regime in connection with the stagnancy of the water situated above the frozen layer, which is poor in oxygen and contains hydrogen sulfide. The soil below the frozen layer is not aerated (A. R. Werner). Not infrequently, seasonal freezing of soils is attended by the formation of cracks, overmoistening of soils and formation of ice sheets.

Frozen soil possesses special properties. It may be water permeable, if it froze at a moisture content below full water capacity. Water may penetrate through the large pores and voids, through the burrows made by earthworms and grubs, through the passages left by decayed roots and through cracks. Frozen soil whose voids are occupied by ice crystals may be impermeable.

Freezing and thawing of soil are of great agrophysical significance. Freezing influences the physical properties of soil, causing it to become more friable and more easily penetrated by plant roots. Deep freezing and cooling of soil in arid zones, by favouring moistening, promotes the development of crops. The freezing of soil attended by winter irrigation of land of arid zones may contribute to the desalinisation of the upper soil horizons, due to the freezing out and displacement of salts and solutions downwards, decrease of evaporation and setting up of conditions for a greater condensation of water vapour.

Permafrost. Permanently frozen soil grounds are found where winter freezing prevails over summer thawing. This occurs as a result of a negative temperature balance in places where the mean annual temperature of the air is lower than the freezing point of water and where there is an accumulation of cold in soil. In the Far North, permafrost may penetrate to a depth of 50-100 m and more.

To the south of the line of delimitation of the present-day distribution of continuous permafrost, it exists in the shape of isolated areas, pockets, especially under peaty horizons in bogs, penetrating into north-eastern China. In large river valleys crossing the zone of permafrost, there is no permafrost, due to the warming effect on soil of the water mass. In the valleys of these rivers, fossil ice is found in places, covered up by deposits from migrating water flows.

In summer, the permafrost layer thaws from the surface down to a varying depth. This depth is greater as we go further south. Summer thawing penetrates deeper (1.5-2 m) in sandy soils and less deep (1 m) in clay soils. On the isle of Kolguyev, the thawed horizon does not exceed 40-50 cm, and south of the 55th parallel (in the Baikal region), thawing affects a layer of 3-4 m.

The annual seasonal (winter) freezing of the ground on most of the northern territory of the U.S.S.R. merges with the lower lying permafrost. Sometimes, in connection with weather conditions, a thawed soil horizon remains at a certain depth from the surface.

Permafrost, as investigations have now confirmed, is subjected to changes. In the distant past, the southern boundary of the distribution of permafrost lay considerably more to the south of its present distribution. Rising of the mean annual temperature of the surface of the earth slowly leads to the intensification of thawing from above and decrease of the layer of permafrost from below. Forest fires cause a lowering of permafrost to a depth of 8 m for a number of years, after which it is restored once more, especially under a developing mossy cover. At the present time, in places, permafrost steadily retreats.

The setting in of permafrost is tied with a decrease of the influx of heat and an increase of its losses in the distant past. At the present time, permafrost persists in the Arctic, Antarctic and everywhere on the peaks of high mountains.

Permafrost does not favour agriculture. In places where fossil ice is retreating, we get the formation of "thermokarst",* which

* Thermokarst—subsidence and sinking of grounds, occurring upon thawing of the ice.

causes a marked deterioration of the forms of the earth's surface. In connection with the uneven seasonal freezing and weak thawing of the soil overlying the layer of permafrost, the root system of plants suffers from the low temperatures of the soil and from the deformation of the shallow horizon which harbours the roots. Often, the layer of soil overlying permafrost becomes bogged up. On permeable soils, permafrost may retard excessive filtration of water, creating relatively favourable conditions for the cultivation of agricultural crops.

Over permafrost, when the thawed layer freezes, there is a formation of irregularities in the microrelief and distension on the surface, causing pushing out of piles and foundations. This leads to the destruction of structures, railway tracks, disruption of the profile of drainage channels, rupture of roots, etc. Deformation of soil-grounds and other unfavourable occurrences are tied with the distillation of vapour and its condensation on the surface of underground ice and permafrost. The harmful influence of permafrost on agriculture can be successfully counteracted. Of great significance in this respect are thermal meliorative measures: the application to soil of organic manures, covering up through the winter, reclamation of swamps and bogged up land, ploughing up of virgin land, improvement of the structure and composition of soil, etc.

Heat balance. The heat balance—the incoming and outgoing of heat—is no less important for the development of soil and its fertility than the salt (nutrient) and water balance, with which it is closely connected. It is obtained from a comparison of the heat coming in and going out (utilised and lost) in a given period of time. Heat balance is expressed in terms of cal cm^2 (per second, month, season, year, period of years). It undergoes constant fluctuations, depending on changes of the native conditions and under the influence of man's productive activity.

Due to the fact that many sources and losses of heat cannot be measured with sufficient accuracy, heat balance is determined with greater or lesser approximation.

According to D. G. Vilensky, the heat balance of soil is expressed by the formula:

$$S = B + L + V$$

or

$$S - V = B + L,$$

where S—radiation balance (incoming-outgoing of radiant energy);
B—heat exchange in the active layer (soil+plants);
L—heat exchange in the air;
V—heat exchange connected with the moisture circulation—evaporation and condensation.

In the daytime, in the summer period, the radiation balance is positive. The surface heat (*S*) received goes towards the heating of soil (*B*), of air (*L*) and evaporation.

The heat balance is made up of the following sources of incoming and outgoing heat, of widely varying values:

A. Sources of heat:

1) direct heating of soil by sun-rays; this is the main and most important source of heat;

2) heating of the soil as a result of contact with the air of the ground layer of atmosphere;

3) heat given out by atmospheric precipitations;

4) incoming of heat into soil as a result of the adsorption of gases and water vapour and condensation of the latter (underground dew);

5) heat of wetting;

6) inflow of heat from warmer soil horizons to less warm ones;

7) inflow of heat as a result of the oxidation of organic matter of exothermic, chemical and biological processes;

8) latent heat upon the formation of ice; inflow of heat from the low-lying layers of the lithosphere and other less important sources.

B. Losses of heat from soil:

1) radiation of heat from the soil's surface into the atmosphere;

2) cooling of soil as a result of its contact with the cold air of the ground layer of atmosphere;

3) cooling of the soil due to the evaporation of atmospheric precipitations falling on the soil surface;

4) cooling of the soil due to evaporation of soil moisture;

5) heat surrendered by soil to gravitational water;

6) outflow of heat to less warm soil horizons;

7) loss of heat through endothermic reactions upon physico-chemical and biological processes;

8) heat used up in the melting of snow and ice and other losses.

It follows from the above that the heat balance (*E*) can be expressed by the following simplified formula: $E=A—B$ (inflow-outflow).

The sources of heat reaching soil and its losses vary from soil to soil and the heat balance can therefore be either positive or negative. In the first case, more heat is received by soil than is lost and in the second case, it is the reverse. But the heat balance of the soils of any given zone does not remain unchanged; it changes appreciably with time.

General changes in the heat balance of soils may be brought about by periodical warming up or cooling down of the climate, in connection with natural phenomena (formation of mountains, changes in the coastline, changes in the direction of cold and warm sea currents, nuclear reactions in the earth's crust, cosmic causes, etc.). Particular changes of the balance may be due to changes in the local physico-geographical conditions (changes of the relief in connection with the manifestation of erosion and the accumulation of deposits, changes in the drainage of the territory, marked decrease or increase of the surface of water bodies and reservoirs, changes in the vegetative cover, ploughing up of virgin land, artificial irrigation and reclamation of land, snow retention, retention of melted snow and ice, moisture supply, planting of forest belts, etc.). It is well known that drainage and reclamation of swamps and bogged up land, snow retention, autumn and winter mulching* lead to warming up of soil and that irrigation reduces temperature gradients, i.e., attenuates excessive heat in the layer of air close to the ground and lowers high soil temperatures which, in both cases, has a favourable influence on soil fertility and yields.

The heat balance of soils of native zones can be regulated not only through hydromelioration but with the help of corresponding agromelioration and forest reclamation and certain agrotechnical measures. A vegetative cover has the general effect of equalising the temperature of the soil, reducing its annual heat circulation, contributing to the cooling of the layer of air close to the soil, due to transpiration and radiation of heat. Large water bodies and reservoirs exert a similar beneficial influence, but indirectly, on the heat balance of soil. The same goes for mulching, a fairly effective agrotechnical measure, which reduces evaporation of moisture from the soil and regulates its heat. The overall heat balance of soil depends on the character of the mulch: its colour, degree of friability, compactness and composition. A black mulch increases warming of the soil. Various types of mulches are applied to soil to raise its temperature: black bitumens, sapropel, earthy brown coal of low calorific value, gumbrine,** lowland peat, black clays and others. Mulch also protects soil from erosion, improves its biological conditions, contributing to an increase of the nitrates. By increasing the CO_2 content in the layer of air close to the soil, organic mulch favours photosynthesis. Mulch used for darkening the surface of the soil acts at the same time as a fertiliser (peat, sapropel, gumbrine). The materials used to darken soil in order to warm it up or to accelerate the melting of snow may be ash, black bituminous clays, brown coal and others. Mulching of soil with a fairly thick layer (several tens of centimetres) may also be adopted in order to accumulate cold in the soil (M. M. Krylov), for forming ice sheets,

* Mulching is the spreading of loose material on the surface of the soil.

** Gumbrine is a fine-dispersive clay which, once it has been freed from petroleum oils, acquires the aspect of a black powder containing up to 40% of organic matter.

to intensify thermal condensation of water vapour within the soil, for winter irrigation and colmatage of sandy soils of semideserts.

The thermal balance of soil is amenable to regulation within varying periods of time: 24 hours, one season, one year, a number of years, which makes it possible to create a more favourable thermal regime of soils and to augment the area of land that can be brought under cultivation. The regulation of the heat balance of soil has not yet become the object of exhaustive studies but it is a promising field of research.

Fairly simple measures, such as mound planting and ridge culture, contribute to the creation of favourable conditions for the right heat, light and water-air regimes in the Far North. On sunny days, the average daily temperature in the root zone is several degrees higher when the soil is in mounds than when it is in flat beds. A promising field for thermal melioration is the use of electric, water and vapour heating of open fields for growing early vegetables.

The regulation of the heat regime and heat balance of soil together with the water-air ones is of very great practical and scientific significance. We are confronted with the task of regulating not only the water-air but also the heat regime of soil, in particular with regard to reducing freezing up and accelerating thawing of soil.

Soil Air and Air Regime

Soil air, or the gaseous phase, is a most important component of soil. It is a source of oxygen for the plant roots and aerobic microorganisms. Air is found in soil in three conditions: a) free—that which occupies the pores free from water, or part of these pores; b) adsorbed—concentrated on the surface of soil particles; c) dissolved in soil water.

The composition of soil air is not constant. It depends on the amount of it present in soil and on its mobility, on biochemical and other processes and differs markedly from the composition of atmospheric air. Particularly subjected to changes is the oxygen content during the period of intensive development of plants and bacterial processes, such as nitrification. During a sharp increase in the production of nitrates, there occurs a marked decrease of the oxygen content, particularly in overmoistened soils.

The difference between soil and atmospheric air is due to the processes going on in soil, which involve the fixation of O and the production of CO_2, the respiration of roots and the activities of microorganisms which bring about the decay of organic remains. The oxidation of organic compounds upon decay is attended by an appreciable enrichment of the soil air with carbonic acid. The oxygen content of the soil is all the lower as it is more intensely used up, as the plants develop better and as the microbiological processes are more intense. Investigations conducted by the Russian scientist N. G. Kholodny have shown that soil air and the layer of air close to the surface contain a number of volatile organic compounds which serve as nutrients for bacteria and higher plants.

The composition of soil air changes as we go deeper down into soil. The carbonic acid content increases with depth, due to its high density and other causes and the oxygen content goes correspondingly down. At a depth of 150-200 cm, the carbonic acid content sometimes reaches 10-12%. The presence of a large amount of carbonic acid in soil is detrimental to plants. Soil air always contains more water vapour than the atmosphere. Its relative humidity is usually around 100%. The soil air of swamped and boggy soils contains hydrogen, hydrogen sulphide, methane and other gases.

That part of the soil air which is in the bound, absorbed condition is represented principally by N and CO_2. It may be devoid of oxygen. Soil air is more or less radioactive and ionised. The radioactivity of soil air is connected with the emission in soil of radioactive radiations, due to the fact that it contains radioactive elements: uranium, radium, thorium, actinium, potassium, rubidium, samarium, carbon. Radioactive elements play an important role in the life of soil organisms. In the absence of radioactive elements, there is no development of nodule bacteria on the roots of leguminous plants. Radioactive elements stimulate growth, accelerate flowering and ripening of plants.

The gases and vapours are absorbed by the solid parts of the soil with varying energy and there is a production thereupon of heat of sorption or heat of wetting. According to the degree of their absorption by the solid phase of soil, gases are arranged in the following order: H_2O (vapour) $>NH_3>CO_2>O_2>H_2S>CH_4$. Absorption goes up with an increase of the soil's humus content and especially of its content of iron hydroxide. As the temperature of soil goes up, its capacity to retain gases goes down.

Soil water dissolves gases, whose solubility increases with a fall in temperature and decreases with a rise in temperature (Table 26).

Table 26

Solubility of Air and Gases in Water in 1 cm³ to 100 ml of Water

Gas	Temperature in °C			
	0	10	20	30
air	3.0	2.3	1.9	1.6
CO_2	171	119	88	66
O_2	4.9	3.8	3.1	2.6
N_2	2.4	—	1.5	—

With a rise in pressure of the gases, their solubility goes up.* Upon an increase in the concentration of the solution in salined soils, the content of gases in soil goes down.

The composition of soil air changes constantly as a result of gas exchanges with the atmosphere in connection with the daily and annual temperature fluctuations, the speed and direction of the wind, infiltration of water, etc. Changes in the composition of soil air depend on the rates of the biochemical processes. Gas exchanges between soil air and the atmosphere take place also in connection with the diffusion of gases, with fluctuations of the atmospheric pressure, with changes in the level of the ground water, upon a difference of temperature between the layer of air just above the ground and soil air. With a rise in the temperature of the soil, the diffusion of gases goes up. In the daytime, upon warming of soil, air increases in volume and part of it is expelled into the atmosphere together with carbonic acid. At night, on the contrary, atmospheric air enriched with oxygen is drawn in. Gas exchanges take place principally in the noncapillary pores.

Changes in soil air depend on the composition and character of the vegetative cover, the weather conditions, the manures applied to the soil, etc. The carbonic acid content of soil air is higher in summer than in winter. In spring and in the first part of summer, overmoistening and shortage of oxygen encourage anaerobic processes. Excessive watering of the soil may lead to the same result. This, in the main, is also the reason why soddy-meadow soils are richer in carbonic acid than ploughed up land.

An increase of the carbonic acid content exerts a substantial influence on the course of soil formation, by affecting the reaction of the medium. A shortage of oxygen in soil promotes denitrification and accumulation in soil of incompletely oxidised compounds, and in soil air, of harmful CH_4, H_2S, etc. An increase of the carbonic acid in the solution is accompanied by an increase of the concentration of hydrogen ions. This leads to an increase of the solubility of $CaCO_3$ and other salts.

An increase in the concentration of hydrogen ions in the soil solution disturbs the equilibrium between the solution and the soil's colloidal complex. Part of the hydrogen ions pass into the colloidal complex in the exchange condition, displacing into solution an equivalent amount of exchange ions, such as Ca, Mg and others.

The air regime of soil is no less important than the heat, water and nutrient regimes. "Soil taken without the gases is no soil,"

* According to Henry's law, the solubility of gases is determined by the ratio of the concentration of the dissolved gas to its pressure. The concentration of gas in solution is directly proportional to the concentration of gas in air over a liquid.

said Academician Vernadsky. The significance of the gaseous phase of soil has been proved by agricultural practice. The plants growing in soil and the organisms which live in it need constant aeration—the exchange of soil air and gases contained in it. Aeration or the exchange of soil gases, in the broad sense of the term, depends on the air capacity, air permeability and air exchange of soil. Air capacity, i.e., the amount of air it can contain, is a dynamic quantity, which depends on the soil's pore-space and its moisture content. It changes with changes in the soil's moisture content. Maximum air capacity is exhibited by dry soil in which the content of air is equal to the overall volume of the pores. Air capacity may be at its maximum, whereupon it is equal to the pore-space; it may be at its minimum and corresponds then to the condition of trapped air and it may possess its field value, specific for the given conditions, or constitute mere aeration, in the narrow sense, meaning that soil contains air as distinguished from aeration or the process of gas exchange between soil and atmosphere. Soil exhibits minimum air capacity when all the interstices (pores) are filled with water and when soil contains a minimum quantity of gases. Field air capacity (aeration) is a variable quantity and depends on weather and agromeliorative conditions. On the whole, aeration fluctuates with changes in the soil's moisture content. Upon an increase of the soil's water content, there is a corresponding decrease of the volume of the air present in soil or of the number of pores free from water, in other words aeration decreases. A lowering of the moisture content raises the aeration of soil and the supply of oxygen to plants and soil organisms. Improved aeration depresses anaerobiosis and intensifies aerobiosis.

The amount of air present in soil drops to its minimum in the early spring when the snow melts or immediately after rains and watering and rises to a considerable extent in the relatively dry period in summer.

From the air capacity or aeration of soil depends its air permeability or its capacity to let air pass through. Air permeability is directly dependent on the structural condition and degree of compactness of soil. It is all the higher as soil has more structure. Air permeability increases upon liming of acid (podzolic) soils and gypsuming of alkaline (solonetzic) soils. Applications of farmyard manure and other organic manures have the same effect.

The dynamics of soil air permeability upon wetting in field conditions give an indication of the soil's structural composition. In a soil possessing a water-stable structure, water is absorbed through the capillary pores of the aggregates and air circulates in the noncapillary interstices. The absence of noncapillary porosity substantially lowers air permeability, which reaches its minimum in a structureless soil. Good air permeability is ensured by the setting up of noncapillary porosity reaching up to 10% of the soil's volume.

The permeability of soil to air is calculated by various methods: the potetometric method (by measuring the speed of motion of air bubbles under pressure in a horizontally disposed curved graduated glass tube), the manometric method (from the speed with which air passes through soil and a rarefied space), the rheometrical method (by direct measurement using a rheometer of the speed with which air passes through soil).

Soil is the seat of a constant air exchange or bilateral movement of air into and from soil. This movement is connected with the temperature gradient, with fluctuations in barometric pressure and with changes in the soil's moisture content. All other conditions being equal, aeration somewhat deteriorates in a warmed up soil, if the moisture content does not decrease thereupon. In a cooled soil, aeration increases.

Circulating water may exert a mechanical influence on the mobility of soil air. Thus, upon an excess of water on the soil's surface, we get the formation of a zone of temporary saturation, which, as it slowly sinks lower, squeezes soil air in front of it and sucks it up after itself. The reverse happens upon evaporation and pulling up of water to the surface.

Changes in the volume of the gases and constituent parts of the soil, in connection with the daily temperature fluctuations, cause soil air to become mobile. The same changes in atmospheric pressure bring about a bilateral movement of water and air in soil. Winds of varying force and speed cause evaporation of the water and alterations in the condition of the air. The latter occurrence is also tied with the diffusion of gases, their utilisation by soil organisms and their formation in soil (absorption of oxygen, production of CO_2). The rate of the gas exchange between soil air and atmospheric air is conditioned by the structure of the soil and is ensured mainly through the diffusion of gases.

Gas exchange between soil and atmosphere is sometimes figuratively referred to as the soil's respiration. Maximum exchange of the soil air gases takes place in the upper horizons of the soil, in connection with its loosening and the various agricultural operations to which it is subjected. The exchange affects the soil to a depth of 20-30 cm in a matter of hours. Over a period of 24 hours and more, the gas exchange affects a greater depth of soil, but to a much lesser degree. Obstructions to normal gas exchange in soil lead to oxygen starvation, to enrichment of soil air with carbonic acid and even to the death of plants, which does take place in very compact and structureless soils. The same thing happens in soils on whose surface forms a compact crust. Breaking up of the crust and loosening of structureless soils are of great significance from the agricultural point of view. It leads to increased yields.

An improvement in gas exchange leads to an intensification of the biological aerobic processes.

The degree to which the medium is supplied in oxygen (aerobic conditions) may be characterised by the value of the oxidation-reduction potential, or *Eh* in volts. The difference of potential on different electrodes (platinum and calomel) in soil depends on the properties of the reducing and oxidising agents which it contains and on their concentration. The correlation between the oxidising and reducing substances gives an indication regarding the amount of oxygen available in soil. The value of the oxidation-reduction potential corresponds to the degree of aeration. A lowering of the *Eh* points to a higher reducing and lower oxidising tension of the medium. During the vegetative period, the soil's *Eh* is subjected to considerable fluctuations. The air regime of the soil, which depends upon the moisture content, determines the oxidation-reduction potential and the trend of the biological processes on the whole and nitrification in particular.

The requisites for a normal development of plants are an optimum air regime and a favourable balance of the soil air. The gaseous phase of the soil is subjected to substantial changes during the annual cycle and from season to season. Of great significance with regard to the overall balance of soil air is the gas balance of the components of air: O, N, H_2O, CO_2, etc.

Chapter XI

CLASSIFICATION OF SOILS AND TYPES OF SOIL FORMATION

Classification of Soils

The soils found on the earth's surface present considerable diversity; a good knowledge and proper utilisation of these soils can only be achieved on the basis of their classification. The classification of all the diverse soils is their division into corresponding groups (types, species). This division of soils rests upon the chief points of difference and resemblance between them. Soils are classified according to a complex of characters and properties; they cannot be classified according to some one single character, such as mechanical composition or humus content, for example, owing to the fact that soils of different properties and origin may nevertheless possess the same mechanical composition and the same humus content. The same goes for structure, water, chemical and other properties of soil.

The basis for the classification is afforded by forms of movement, viz., the stages of development of soils and the changes which they undergo under native and production conditions. Such a classification of the soils reflects the genetical sequence of their development as a series of interdependent phases of one single soil-forming process and their spatial distribution on the earth's surface.

To suit production purposes, the classification of soils reflects their properties, inasmuch as they form the basic means of agricultural production. The following classifications are known today:

a) natural-historical (native)—genetical;

b) technical or applied: agronomical, meliorative, road construction, forestry, sanitary, etc.;

c) economic—the soils being grouped according to fertility, productivity, etc.

Genetical classifications can be traced back to the soil classifications worked out by V. V. Dokuchayev and N. M. Sibirtsev, who were the first to define the main types of soil formation. P. S. Kossovich and K. D. Glinka devised a system of soil classification based on moisture content, internal properties of soils and particularities of the soil-forming process. K. K. Gedroits proposed to classify soils according to their colloidal complex and their degree of saturation or unsaturation with bases. He distinguished: a) soils unsaturated with bases (acid): lateritic, podzolic; b) soils saturated with bases: chernozemic (neutral) and solonetzic (alkaline); bog soils being also subdivided into unsaturated (northern) and saturated with bases (southern).

A chart illustrating soil classification will be found in Fig. 37.

In the classification of soils elaborated by the Soil Institute of the Academy of Sciences of the U.S.S.R., the following taxonomic units are distinguished:

I. *Soil type and type of soil formation*. Soil type is a summary concept embracing all the various soils linked by a common origin and retaining for a prolonged time the common features of a type of soil formation. A type of soil formation manifests itself in space over a vast territory, corresponding to a native soil zone. It is governed by a complex of physico-geographical conditions of soil formation; reflects a period of development of one single soil-forming process on earth. A soil type possesses its own inherent features and structure, the characteristic composition and properties of the soils of which it is composed and reflects a type of soil formation. It reflects the characters and properties of many concrete soils, genetically related. A type of soil formation can be represented by several soil types.

II. *Soil subtype*. Each soil type is divided into subtypes according to the degree of expression of the main and superimposed soil-forming processes. It reflects some stage of development of a main type of soil formation, being an intermediate stage from one type or subtype to the next one. Some also distinguish soil provinces of one or the other type of soil formation.

III. *Soil genus*—includes a set of soils within the limits of a subtype. The genetical particularities of a genus of soils are governed by modifications in the soil-forming process and the influence of the complex of local conditions: the composition of the parent rocks, the chemism of the ground water, changes in the forms of the earth's surface, biological particularities, etc., including the residual soil properties acquired in the course of previous phases of soil formation.

IV. *Soil species*—includes a set of soils within the limits of a genus, corresponding to a certain less marked degree and stage of development of a soil-forming process. Soils are grouped into species according to degree of podzolisation, leaching-out, stage reached by the sod formation process, degree of salinisation, depth and degree of humus formation, swamping, etc. Not infrequently, within one species are distinguished further subdivisions (subspecies) according to the character of the soil-forming rocks, the chemical composition, humus content, character of the illuvium, structure, peat formation, thickness and degree of expression of the genetical horizons, degree of development of certain soil properties, etc.

V. *Variants*. They are distinguished according to the main particularities in petrographic composition connected with the parent rock, according to the degree of taming, the effect of meliorative measures, etc. Further subdivisions are conceivable in accordance with the scale of the soil survey and plans for their utilisation.

The classification of soils serves as the basis for making soil and soil-melioration maps of various scales, from rough small-scale maps to detailed large-scale ones, on which are indicated the species, varieties and variants of soils, as well as finer subdivisions and meliorative particularities. Soil and soil-meliorative maps serve as the basis for soil-meliorative division into zones, which is of particular importance with regard to agricultural production. Without dwelling on a detailed examination of the taxonomical units involved, we shall pass to a brief description of the main types of soil formation.

Types of Soil Formation

Extending in a latitudinal direction over the vast expanse of level land stretching across Europe and Asia in the form of broad bands or belts, we find the tundra, forest, forest-steppe, steppe, semidesert, desert and partly subtropical native soil zones. This zonality of soils is bound up with the zonality of the factors and conditions of soil formation themselves, in the first place with the zonality of climates—the main zonality of the whole complex of natural processes.

The native soil zones are the forms of distribution in space, of the types and in time, of the periods of soil formation. On the territory on which are distributed the soil types, in the form of separate subzones—subtypes, areas of soil species and contours of subdivisions, we find reflected all the stages of development of soils, which are bound up with the age of the land.

The types of soil formation are the result of a physico-geographical process, which does not follow the same course in the various separate zones of the earth's territories. The soil formation typical for each native zone manifests itself by deep changes in the parent rock under the influence of the corresponding combination of factors and conditions of soil formation. It is characterised by the exchange of matter and energy inherent to the given native soil zone. To the types of soil formation correspond the soil types which develop in interdependence with the conditions of the environment and which have their own morphological features, composition, structure and properties. The soil types are distinguished one from the other by grouping together the diverse concrete soils related from the genetical standpoint. Each representative of a genetic type of soil formation possesses its main features, showing differences in details in accordance with modifications in the soil-forming process.

The following types of soil formation are distinguished:

1) taiga-podzolic; 2) meadow soddy-podzolic; 3) boggy: a) hydrogenous (acid), b) hydrohalogenous (alkaline); 4) steppe: a) chernozemic, b) chestnut; 5) solonetzic; 6) desert; 7) lateritic (subtropical).

If the types of soil formation given above have, on the whole, been established and have received universal recognition, the same cannot be said of the soil types, which are still only in the process of becoming established and have not yet received their full expression in soil classification. That is why we shall enumerate the soil types in the form of a systematic list: 1) tundra-gley, 2) podzolic, 3) soddy-boggy, 4) podzolic-boggy, 5) boggy, 6) humus-calcareous, 7) soddy-gley (dark coloured), 8) meadow-boggy, 9) grey forest, 10) grey forest gley, 11) chernozems, 12) meadow-chernozemic, 13) chestnut, 14) meadow-chestnut, 15) brown desert-steppe, 16) brown meadow-steppe, 17) solonchaks, 18) solonetzes, 19) solods, 20) sierozems, 21) meadow-sierozemic, 22) grey-brown desert, 23) takyr-like, 24) takyrs, 25) brown forest, 26) mountain-meadow, 27) mountain meadow-steppe, 28) red soils, 29) yellow soils, 30) soddy-alluvial.

Soil formation types are bound up with soil-climatic and vegetation zones.

The boggy (hydrogenous) type is, in the main, confined to the tundra region, although soils of this type can be found in all the other zones and in the taiga zone in particular. The taiga-podzolic type is distributed in the coniferous taiga belt. The meadow soddy-podzolic type is confined to the coniferous-deciduous taiga. The steppe type is inherent to the steppes of southern regions, including semideserts. The solonetzic type of soil formation also belongs to the steppe zone. In deserts are formed soils of the desert type of soil formation. The lateritic type of soil formation (subtropical) is inherent to the tropics and subtropics.

The formation of soils of the podzolic and boggy type of soil formation proceeds under the conditions of a moderately cold climate with a prolonged winter. During the period of low temperatures, when the biological activity is checked, there is an interruption in soil formation. Soils of the podzolic type of soil formation are formed under a cover of coniferous forest taiga vegetation, which gives rise to a forest floor. Atmospheric precipitations, which here reach 350(500)-550(650) mm, pass through the forest floor and a considerable part of the water percolates downwards, leaching the soil. Intensive fungous decay of the woody remains of the forest floor is attended by the formation of strongly dispersive organic acids and their salts, which are readily soluble in water. Under the influence of the organic acids (mainly fulvic acid: crenic and apocrenic) as well as of the microorganisms and CO_2, the mineral part of the soil undergoes pronounced destruction.

The solutions of acids, which circulate through the soil during a prolonged period, destroy the silicates of aluminium and colloids, with the formation of a chemically inert quartz, lagging behind in the cycle of changes. In the absence of coagulates (Ca, Mg), the solutions of hydrogels which are formed are leached from the upper soil horizons down to a certain depth where a well-defined washed in horizon (illuvium) is formed. In the illuvial horizon, the pores become filled with displaced colloids, which is the first stage towards a sharp decrease in water permeability and the setting in of swampy conditions in podzolic soils.

The podzolisation of soils lowers their productivity and the stand of trees gets thinner. Thinning out of the forest favours the growth of a grassy vegetation and under it, on the basis of the podzolic soils, are formed soils of the meadow soddy-podzolic type of soil formation.

An important role in the process of podzol formation is played by the surplus of moisture in soil from the surface, which causes temporary anaerobiosis and high dilution of the soil solutions. Newer and newer portions of substances coming from the solid phase of the soil are inevitably drawn into the solution. A lowered coefficient of filtration of soil-forming rocks does not hamper podzolisation and even promotes it, prolonging the duration of the leaching effect. As the forest vegetation, which is not rich in ash elements, dies out, it does not ensure the neutralisation of the humic acids and is therefore unable to stop the leaching out process. With thinning out of the forest and the establishment of a grassy vegetation, more and more bases are drawn into the biological cycle of changes. Being set free upon the decomposition of the grassy vegetation, the bases slow down podzol formation, which passes into a soddy-podzolic soil-forming process.

Podzol formation sets in as a result of the relative prevalence of the geological eluvial process over the accumulative biological process; it is not indispensable, in this connection, that atmospheric precipitations should sharply prevail over evaporation, the only important condition being the prevalence of a permacidous leaching type of water regime.

Podzol formation is attended by soil gleisation—the process of translocation of reduced substances. As this process grows in intensity, it displaces the podzolic process and passes into a boggy process. To begin with, gley formation is accompanied by peat formation but later, the latter becomes prevalent. The reason for the boggy process of type formation may be afforded by the succession of plants and the accumulation of plant remains capable of retaining water, a surplus of moisture being the consequence. But subsequently, the surplus of moisture becomes the main reason for the boggy soil formation. In bogs proceeds a process of intensive accumulation of slightly decomposed organic matter—peat. The stifling of podzol formation by a soddy process of soil formation proceeds in nature at a fairly slow pace and with uneven progress. Under productive conditions, podzol formation can be checked within a very short period.

Soils of the steppe type of soil formation are formed under conditions of a negative water balance under grassy steppe vegetation. These soils are characterised by saturation of the soil absorbing complex with bases (chiefly Ca and Mg). Translocation of the sols in sesquioxides and silt-like fractions in the soil section is either absent or very slight. The wetting of soils by atmospheric precipitations is light, which conditions the relatively weak differentiation between the genetical soil horizons.

The steppe zone occupies quite a broad latitudinal band of territory, hence the variety in soil formation environments. As we move from north to south, the climate becomes drier and, consequently, the character of the vegetation and soils changes. The meadow steppe of the northern zone passes, to the south, into dry steppe and still more to the south, into semidesert and desert steppe. Accordingly, the soils change too. Under the lush vegetation of meadow steppes and conditions of an optimum water-air regime, are formed northern, typical, high in humus and southern chernozems. Further south, the southern chernozems pass into the chestnut soils of dry steppes and then into the brown semidesertic soils and the sierozems of desert steppes. Passing from north to south, the alkaline reaction of the soils goes up and with it also, to a certain extent, the concentration of the soil solution. A carbonate illuvium forms at a certain depth from the surface. In the steppe zone, the amount of atmospheric precipitation is barely sufficient to remove the readily soluble salts and ensure the leaching to a certain depth of calcium and magnesium carbonates. There is little translocation of silica and sesquioxides through the profile.

The chernozemic type of soil formation is characterised, on the whole, by a process of carbonate enrichment with some leaching in the northern part of the zone. As we move towards the southern regions of the chernozemic zone, we note signs of a shortage of moisture (dry conditions) and an increase in carbonate enrichment (Fig. 38).

Soils of the zone of dry steppes (chestnut and brown) are formed under conditions of an impermacidous leaching water regime with a pronounced moisture deficit. The soil is slightly moistened to a depth of not more than 100-150 cm. The soils of the zone of dry steppes (chestnut type of soil formation) have a lower organic matter status, little structure and are characterised by a certain salinity. The relatively stable water regime of meadow steppes is replaced by an intermittent one. As the climate becomes drier, meadow steppes pass into dry steppes with a sparse stand of grass. The humus accumulated earlier on is intensely decomposed, with the formation of salts of monovalent metals, until they prevail over the others. In this connection, the absorbed calcium is partly displaced from the humus horizon, being replaced by magnesium and sodium. The carbonate illuvium moves upwards.

The soil effervesces with hydrochloric acid from a depth of 15-25 cm. Below the carbonate horizon, and here and there partly combined with it, arise a sulphate (gypsum) and a salt horizons.

Chestnut soils are considerably more packed in profile, in connection with the fact that the role of the mineral part of the soil takes progressively precedence over that of the organic part. Towards the south, chestnut soils pass into still more packed, containing less humus, more calcareous and saliferous brown soils. To the south-east, in the Transcaspian region and in Trans-

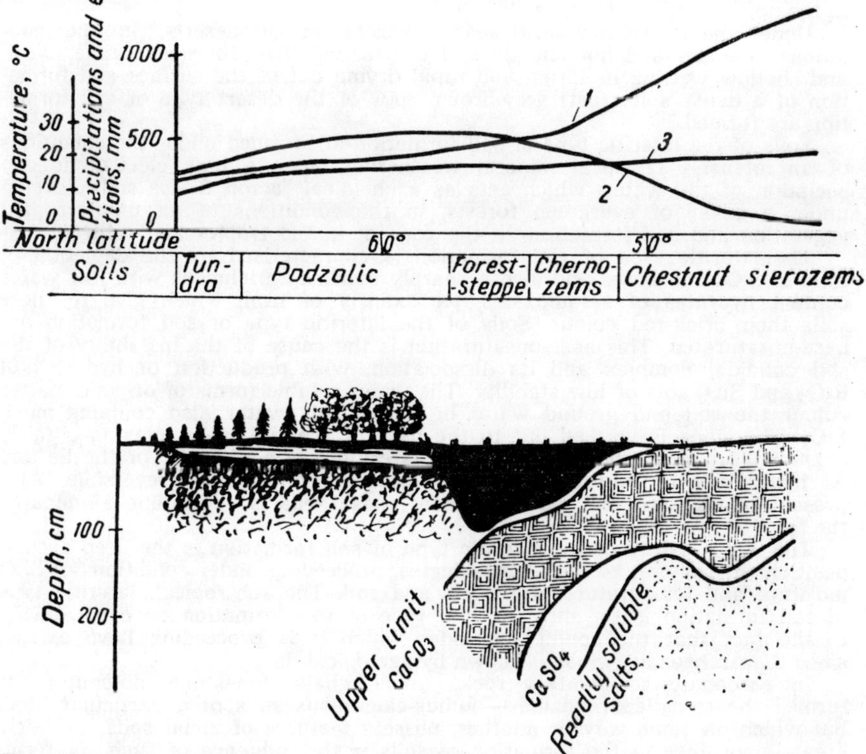

Fig. 38. Diagrammatic meridional profile of an area in the European part of the U.S.S.R.:

1. evaporation; 2. precipitations; 3. mean July temperature

caucasia, brown soils pass into sierozems and yellow soils, characterised by absence of translocation of R_2O_3. The humus content of these soils falls to a minimum (less than 1%); the carbonates and soluble salts reach the surface, owing to the fact that their leaching is excluded. In connection with the extremely low activity of the water and in spite of the prolonged warm period, the soil-forming processes cause little change in the soil-forming rock. The profile of sierozems is still less differentiated into horizons. The absorbing complex of sierozems is saturated with Ca, Mg and partly Na. The mineral part of the soil prevails over the organic part so that in appearance, the sierozems are close to the parent rocks. Salt accumulation is greater in the sierozemic zone than in the neighbouring zones, which explains the wide distribution here of solonchakous soils and solonchaks.

To the steppe zone belongs the solonetzic type of soil formation, arising as a result of the high concentration of salts in the soil solution, which ensures the displacement of calcium by sodium in the soil absorbing complex. The accumulation in soil of free silicic acid is connected with the saturation of the absorbing complex with sodium and the instability of the latter. Solonetzic soils are formed where the halohydrogenous* processes of soil formation prevail over the organogenous ones. When passing from the chestnut zone to the brown one, alkalinity decreases and towards the sierozemic zone, it falls to a minimum, in connection with complete absence of leaching out in the upper horizons of the soils.

Under the thistly-low bush sparse vegetation of deserts, in the conditions of a dry and hot climate and contrasting hydrothermal regime (weak and shallow wetting in spring and rapid drying out of the surface and formation of a dense soil crust) grey-brown soils of the desert type of soil formation are formed.

Soils of the lateritic type of soil formation are formed under the conditions of an intensive chemical mineral weathering, upon a high electrolytic dissociation of the water, which acts as a chemical factor. These soils develop under a cover of evergreen forests, in the conditions of an uninterrupted vegetation and soil formation, in the zone of humid tropics and subtropics.

The lateritic type of soil formation is characterised by the depletion of soils in SiO_2 and bases and also a sharply marked enrichment with low water content hydrates of sesquioxides, particularly of iron, which give to these soils their brick-red colour. Soils of the lateritic type of soil formation are base-unsaturated. This base-unsaturation is the cause of the instability of the soil colloidal complex and its dissociation, with production of hydrates of R_2O_3 and SiO_2 sols of low stability. The water-soluble forms of organic matter colour the soil and ground water in black. This water also contains much SiO_2. The silica is washed out to the base of the soil layer, playing a fairly appreciable part in the lithogenesis of lateritic rocks. The gels of the Fe and Al hydroxides of lateritic soils age, lose water, become irreversible. Also present in the laterites and typical for these soils is crystalline alumina in the form of hydrargillite.

The main feature of the lateritic type of soil formation is the deep decomposition (hydrolysis) of the alumosilicates, proceeding under conditions of high moisture and temperature of the air and soil. The subtropical, lateritic type of soil formation is the most ancient type of soil formation on earth, owing to the fact that the conditions under which it is proceeding have existed since the Archeozoic, which is shown by geological data.

In carbonate sedimentary rocks (marl, chalk, limestone, dolomite) are formed the so-called rendzinas—humus-calcareous soils of a particular kind, but which, in some way or another, possess features of zonal soils. Of fairly great importance in the formation of soils is the influence of such soil-forming rocks as the friable quartziferous sands of the dense sandstones and quartzites, granites, basalts, porphyries and others, on which immature soils are formed.

Standing apart are the soils of flood-plains, which are formed under the influence of the fluviogenic** factor of soil formation, which smoothes down but does not altogether erase the zonal differences between flood plain soils.

For mountainous regions, the soil formation types will be approximately the same, reflecting a vertical zonality possessing its particularities of soil formation, tied with orography, more insolation and other specific conditions. But zonality is not a universal law. In the first place, horizontal and all the more so, vertical zones, do not possess regular geometrical outlines. One zone passes imperceptibly into the next one and the boundary between them forms a very irregular line, whereupon some zones may be enclosed within others or be present as isolated areas or "pockets" of one zone inserted into another. This concerns also the distribution of one type of soils in another.

* From the Greek words meaning salt, water, beget.

** From a Latin word meaning flow.

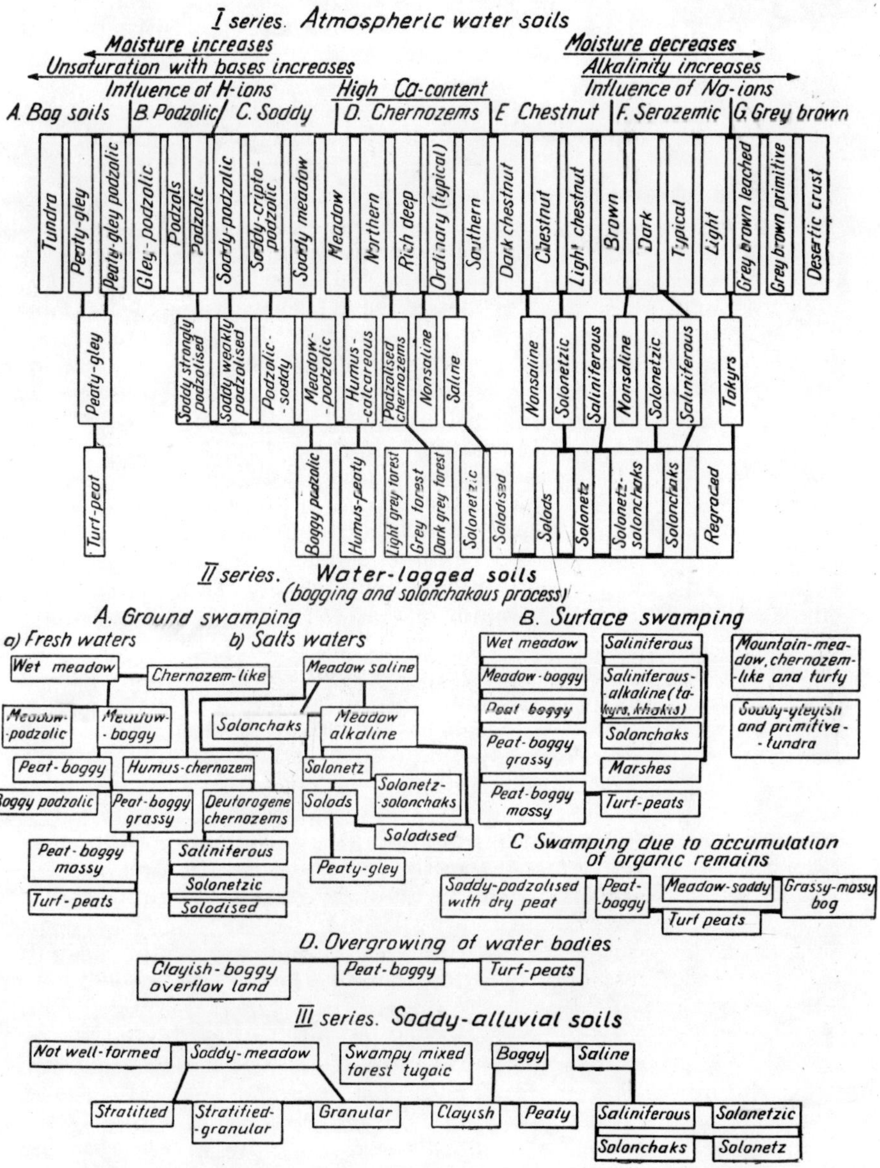

Fig. 37. Chart illustrating the genetical classification of soils (after A. N. Sokolovsky)

Such insertions are directly related to the local conditions, i.e., to the irregularities of the earth's surface, to the sharp differences between the moisture contents of soils, to changes in parent rocks, to climatic and other particularities. Sometimes, this causes the insertion of soil types of different zones, which are not even contiguous. Thus, in the zone of semideserts, we occasionally meet sandy areas with birch and aspen woods and sphagnum peat. In the forest zone, on steep slopes exposed to the south, on carbonate rocks, occur inserted areas of soils of the steppe type of soil formation. Saline soils, which are widely distributed in southern regions, are found inserted, here and there, as isolated areas on gypsum-bearing rocks, in the form of sulphate solonchaks, far into the north (Yakut A.S.S.R.).

Still greater encroachments or divergences from zonality are met with in the mountains, in connection with the exceptional influence of topographic, climatic, lithological and other conditions. Only on the flat tracts of land of watersheds does the type of the mature most widespread soil correspond to some or other natural zone and the rest of the soils, which are not related to the main zonal type of soil formation, have but a secondary distribution.

Within the boundaries of any soil zone, when the leading role belongs to the biological factor of soil formation, the geographical distribution of the soils reflects principally the influence of the climate. And within the boundaries of regions with the same climatic conditions, the soils are formed in accordance with the forms of the earth's surface and the distribution of the moisture.

Inside the zones, these differences lead to the formation of soils of the subzones of the corresponding subtypes of soil formation and of the various species of soils. In accordance with the local (provincial) conditions of soil formation, natural soil zones are divided into parts corresponding to so-called provinces (after L. I. Prasolov).

The regular passage of some soils into others and their strict distribution over the main elements of the relief: watersheds, slopes, terraces and flood plains of river valleys correspond to the so-called soil scales or combinations of soils of the given zone. In some zones, these combinations may possess a relatively greater amplitude (in forest-steppes, for example) or smaller (in the steppe zone, for example). This amplitude exhibits a relative decrease to the north and south of the forest-steppe, in connection with the steady humid conditions to the north and relative dryness to the south.

In order to establish the zonal differences between the soils of the various types of soil formation, we must compare soils of identical elements and forms of relief and should avoid comparing the soils of the watersheds of one zone, for example, with the soils of the slopes of another zone; the same goes for convex and concave forms of the surface.

The combinations of soils, i.e., the scales and complexes of soils are of great practical meliorative significance owing to the fact that it is in accordance with them that the various combinations of chemical, hydrotechnical and agrotechnical meliorations, corresponding to the natural soil zones are adopted. In the north, the problem is to check the excess of moisture and the acidity of the soils. In the south, on the contrary, the problem is to conserve and accumulate moisture in the soils, to neutralise the alkaline reaction, etc. All these measures bring about radical changes in the soil aimed at improving their agricultural value. These measures and the related problems of soil improvement are discussed in the second and third parts of this book.

The percentage distribution of the various soils over the continents is given in Table 27.

Table 27

Percentage Distribution of the Soils over the Continents

Soils and snow	Eurasia	Africa	America		Australia	Total dry land
			North	South		
Tundra	3	—	17	—	—	4
Podzolic	16	—	23	—	—	9
Grey and brown forest soils	7	9	6	8	7	7
Chernozems and black tropical	6	7	7	5	4	6
Chestnut and red brown soils	7	9	7	6	10	7
Desert	15	37	7	3	44	17
Red soils and laterites	9	29	10	59	25	19
Flood-plain soils	4	6	1	7	—	4
Mountain soils	33	3	14	12	10	16
Permanent snow and ice	—	—	8	—	—	11

The most widespread soils of the earth are the red soils and laterites (19%) and the desert soils (17%). An enormous area is occupied by mountain soils (16%).

So far, the utilisation of the above-named areas of the earth for agricultural purposes is most inadequate, since even the most cultivated soils, viz., the chernozems, are utilised to only 35% of their surface. This shows that, up to the present time, the surface of the earth has been in a stage of extensive utilisation, due to the fact that man, deprived of the necessary resources in machinery, has not been in a position to direct the powerful forces of nature towards its transformation. That is why the reserves of potentially useful land are enormous. The advent of electrical and atomic energy opens up boundless possibilities for the development of agriculture.

PART TWO

ELEMENTS OF SOIL GEOGRAPHY

Soils of the Earth and Their Utilisation

The variety exhibited by the soils of the earth and their distribution can be seen on the world soil map. An analysis of the map shows that the soil types form soil zones on the surface of the continents.

In the northern hemisphere, we note the widespread development of the tundra zone and tundra soils. South of the tundra are distributed gley-podzolic soils and still further to the south, podzols, soddy-podzolic and boggy soils. This is followed by a narrow band of grey forest soils of forest-steppes; to the south of these are the chernozems, which develop in the central parts of the continents, without reaching the oceans. Chernozems are mostly found in Europe and Asia. The direction in which they are disposed in North America shows a considerable deviation from the longitudinal direction. The chernozems of North America are not so deep as those of Europe and contain less humus (3-7%). South of the chernozemic zone, are the chestnut and brown soils, which are also distributed in the interior of the continents.

In the southern hemisphere, the chernozems, chestnut and brown soils occupy a much smaller territory. Brown soils (brown forest soils) of broad-leaved (beech) forests, as well as cinnamonic soils of dry subtropical forests are distributed in the maritime regions of continents.

Vast areas of the continents in the northern and southern hemispheres are covered with grey-brown soils and sierozems. Small areas of the humid subtropics are covered with red soils. In the tropical zone, the main soil types are the laterites and red earths.

Podzolic-red soils and podzolic-lateritic soils are distributed under the humid forests (rain-forests) of the tropical zone. In the same zone are also found black soils of dry tropical savannahs. Their black colour is linked with the circulation in them of black alkaline soil solutions and is due to the presence of organo-mineral compounds of iron, manganese and titanium. The humus content of these soils does not exceed 0.5-3%.

A considerable portion of the earth's territory (4%) is covered by soddy-meadow alluvial soils. Soils of mountain regions compose vertical soil zones on the slopes of mountains of vast territories.

Virgin lands occupy enormous areas of the earth's surface. In Australia, Canada and Brazil, the surface of land brought under cultivation does not exceed 2.3%. There is much virgin land also in Alaska and other countries. As a result of the agricultural activity of man, which in places has been going on for thousands of years, the cultivated soils have undergone more or less pronounced changes. In some cases, human interference has led to the formation of swampy (gley), boggy and saline soils, whose fertility has either been considerably impaired or has been lost altogether and, in other cases, to the formation of more or less tame (in most cases to a slight degree) soils. In all cases, the utilisation of land has been haphazard. It has given adverse and, to a lesser extent, favourable results.

Up to the present time, the soils of the earth have not yet been properly tamed, which strongly hinders the economic development of the countries of the world.

In this respect, meliorative soil science opens up enormous prospects with regard to the taming of the soils of the earth. Humanity will turn to the realisation of these possibilities immediately following general and complete disarmament, on the basis of solidarity among the nations, using modern machinery and sources of energy.

Here is a brief description of the soils of some soil zones and regions of the northern hemisphere.

Chapter XII

SOILS OF THE TUNDRA AND FOREST ZONES

Soils of the Tundra and Forest-Tundra Zone

The tundra zone is divided into subzones: 1) arctic desert; 2) arctic tundra; 3) lichen-moss (spotty) tundra; 4) brush tundra; 5) forest-tundra.

In the polar (arctic) zone a few centimeters thick primitive soils are formed, due to the poorly developed biological activity. The work of running water is almost nonexistent. On the other hand, glacial denudation is pronounced. On denudated rock peaks sticking up through the ice, or nunataks, weathering due to frost action (causing physical breakdown of rocks) goes on almost uninterruptedly all the year round. Here and there modern moraines are formed. Chemical weathering in the tundra is weakly developed. The combination in the tundra zone of permanent frost with soil frost leads to the bulging out of the surface, with the formation of mineral and peat hillocks in the form of patches. Here and there occurs solifluction and the formation of stone polygons in the shape of polyhedral patches of denudated soil, bordered with stones. Fossil ice and hydrolaccoliths (underground ice bodies) are met with. There is some peat formation in its early stages.

The forest-tundra is distributed in the transitional band between the tundra and the taiga. It is characterised by areas of isolated, thin, stunted forests, which disappear further north and merge further south with the forest of the taiga zone. The vegetation of forest-tundras consists of stunted birches, spruces, larches (in Siberia), occasionally firs, poplars and willows. The forests here are characterised by the spareness of the stand of trees and stunted growth. The trunks are low, bent in the direction of the prevailing wind. The spareness of the tree stand favours the development of a brush-lichen vegetation in the lower stories. 200-300 years old stunted trees have a diameter of 5-8 cm. There are also flood plain marshes and high bogs in the forest-tundra.

Along river valleys, the forest-tundra penetrates far into the north, into the depth of the tundra, due to the fact that the warm water flowing from the south moderates the severity of the climatic conditions.

Here and there, the forest-tundra penetrates into the tundra in the form of advance forest projections and the forest-tundra cedes the ground to the taiga. This is manifested by an increase in the size of the forest projections and in the height of the tree trunks, a denser stand of trees, etc.

The climate of the tundra is quite severe. The soil of the tundra is permanently frozen down to great depth, which influences the whole course of soil formation. The thin horizon of thawing limits the accumulation of ash plant nutrients.

Along the river valleys of the tundra develop soddy-meadow soils with an A_1 humus horizon up to 20-30 cm thick. On the greater part of the territory of the tundra, outside the river valleys, the soil gets little warmth. Below persists an unmelted layer of permafrost and this source of cold hinders the development of the soil. In the frozen horizon, biological processes are halted. This horizon hinders the penetration of plant roots down to the required depth and promotes swamping of the soils. The thickness of the horizon affected by thaw (30-80 cm) is not quite enough to provide plants with ash nutrient elements. Moses and lichens, on the other hand, develop normally. Waterlogged peaty soils thaw down to a depth of only 30 cm and only a mineral dry sandy soil can thaw down to a depth of 200 cm. Below that, everywhere reigns permafrost, which conditions overmoistening of the soils caused by thaw and the impermeability of the permafrost layer. Under such conditions prevails anaerobiosis.

In the permanently frozen ground of the tundra, here and there, are found fossil remains of woody vegetation and the bones of now extinct animals (mammoths, etc.), which are evidence of the existence in the past of higher temperatures and a displacement of the boundaries of the tundra. Here and there, in the tundra, fossil ice is found. Soil formation in the tundra possesses its particularities, linked with low temperatures, rather high humidity of the air and the influence of permafrost.

The tundra is a zone of wide distribution of swamped soils. Swamping is due to the impermeability of the permanently frozen ground and particular climatic conditions: low temperatures, high relative humidity of the air and weak evaporation from the surface of the soil. On the surface of the frozen horizon occurs a condensation of the soil moisture, which also raises the moisture content of the layer above the frozen ground. The biological and biochemical processes are slowed down in the tundra, in connection with the insignificant annual litter-fall (0.5-1.0 t/ha), which also explains the low carbonic acid content of the soil. The microflora of the soil is uniform and poor. Azotobacter is absent from the tundra soils in connection with the absence of the necessary nutrient elements and the otherwise unfavourable conditions. The soil solutions contain little soluble substances; the conditions for humus formation and mineralisation of plant remains are unfavourable. Organic matter is accumulated in the soils as raw semipeaty acid humus. Biologically absorbed elements only slowly return to the soil. The absorption capacity of the soils is insignificant, reaching 9-16 m-equiv. to 100 g of soil.

The soils of the tundra are base-unsaturated and possess high exchange acidity. In connection with the weakening of the biological activity, the soils here do not reach a high degree of development and are characterised by the possession of thin genetical horizons and an overall shallow profile.

The soil-forming rocks of the level part of the tundra are moraine formations, fluvio-glacial, lacustrine and river deposits of various mechanical compositions. In places, the soils are formed on deposits of late quaternary boreal sea transgressions. In mountainous regions, the soil-forming rocks consist of crystalline and sedimentary geological formations of various ages. In the tundra the gley-boggy type of soil formation predominates, which leads to the development of peaty-gley and peat-gley soils. Of secondary significance is the podzolic type of soil formation on sandy rocks.

The following soils are distributed in the tundra zones:

a) polygonal, ochre-gley and gleised primitive, diminutive soils, distributed mainly in the arctic tundra subzone;

b) boggy, typical tundra-gley, peaty-gley and peat-gley soils, distributed mainly in the lichen-moss and brush tundra;

c) boggy-podzolic, crypto-podzolic, weakly podzolic-gley, peat-podzolic-gley and podzolic soils distributed in the forest-tundra subzone;

d) soddy-meadow, weakly soddy-gley soils, distributed in river valleys.

Large areas of the tundra are occupied by shallow sphagnum turf-peats. The thickness of the peat rarely reaches 1 m.

According to data obtained from their analysis, polygonal primitive soils contain up to 1.5% of humus; the total of the absorbed bases in the A horizon is approximately 15-16%, of which Ca represents approximately 63% and Mg approximately 37%; the pH being close to neutrality.

The gley soils of the tundra have an insignificant peaty horizon (2-3 cm). They constitute underdeveloped primitive soils. Peaty-gley soils have a peaty horizon approximately 8 cm thick, under which usually lies an ochre stratum and, below, a bluish gley horizon with ochre spots. The humus content=2-3%, the pH of the water extract=5-5.8, of the salt extract=4-4.25.

Soddy-meadow soils are the best soils of the tundra. They possess a humus horizon 25-30 cm thick, with a humus content of up to 10%. These soils are distributed in the flood plains of rivers and on well-drained slopes.

In the southern part of the tundra, on sands, are found podzols with a diminutive profile 10-15 cm thick. Under the 1-2 cm thick raw humus A_1 horizon, lies a whitish podzolic A_2 horizon 2-4 cm thick and below that, a shallow illuvial B horizon.

Podzolic-gley soils possess the following shallow profile: A_1—3 cm, with a humus content of 1-3%, A_2—7-10 cm, B—20-26 cm. These soils are found in the forest-tundra region. Podzol formation in the forest-tundra region can be traced not only on sandy but also on loamy soil-forming rocks. The development of soil formation is hampered by ground-soil and surface swamping. The stumps and trunks found in the soils of the tundra are evidence of the past distribution of forests far up to the north, beyond the present northern forest-line. At the present time, there is a trend towards the advance of the forests at the expense of the tundra and a retreat of permafrost.

The obstacles hindering the development of agriculture in the tundra are the short vegetative period, low summer temperatures, permafrost, excess of moisture and poor aeration. Of great significance with regard to the overcoming of these obstacles are the specialised subpolar agrotechnique and melioration, aimed at improving the thermal and water-air conditions and at eliminating the harmful effect of permafrost. It is indispensable to increase the energy of the nitrification process in the soils, which is quite low in the tundra region. This is attained through the application of organic manures (muck), mineral fertilisers (NPK) and bacterial inoculations. The latter favour the intensification of the microbiological processes. A particularly important measure for improving the soils of the tundra is their warming up and the improvement of their aeration through taming, which begins with drying of the land, afforestation, snow retention, liming of acid soils, sowing of leguminosae, structurisation, acceleration of spring thawing, etc.

Soils of the Forest-Meadow Zone

The predominant vegetation of the zone consists of taiga coniferous and mixed forests. The forests alternate with dry meadows.

The taiga-forest belt is divided into subzones:

a) northern taiga: a belt of thin forests, lichen-moss coniferous forests with some birch. The underwood consists of shrub. The subzone contains many swamps and there is a predominance of gley-podzolic soils;

b) central taiga: dense spruce and fir forests, ground cover of green mosses. The soil surface is covered with dead forest litter. Here and there are distributed small-leaved forests consisting chiefly of birch. The soils are podzolş and podzolised and here and there podzolic gley;

c) southern taiga: coniferous-broad-leaved forests; where the forests border upon fields, the species are spruce, lime, oak, maple with a grass cover. Pine grove areas are found here and there, and secondary birch groves and asp forests in the felling areas. In the Asiatic part, broad-leaved varieties are almost nonexistent. The soils are podzolic and soddy-podzolic.

The soils of the forest-meadow zone are formed in a moderately cold continental climate under the conditions of a positive water balance and systematic moistening. Prolonged winters cause long interruptions in the biological processes of soil formation.

The relief of the forest-meadow zone is relatively level and presents the aspect of a typical moraine landscape. The main soil-forming rocks here are moraine deposits. On the moraine occur eluvium of watersheds and deluvium of slopes. In river valleys, the soils develop on alluvium.

The podzolic soil-forming process proceeds under coniferous and mixed forests, the conditions being: a flat relief, systematic moistening and leaching of soils. The formation of podzolic soils is due to anaerobic microbiological processes proceeding intermittently in the upper part of the soil in connection with periodic overmoistening in the A_2 horizon. Under anaerobic conditions occur reduction processes, which lead to the formation of mobile aluminium, intense leaching, and gleisation. Of paramount importance is the role played in podzol formation by forest litter, which consists of dead parts of plants, viz., needles, branches, stems, spruce and pine cones and other plant remains. The forest litter possesses good water permeability and high water capacity. The forest litter protects the surface of the soil from erosion and from losses of soil moisture through evaporation. It is rich in tannic substances and possesses acid properties, which hinder the development of bacteria. At times, the forest litter causes overmoistening, which leads to anaerobiosis. In summer, when rainfall is relatively low, occur oxidation processes. During that period there occurs a rapid fungal decay of woody plant remains, with the production of crenic and other organic acids. Crenic acid forms soluble salts, which are removed with water, and this attenuates the harmful effect of the acid medium. Among the fungi which attack the woody remains are the blue, brown and red rots, as well as various low mould fungi (*Penicillium, Aspergillus, Cladosporium*), ray fungi (*Actinomycetes*) and higher fungi (*Basidiomycetes*). The fungal flora of forest soils constitutes one of the important factors of soil formation and nutrition of plants.

In connection with the considerable amount of decaying organic matter present in soil, periodic overmoistening and weak aeration, denitrification processes are very pronounced. Nitrification is either depressed or altogether absent, in view of the scanty distribution of azotobacter. Forest-steppe soils contain ammonia-producing bacteria. They contain but scanty numbers of organisms responsible for aerobic decay. The decay of plant remains in podzolic soils is accompanied by their mineralisation, with the liberation of bases, which react with the acids. But there are not enough bases for the full neutralisation of the acids, due to the fact that the ash of coniferous vegetation contains but small amounts of them (50-100 kg/ha).

The reaction of podzolic soils, particularly under coniferous forests, is markedly acid. The acidity is less pronounced under deciduous forests.

Forests in the taiga zone constitute the major factor of soil formation. They intensify and accelerate the cycle of changes. The podzolic soil-forming process is characterised by a marked breakdown of primary and secondary minerals and the removal of the products resulting from their breakdown.

Podzol formation leads to unsaturation of the soil with bases, intense leaching of the soil and progressive displacement of exchange cations by hydrogen ions. As a result of prolonged leaching of the soil, in the podzolic soil forms a clearly marked, bleached A_2 horizon completely devoid of soluble salts.

Podzol formation is characterised by decay of the organic and mineral part of the soil, by progressive leaching out of the bases from the soil, depletion of colloido-clayey elements and relative enrichment with SiO_2. The solutions and dispersions are carried with the downward flow of water to the lower sections of the soil, where they are retained as a result of mechanical absorption and coagulation, which cause the formation of an illuvial horizon (B). Podzolisation begins immediately below the forest floor and progresses gradually downwards.

Podzols reach their maximum development under the firwoods and pine forests of the zone. The soils formed under the thin forests of the northern taiga are thin swamped podzols with a podzolic (A_2) and a gley (G) horizons. The humus-accumulative

(A_1) horizon is almost nonexistent. Under the coniferous-broad-leaved forests of the southern part of the taiga, we get the formation of soddy-podzolic soils with a relatively thick soddy (A_1) horizon. Deciduous trees are capable of reducing the unsaturation of soils with bases and of enriching it with ash elements, provided there is no marked leaching out. Deciduous trees slow down podzol formation.

Podzolisation is more intense in periods of higher moistening of the soils, especially in spring and autumn, so that it proceeds intermittently, being either slowed down or accelerated, according to the season of the year. This intermittent character may possess greater amplitude if it corresponds to a series of several wet and dry summers. In years when permacidous leaching becomes unstable or is replaced by impermacidous leaching, the progressive development of podzol formation is slowed down, in connection with the relative predominance of the accumulation of ash nutrient elements. The latter occurrence may lead to the formation of soils of other types of soil formation.

Under a deciduous forest, brush and meadows, podzol formation does not take place, on the contrary, here the soil becomes systematically enriched with ash nutrient elements, the humus horizon becomes thicker, its content goes up, the soil becomes enriched with bases and acquires structure. In the taiga zone, we frequently get a lowering of productivity of the standing timber and a deterioration of the tree-growing capacity of the soil. The inclusion of deciduous trees into the area leads to an improvement of the situation. Favourable conditions are gradually brought about which allow the development of more exacting forest cultures and grassy vegetation.

The process of podzol formation proceeds quite differently depending on the conditions of the environment. This process leads to the formation of more typical podzols on loamy and heavy-loamy parent rocks. On relatively lighter parent rocks, under conditions of podzol formation, there occurs a kind of colmatage or packing of the B horizon to the condition of ferruginous sandstone (ortstein)*. In clayey soils are formed numerous ferruginous concretions. Below the ore horizon is often formed a bluish-grey horizon of gleisation, characterised by the presence of ferrous iron compounds, including vivianite. In appearance and viscosity, the gley horison can be compared with rich clay oversaturated with water. In contradistinction to the podzolic one, this horizon is formed in the zone of permanent overmoistening and anaerobiosis. Gleisation may eventually involve the whole of the soil profile, replacing podzol formation. Gley formation is caused by stagnant

* Colmatage—filling up, deposition of silt. Ortstein, a German word—hardpan, a ferruginous-stony formation.

soil water and subsequently becomes itself the cause of its formation. At certain stages of podzol formation, under natural conditions, there is a self-thinning of the forest and coniferous trees are replaced by deciduous ones.

In mixed (thinned) forests with an undergrowth and grasses, there forms a sod. Under these conditions, the podzolic process is complicated by the superimposition on it of a soddy soil formation, and podzolic soils pass gradually into soddy-podzolic ones.

Soddy soil-forming process. Under natural conditions, the soddy and podzolic soil-forming processes often proceed simultaneously, affording a clear illustration of the struggle between opposites (mineral versus organic) and between rising contradictions (losses of substances through leaching out and accumulation of ash elements).

The soddy process begins under the canopy of a thinned forest with grassy vegetation. The litter layer is gradually replaced by a live cover and disappears. The formerly poorly developed A_1 horizon of podzolic soils gradually increasing in thickness, acquires a granular structure and becomes fully expressed in the shape of a humus-accumulative horizon. In this horizon occurs a synthesis of clayey minerals and the humus content goes up. At the same time, the amount of ash nutrient elements and exchange bases, including Ca and R_2O_3, increases. There is also an increase of the base exchange capacity and of the saturation of the soil with bases. The reaction of the soil solutions approaches neutrality.

The grassy vegetation gradually brings about more favourable conditions for its development, successfully competing with the forest. From a biological point of view, the grassy vegetation is characterised by the fact that it dies out every year with the onset of negative temperatures and is decomposed by bacteria in spring under anaerobic conditions, with the formation of ulmic acid. In the A_2 horizon and immediately above it, there is a progressive accumulation of organic remains and amorphous humus, which are all the more abundant as the grass stand is more developed. The roots of the grassy vegetation permeate this horizon and break up the mineral mass into separate crumbs. The crumbs agglomerate, become impregnated with active humus, after which, as a result of the coagulation of the humus by the bases, they acquire a fair resistance to water. The podzolic A_2 horizon merges with the soddy horizon and acquires structure, water permeability and water capacity.

Under the influence of the soddy process of soil formation, podzolic soils lose their podzolic character and all the faster as they are less podzolised.

The role of the various elements of the relief and their influence upon the speed with which the soils of the podzolic type of soil formation pass into the soddy type and the latter into the boggy type have only been roughly outlined so far.

The influence of the parent rock upon the course of the podzolic and soddy-podzolic soil formation is quite important. On calcareous rocks, i.e., loess, loess-like loams, marls and calcareous moraine, the process of podzol formation proceeds at a slow pace, in view of the high calcium content of these rocks. Podzol formation brings about only weak changes in these soil-forming rocks and affects them to a lesser depth. Broad-leaved trees and the grasses which penetrate into the taiga forests and grow with them, thrive better on calcareous rocks. The forest becomes relatively rapidly overgrown with meadow vegetation. Under the influence of the superimposition of the soddy-soil formation, podzolic soils pass into soddy-podzolic ones, completely losing their podzolic character.

On calcareous soils, where calcium neutralises fulvic acids, favourable conditions are created for the development of the microflora and for the accumulation of amorphous humus.

On rocks such as chalk, limestone, dolomite, marl and others, the A_1 horizon reaches a thickness of 20-25 cm and more and the humus content goes up to 10-12%. Soils of this kind possess a high absorbing capacity (50-60 m-equiv.), full base-saturation and a pH of 7 to 7.5. Under these conditions, the humus accumulative horizon acquires a cloddy-granular structure, the water capacity goes up and so does the depth of the roots zone. We thus get the formation of deep humus-calcareous soils known as the rendzinas.

Soddy-podzolic soils possess a deeper profile than podzolic ones in view of their greater absolute age and of the fact that they get more uniformly wetted by atmospheric precipitations.

There is a certain equilibrium between the podzolic and soddy soil formations. Although the prevailing trend of soil formation goes from podzolic to soddy, this does not mean that the soils of the soddy type of soil formation are not capable of changing towards podzolisation. Should coniferous forests occupy the areas previously under a grassy vegetation, the process of soil formation may become directed towards the formation of soils of the podzolic type. However, this process should, at the present time, be viewed against the background of a gradually developing soddy-podzolic soil formation.

Classification and description of the soils of the forest-meadow zone. The basis for the classification of podzolic and soddy-podzolic soils is afforded by the degree of podzolisation and the degree of expression of the soddy soil formation, i.e., the thickness of the podzolic and humus horizons, the degree of gleisation and the thickness of the illuvial horizon, their age, etc.

The classification is as follows:

I. Podzols: 1) shallow podzols ($A_2<15$ cm); 2) medium podzols (A_2—15=25 cm); 3) deep podzols ($A_2>25$ cm).

II. Soddy-podzolic soils: 1) weakly soddy, strongly podzolic ($A_1<10$ cm, A_2 clearly marked); 2) soddy-medium podzolic (A_1—10=20 cm, A_2 also marked); 3) deep-soddy-weakly podzolised (A_1 clearly marked, more than 20 cm thick, A_2 is represented by isolated spots or is absent altogether).

III. Soddy soils: 1) soddy soils ($A_1<40$ cm); 2) soddy-meadow soils ($A_1>40$ cm); 3) soddy-humus soils; 4) humus-calcareous soils.

According to their degree of gleisation, podzols and soddy-podzolic soils are grouped as follows: a) weakly gleyed (weakly marked ortstein horizon and concretions. Presence of bluish gleisation spots in the illuvial horizon); b) medium gleyed (more concretions; more clearly marked gleisation; merging together, the gleisation spots partly invade A_2); c) strongly gleyed (complete gleisation affects all the horizons. The soil acquires a bluish-grey colour).

The soils most widely distributed in the forest-meadow zone are the boggy soils, of which a classification and description are given below.

Podzols possess a clearly marked typical profile, on which all the genetical soil horizons are clearly seen, beginning with the litter layer, which reaches 5-7 cm. Sometimes, the bottom part of the litter layer acquires a peaty character and increases in depth. Under the litter layer lies a shallow (3-5 cm) grey, coloured by humus, A_1 horizon.

Particularly clearly seen in the soil section is the podzolic proper, strongly leached, loose, structureless or horizontally stratified (laminated on clays) A_2 horizon of a whitish colour, floury, loess-like, enriched with quartz. This horizon often lies immediately below the litter layer. Its thickness varies from a few to 50 cm and more, which gives an indication as to the degree of podzolisation. Under A_2 lies a no less characteristic packed illuvial washed-in horizon (B), enriched with the products of leaching. This horizon is usually reddish-brown, has a nutty structure and is viscous. Its bottom part (B_2) is sometimes packed (ortsand), ochre-brownish in colour.

Chemical analyses reveal a low humus content in the A_1 horizon. The A_1 and especially the A_2 horizon are characterised by relative enrichment with silica, which is due to leaching out. The leaching out process is also revealed by a decrease of the gross amounts of sesquioxides in the surface horizons of the soil. The upper horizons become depleted of their silt, whereas the silt content of the lower horizons goes up due to inwash. In this connection, the lower horizons become packed and weakly water permeable. Podzolic solis are characterised by base-unsaturation.

Among the absorbed cations of the absorbing complex, we find hydrogen and aluminium. Maximum unsaturation can be observed in the upper podzolised horizons. In the A_2 horizon, the absorbed hydrogen may sometimes prevail over the absorbed Ca and Mg. In the C horizon, the absorbed hydrogen goes sharply down or even disappears. The presence of exchange hydrogen confers acid properties to the soil. The greater the podzolisation, the more pronounced the acidity, the pH going down to 4. Base-saturation goes up in soddy-podsolic soils.

Podzolic soils possess an illuvial horizon enriched with colloids, which manifests itself in an appreciable increase of the apparent density with depth (from 1.2 in A_1 to 1.6 in B_2). The pore-space falls from 55% in A_1 down to 40% in B_2. Packing of the B_2 horizon causes overmoistening of the soil in the A_2 horizon and swamping.

Typical soddy-podsolic soils possess the following morphological features:

A_0—2-3 cm thick. The litter layer consists of the needles of conifers, the leaves of deciduous trees and the remains of the grassy vegetation. Not clearly marked. Melts into the humus-accumulative horizon.

A_1—0-15(20) cm, soddy, dark grey at the top and even brownish-black, loose. The upper part of the horizon consists of stable clods and grains. The structural aggregates are bound together by the plant roots, forming a root mat.

The transition downwards is gradual in connection with the decrease in the contents of organic remains and soil humus, in the intensity of the colour and with weakening of the soil structure.

A_2—15-45(55) cm—podzolic horizon, whitish-grey, slaty-laminated structure, loess-like. Passes gradually into the underlying ore horizon.

The white coloration of A_2 is dulled due to the presence of humus, especially in the upper part of the horizon.

B—45-130(180) cm—ore horizon, packed, cinnamon brown, sometimes yellowish, grey, reddish-brown, nutty structure.

Along cracks and passages left by roots are sometimes seen whitish podzolisation spots and glassy powdering. Sometimes found in sandy soils are rusty-brown veins and pseudofibres in the form of benches and a tongue-like packed fringe. Ore grains are met with—ferruginous and ferromanganic concretions. Concretions, veins and pseudofibres contain R_2O_3, manganese oxide, P_2O_5 and organic matter. Microorganisms and among them, iron bacteria play a role in the formation of the mineral part.

C—soil-forming rock, weakly affected by soil formation, bears traces of inwash. As a result of gleisation, a grey gley horizon (G) is formed at the expense of the lower part of the B horizon and the upper part of the C horizon. Breaks down to a slight extent to give nutty units, gleised on the outside. Inside, they retain the colour of the soil-forming rock or of the units of the podzolic soil of the previous stages.

In some soddy-podzolic soils, there is sometimes something like a second humus horizon in the upper part of the illuvium, at a depth of 25 cm and more, which arises as a result of inwash.

The mechanical composition of soddy-podzolic soils changes with age, under the influence of the podzol formation process, in connection with the removal from the upper section of the soil of the silty particles (<0.001 mm) and their transportation to the lower section of the soil.

In time, in the upper part of the soil, the quantity of sandy particles (>0.01 mm) goes up. More seldom, it is the quantity of the relatively fine fractions of mechanical composition which goes up in sandy soils, on account of the podzolic process of soil formation.

The absorption capacity is very low, the soil is base-unsaturated and the reaction is acid. The upper horizons of podzolic soils are poor in fine mechanical elements. The amount of water-soluble substances in soddy-podzolic soils is unsignificant. The dense residue of water extracts from soddy-podzolic soils does not exceed 0.1%.

Under natural conditions, soils of the forest-meadow zone vary greatly in fertility, this being governed by the degree of podzolisation, the amount of ash and nitrogenous nutrients present in the soil, the degree of unsaturation with bases, the acidity, etc. In the podzolic zone, on soils with a pH of 4.5-5, all crops suffer from the excess of acidity and the unfavourable physical properties. The defects of forest-meadow soils are easy to remedy. They form useful agricultural soils owing to their positive water balance, systematic moistening, relatively adequate amount of heat and light, with days of long duration; the parent rock is mostly moraine, which is one of the best parent rocks; the relief is relatively level; the soil is highly responsive to applications of manures, etc.

The most important measures apt to raise the fertility of soddy-podzolic soils are: the creation of a deep structural plough layer (A_p), neutralisation of the acidity, removal of the excess of moisture, raising of the humus and ash elements contents. This is achieved through a high level of agrotechnique, the application to soil of lime, farmyard manure, peat, compost and mineral fertilisers, as well as green manuring (sowing and ploughing in of lupins and other plants), reclamation of swamped land, removal of the boulders, etc. When the soil is ploughed to a depth of 25 cm and more, the A_2 horizon is almost always drawn into the plough layer (A^p) and it therefore dilutes the A_1 horizon. That is why the depth of the plough layer should be increased gradually and the soil generously enriched with organic matter. Soddy-podzolic soils should be subjected to taming.

The natural factors remaining relatively unchanged, the conditions in tame soils favour the retention of a reserve of humus, owing to the fact that the podzolic process is slowed down (stifled), and in very tame soils, it is totally suppressed. The podzolic soil-forming process is eliminated as a result of man's

Fig. 39. Taming of podzolic soil

productive activity, through the cultivation of agricultural crops, which contribute to the development of the soddy process of soil formation. When the soil is being tamed, at first, A_2 decreases and then it disappears. Highly tame soils should be included in the group of soils referred to as the cultivated soils of the soddy-podzolic type (Fig. 39).

Taming of the soils brings about an increase in the humus content and thickness of the humus horizon, in the amount of nutritive elements, absorption capacity, degree of saturation, structure and its water-resistance, the acidity going down. The availability of the phosphoric acid goes up and nitrification is accelerated.

With the disappearance of forest and ploughing up, the soil is no longer shaded by a canopy and receives no more litter-fall, the forest floor is destroyed as well as the surface vegetation. Ploughing up brings about changes in the biological, chemical and physical regimes of the soil. Rapid taming of the soil requires a high level of agrotechnique. A low level of agrotechnique may bring about a deterioration of the chemical composition, which is radically improved through the application of manures and lime. With a low level of agrotechnique, the atmospheric precipitations which penetrate into the soil continue to leach the soil to no lesser an extent than under forest.

Of great importance is the application of lime to acid soddy-podzolic soils. Lime eliminates the harmful effect of mobile aluminium compounds, unsaturation and acidity. Liming lowers the solubility (losses) of humus, improves the physical properties and stimulates the microbiological activity, which leads to an increase of the accumulation of plant nutritive elements. But excessive dressings of lime bring about an alkaline reaction and may depress the fertility of the soil. The role of liming is enhanced when it is combined with the other measures adopted for the taming of soddy-podzolic soils, such as deepening of the plough layer, applications of organic and mineral fertilisers. The positive effect of liming lasts for several years and it is an important means of improving the physical and biological properties of podzolic and soddy-podzolic soils, but it cannot replace fertilising.

Of great importance in the taming of soils of the zone is the application of potash and especially of phosphate fertilisers combined with generous applications of peat, compost and farmyard manure. The taming process is intensified through the sowing of mixtures of grasses and legumes (clover+timothy and others).

The plough horizon A_p of highly tame soddy-podzolic soils has a thickness of more than 22 cm. It has a good cloddy structure. The humus content reaches 5-6% and more, the pH of the salt extract approaches 6-7, saturation exceeds 75%. The plough horizon of medium tame soddy-podzolic soils has a thickness of nearly 20 cm, with a well marked precariously cloddy structure. The humus content in A_p reaches 3%, saturation with bases 55-75%, the pH goes up to 5-5.5. In weakly tame soddy-podzolic soils, the thickness of A_p does not, as a rule, exceed 20 cm, the structure is not well marked. The humus content is about 2-2.5%, the saturation of the soils does not exceed 55%, the pH of the salt extract reaches 4-5.

The productive capacity of soddy-podzolic soils is sharply lowered by swamping due to the fact that owing to the inevitable delay with which the agricultural operations start in the spring, the length of the vegetative period is shortened. In summer, under anaerobic conditions the root system of crops is apt to rot. In autumn, when the soil is water-logged, autumn-sown crops may perish towards the beginning of winter, due to their becoming soaked with water and rotting. All swamped soils are in need of a radical improvement of their water-physical properties through the regulation of the surface and internal runoff by drainage and the elimination of the impermeability of the illuvial horizon. Drying the soils brings about an improvement of the physical properties not only of that land which is actually subjected to drying but also indirectly of the soils of the adjacent territory.

Drainage promotes the aeration and oxidation processes and leads to a gradual increase of the filtration and water capacity and eventually to the elimination of the ortstein.

Recent research has shown that the fertility of podzolic and soddy-podzolic soils can be greatly raised by thoroughly working down the packed B_1 horison to a microstructural condition and conferring stability to this structure through the application to the soil of high-molecular glues (polymers, metacrylamide and metacrylic acid) in small quantities, not exceeding 0.05% of the weight of the soil.

On the example of soddy-podzolic soils, where the process of soil formation proceeds according to the podzolic and soddy types, which are in contradiction, we see that the fight between opposites leads to different results. The process of podzolisation periodically prevails over the accumulation of humus and ash nutrient elements, but in the long run, it is ousted by the soddy process. The latter exerts a favourable influence upon the accelerated taming of the soils of the forest-meadow zone.

Chapter XIII

SOILS OF FOREST-STEPPES AND CHERNOZEMIC STEPPES

Soils of Forest-Steppes

The forest-steppe forms a physico-geographical landscape, a developing aggregate of interrelated natural phenomena. Forest and meadow-steppe compete with one another, coexisting on the same territory. The soils develop under the influence now of the forest, now of the grassy vegetation.

As a landscape, forest-steppe formed in the preglacial epoch on Tertiary savannahs for basis. The outline of its boundaries and its general character underwent alterations during the Quaternary Period.

Modern forest-steppe bears the deepest traces of the influence exerted by man on soil formation. Even within historical times, dense forests spread substantially further south than their present boundary. As a result of continuous clearing and burning lasting for centuries, the forests retreated northwards to their present position.

The formation of soils and forest-steppe landscape is tied, on the whole, with climatic particularities: deviation towards potentially negative water balance, contrasts between the seasons of the year and changing climatic elements. Dry periods alternate with humid ones.

The vegetation of the forest-steppe zone is represented by broad-leaved forests and meadow steppes. The forests consist of oak-grooves with some lime, maple, ash, elm and hornbeam. The undergrowth consists of hazel, spindle-tree, buckthorn.

Paleobotanical data confirm the antiquity of the forest-steppe landscape. As a zone, the forest-steppe was never entirely destroyed, even in the period of maximum glaciation. There persisted peculiar forest-steppe areas consisting of birch, pine and larch; broad-leaved forests were also found on isolated heights of the forest-steppe zone.

As a zone, the forest-steppe acquired its modern features earlier than the forest zone, which was freed from the continental ice sheet at a later period. The distribution of the forest areas and meadow steppes is tied with the differences in land forms, heat, moisture, rocks and soils. This distribution changes all the time.

The dismembered relief and the diversity of the soil-forming rocks condition the diversity exhibited by the soils of the forest-steppe.

Forest-steppe serves as an example of an intrazonal vertical differentiation of landscapes. This vertical differentiation bears some resemblance to vertical mountain zonality but it is less tied with substantial changes in absolute height, which does not involve radical changes in the landscape. In the south of the zone, this vertical differentiation is more pronounced.

The formation of the soils of forest-steppes is governed by processes proceeding in opposite directions: removal (leaching out) on the one hand, and on the other hand, biological accumulation. The latter prevails appreciably over the former, especially in the modern period.

Soil formation in the forest-steppe corresponds to complex climatic conditions. In the north of the zone, with a neutral water balance, develop grey forest soils, forming under broad-leaved forests; in the south, with a negative water balance, northern chernozems arise under the grassy vegetation of meadow steppes. Modern grey forest soils are formed under the influence of a soddy process, under sparse forests.

The forest-steppe occupies a certain intermediate position between taiga and steppe proper. The soils of forest-steppes are characterised by a greater dynamism than the soils of the other natural zones. This is tied with the struggle which is still going on with varying results between forest and steppe. During the postglacial epoch, in the period of overmoistening and predominance of forests, the soils were subjected to steady podzolisation. Subsequently, with the onset of warmer and drier conditions, podzolisation began to decline and, to counterbalance it, there was an increase in the saturation and carbonate content of the soils. There was also an increase in the erosive dismemberment of the surface and drainage of the territory. Podzols were replaced by grey forest soils.

Grey forest soils occupy a special intermediate position between soddy-podzolic and chernozemic soils, spreading as a narrow band along the southern boundary of the former and projecting as tongues into the zone of chernozems.

The grey forest soils of forest-steppes are not yet characterised by a high humus content. Unlike the chernozems, they do not possess a black-grey coloration (Fig. 40).

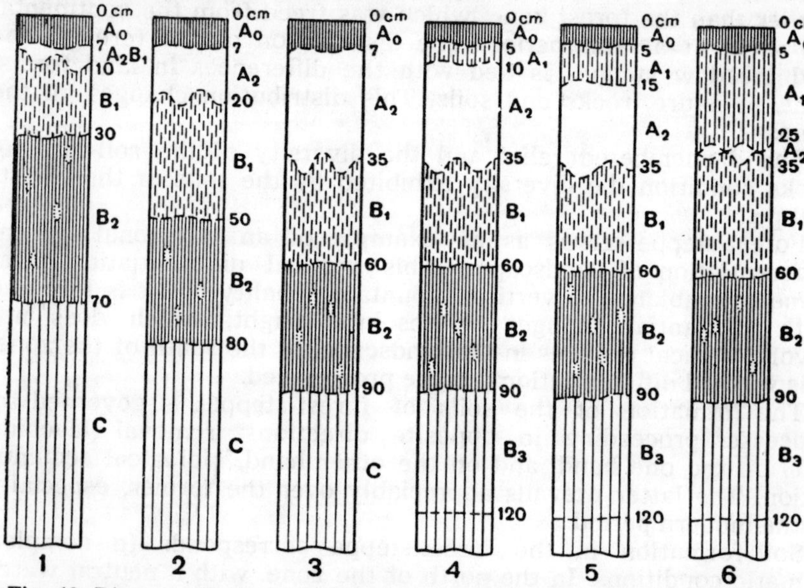

Fig. 40. Diagram illustrating the relationship between degree of podzolisation and soil structure:

1—shallow podzol; *2*—podzol of medium depth; *3*—deep podzol; *4*—soddy-strongly-podzolic soil; *5*—soddy-medium-podzolic soil; *6*—soddy-weakly-podzolic soil

This is due to the unstable balance of organic matter, relatively low accumulation of it and comparatively lower absolute and relative age of the soil. But soils which underwent lengthy formation under broad-leaved forests, i.e., dark grey forest soils, are characterised by a significant humus content. The thickness of A_1 in dark grey forest soils reaches 50 cm and more. The A_2 horizon and the upper part of B_1 are coloured with humus. The amount of humus goes up to 6-8%. The overall reserve of humus in grey forest soils reaches 150-300 and 500-600 tons per hectare in dark-grey forest and northern chernozems. The accumulation of humus in forest-steppe soils increases progressively. The nutty structural units of the B horizon of grey forest soils contain more humus towards the surface than inside, where it is not infrequently missing. The horizon of podzolisation (A_2) of grey forest soils bears a residual and not a secondary character.

The present forest-steppe is distributed where, immediately following the glacial epoch, developed the northern taiga, which passed progressively into broad-leaved, mostly oak forests. The chief component of these forests, viz., oak, does not get its leaves until late in the spring and as, until then, the soil is not yet shaded, there is nothing to hamper the growth of the grasses forming the undergrowth. Apart from that, in oak forests, the litter-fall contains a great quantity of ash elements, sufficient to neutralise the humic and mineral acids which form upon the mineralisation of the plant remains. We therefore get a predominance here not of podzol formation, but of a process of soddy soil formation. Under the cover of the grassy vegetation of meadow steppes develop soils of the chernozemic type of soil formation, with a marked water-resistant structure. We can thus see that from the point of view of the soils and in connection with the transitional character of the physico-geographical conditions, the forest-steppe presents a fairly complex picture.

The northern (podzolised) chernozems of the forest-steppe are characterised by the possession of a relatively less thick humus-accumulative horizon than the other (nonpodzolised) chernozems. This serves as an indication of the fact that they were first formed under a cover of broad-leaved forests and later, under a grass stand of meadow steppes.

Of great influence upon the course of the soil formation of forest steppe soils is the varying length and depth of freezing of the soil in winter.

Classification and description of the soils of the forest-steppe. The most typical for forest-steppes are the grey forest soils.

The grey forest soils type is subdivided into subtypes: light grey, grey and dark grey, a further subtype being the grey forest gley soils. The chernozems of forest steppes are differentiated into podzolised chernozems, leached, deep typical and meadow-steppe chernozems, as well as meadow-chernozemic and chernozemic-meadow soils.

In the zone of the grey forest soils, under a forest, the microrelief is often well marked and consists of microheights and microdepressions (Fig. 41). The microheights, in the centre of which grow trees, represent small hillocks where the soil has become raised due to the development of powerful root systems. Microdepressions receive more moisture and become enriched

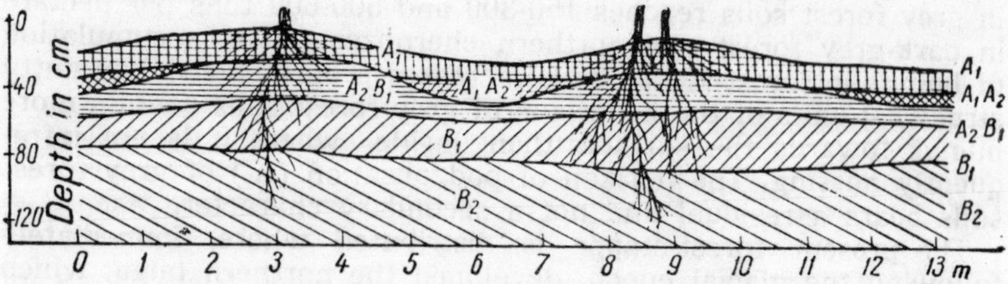

Fig. 41. Diagram illustrating the development of light grey forest soils in accordance with the microrelief

with organic matter on account of litter-fall and dying out of the abundant grass stand. This explains the formation in the depressions of meadow-grey soils with darker-coloured humous horizons of great thickness, with clearly marked gleisation in the lower part of the profile. These soils possess high absorption capacity. Even after the forests have been destroyed and the microrelief has been relatively levelled out due to ploughing, the differences between the soils persist for a long time.

Grey forest soils have the following average profile:

A_0—litter-fall 2-3 cm;

A_1—0-25 cm. Grey fine-nutty, fine-cloddy to dustlike. The coloration gets lighter with depth. Occasional presence of whitish glassy powdering;

A_2—25-45 cm. Grey nutty with whitish glassy powdering. Noticeable laminated-slaty structure;

B—45-90 (100 cm). Cinnamon-brown nutty-prismatic, compact, viscous when wet, illuvial horizon;

C—brownish-pale yellow. Carbonates appear at a depth of 1.5-2 m.

Grey forest soils have a pH of 5.5-6.5, so that they are weakly acid soils. The structure of these soils is not water-resistant. The nitrogen content is 0.1-0.2%. Grey forest soils are the seat of a marked clayisation process.

The apparent density decreases from light grey to dark grey soils. The pore-space increases in the same direction. The aeration of the soils diminishes with depth, going down to 6-7%, the field water capacity being relatively high (over 40%). Noncapillary porosity is represented by the interstices between the flat planes of the nutty and prismatic units.

Dark grey forest soils pass progressively into northern chernozems, their nutty structure gradually changing to a granular one. At the same time, there is an increase of their noncapillary porosity and aeration, without any appreciable rise of the field water capacity.

Grey forest soils are close to soddy-podzolic soils, which can be seen from the high SiO_2 content (82.8%) in A_2, the accretion of sesquioxides in the illuvial horizon, a certain unsaturation with bases and an acid reaction.

Meadow-steppe chernozems are formed under conditions of permacidous leaching and impermacidous leaching, the level of the ground water lying at a depth of 5-6 m and deeper. There are usually no readily soluble salts in the profile of these soils and no signs of alkalinity. The horizon of carbonate accumulation varies in importance. The pseudomycelium is clearly marked. The carbonate concretions do not, as a rule, constitute a subhorizon. Sometimes, at the bottom of the horizon of calcium carbonate accumulation there occurs gleisation.

The meadow-steppe chernozems have a dark-coloured humus horizon and a high humus content. The lower boundary of this horizon is uneven and penetrates as tongues into the underlying horizon. The reaction of the humus horizon is weakly acid, whereas in the lower part of the soil, the reaction may be slightly alkaline.

The water regime of meadow-chernozemic soils is akin to that of meadow-steppe chernozems, but their soil-ground water lies at a depth of 3-4 m, approaching the lower boundary of the soil profile and ensuring constant moistening of the lower portion of the soil. The horizon of calcium carbonate accumulation is gleised. Loose $CaCO_3$ concretions are disseminated in the lower portion of the soil. Sometimes occur silicified $CaCO_3$ concretions in the form of lime nodules.

Chernozemic-meadow soils are characterised by a more pronounced type of permacidous leaching. The soil-ground water lies at a depth of 1.5-3 m, constantly wetting the soil. The thickness of the soil profile of these soils is less than that of the chernozems of meadow-steppe and meadow-chernozemic soils (1.2-1.3 m). Their horizon of calcium carbonate accumulation is strongly gleised. It contains many loose and hard calcareous concretions, but there is no formation of a lime nodules horizon.

Agricultural utilisation of forest-steppe soils. With proper management and a high level of agrotechnique, grey forest soils can rapidly be tamed, whereupon their fertility goes up. This can be achieved through the application of farmyard manure, compost, mineral fertilisers, the sowing of grasses, green manuring, bare, occupied and coulisse fallows, snow retention, irrigation, digging of ponds and reservoirs, planting of field-protective forest belts, etc. Taming of grey forest soils leads to an increase of the amount of humic acids over that of the fulvic acids, i.e., these soils acquire the features of steppe soils.

The soils of forest steppes can easily be transformed into useful highly productive soils, owing to their high potential fertility.

In connection with the weakly acid reaction of the soil solution, the chernozems of forest-steppes possess a greater amount of available phosphorus than the chernozems of steppes.

In view of the shortage of moisture (Table 28) and the dryness of the climate, the soils of forest-steppes need a substantial replenishment of their reserve of useful moisture, which can be achieved through the accumulation in them of autumn and winter precipitations, as well as through water supply irrigation and watering during the vegetative period.

Table 28

Approximate Data Regarding Water Capacity and Useful Water Reserve of Soils of Forest-steppes

Soils	In mm in a one-metre layer		
	at limit field water capacity	at wilting point moisture	useful moisture
Grey and dark grey forest soils . .	300	120	180
Deep chernozem	360	140	220
Ordinary chernozem	320	130	190

The reserve of productive moisture in the soils of forest-steppes are subjected to fairly significant changes, going down during the vegetative period to 50 mm and lower in a one-metre layer.

Soils of the Chernozem-Steppe Zone

Under the herbaceous vegetation of the chernozem-meadow-steppe zone, where biological accumulation reaches its highest level, we get the formation of chernozems, which are the best soils in the world. The word chernozem was coined by the Russian scientist M. V. Lomonosov and this type of soil was first described by V. V. Dokuchayev in his book *The Russian Chernozem*.

The chernozem-steppe zone lies to the south of the forest-steppes.

The climatic conditions of the steppe zone show marked differences in latitudinal and longitudinal directions. Whereas in the north of the zone, the climate is moderately semihumid, in the south it is semiarid to arid.

Not so long ago, the grassy vegetation of steppes was still characterised by a vigorous development spread over a lengthy vegetative period. After the vegetation died in the autumn, there remained a large amount of roots, leaves and stems, which formed the main source of the soil humus. At the present time, the steppe is ploughed up. The natural (virgin) vegetation now covers insignificant areas. The grassy vegetation was, in the past, represented, in the north of the zone, by motley grass-feather grass-sheep's fescue associations. As the aridity became more pronounced, the vegetation changed in the south to feather grass-sheep's fescue associations, and, in the south-east, to gramineae-wormwood associations.

The soils of steppes harbour a large number of earthworms, insect larvae and burrowing animals. Earthworms play a useful role due to the fact that they enrich the soil with organic matter, whereas burrowing animals which cause excessive loosening of the soil and translocate the carbonates upwards, are harmful to agriculture.

Typical for the steppe zone is a level, slightly undulating relief, which promotes chernozem formation.

On the whole, the influence of the various land forms on the course of the soil-forming processes remains in force in the steppe, although it is modified and somewhat smoothed out. The concave parts capture the surface water and there, the soil is wetted throughout and gets leached. Various obstacles found on the surface play a similar role to that of concavities, by retaining the surface runoff or causing the accumulation of snow, obstacles such as the edges of clusters of trees, shrub, a high stand of grass, after-harvest remains, windbreak rows, fences, banks, etc.

On convex land forms a dry steppe water regime invariably predominates, which means that the soil never gets soaked throughout the year. Such areas of soil not subjected to soaking grow larger in the south of the zone and smaller in the north, where they finally vanish altogether.

The question of the origin and age of chernozems remains controversial even today. But M. V. Lomonosov was already aware of the fact that chernozem "occurred as a result of the decay with time of animal and plant bodies". This tallies with the popular view that chernozem originates "from the decay of plants in the earth". In the solution of the problem concerning the origin of chernozem, writes V. V. Dokuchayev, "popular consciousness forestalled science". It is indisputable that chernozem arises as a result of the accumulation of the humus forming upon the decomposition of the grassy vegetation of the above ground part as well as of the root mass in the soil itself. Ruprecht was of the opinion that in origin, chernozems are close to soddy soils, being differentiated from them by their greater age and a different steppe composition of the vegetative cover. According to V. R. Williams, the chernozems originated from soddy-podzolic soils and passed, or did not pass, through a boggy (or rather semiboggy) phase of soil formation. It is quite likely that chernozems originated on the basis of previous soddy-meadow weakly podzolised soils of the same type as the modern semiboggy soils of the Amur region. The major role in the formation of chernozems is played by the accumulation of humus and plant ash nutrient elements.

Chernozems were formed under the conditions of a good drainage and optimum water regime, under the cover of a vigorous grassy vegetation, whereupon there occurred a progressive increase in the quantity of soft humus and the saturation with exchangeable, chiefly alkaline-earth bases went up.

The humic A+B horizon of chernozems reaches a thickness of 1 m and more. This corresponds to the depth of regular moistening of the soil and that of the mass distribution of plant roots. The carbonates lie at the depth of penetration of the main mass of the roots of the steppe vegetation. The falling out of solution of $CaCO_3$ and calcium bicarbonate is conditioned by the liberation of CO_2 when the temperature rises, and the loss of water from the soil through transpiration.

Chernozems reach their maximum development on flat watersheds. Here, the soils are deeply coloured in black, possess a clearly marked cloddy-granular water-stable structure. In places, the chernozem is much dug up by burrowing animals, which causes a worsening of its properties.

The chernozems of slopes possess a humus-accumulative horizon of lesser thickness; as for the chernozems found on the terraces of river valleys, they are characterised by a thick humus horizon of a black colour with a tinge of cinnamon. Here, the soils are more seldom dug up by burrowing animals.

The diversity exhibited by chernozems increases in connection with the diversity of soil-forming rocks. The formation of chernozems is also influenced by the local conditions (hydrology, topography, vegetation, etc.); this is likewise reflected in the morphology and properties of the soil.

The thickness of the humus horizon of chernozemic soils diminishes to the south due to drier conditions and, in places, in connection with more continental conditions (smaller amount of atmospheric precipitations, increase of the depth to which the soil freezes, shortening of the frost-free period, etc.) and with the fact that the main mass of plant roots, which follow the depth of the layer of soil thoroughly wetted, is distributed down to a lesser depth.

Chernozems. General characters. The soil profile of chernozems is characterised by a fairly sequential transition between the genetic horizons. The upper part of the soil stands out most clearly; the humus it contains gives it a black-grey colour, hence the name chernozem (black earth) (Fig. 42).

The average morphological characters of the chernozems are as follows:

A—0-50 (80) cm—black-grey or dark grey, to the south they possess a cinnamon tinge, in general they have a cloddy-granular structure, which also deteriorates in a southward direction. Friable, permeated by plant roots. The transition is progressive.

B—50-110 (160) cm—brownish-grey, irregularly coloured, lighter than A, showing humus tongues in the bottom part, cloddy. At the

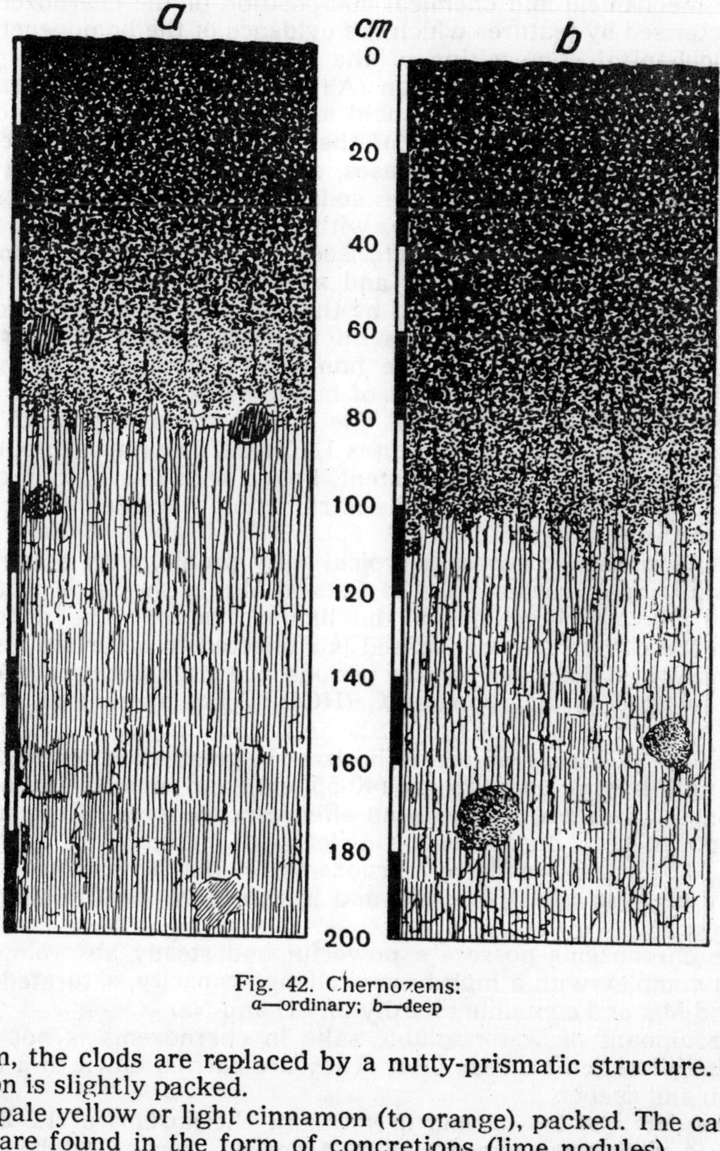

Fig. 42. Chernozems:
a—ordinary; *b*—deep

bottom, the clods are replaced by a nutty-prismatic structure. The horizon is slightly packed.

C—pale yellow or light cinnamon (to orange), packed. The carbonates are found in the form of concretions (lime nodules).

The mechanical and mineralogical composition of the chernozems is connected with the composition of the soil-forming rocks. In view of the fact that the soil-forming loess-like rocks and loesses change from loams in more northern regions, to predominantly clayey rocks to the south, the mechanical composition exhibits corresponding changes.

The mechanical and chemical composition of the chernozems is characterised by features which are evidence of the homogeneity of the mechanical composition of the soil profile and of the great thickness of the humus horizon (A+B)—from 45 to 110 cm. The uniform distribution of silicic acid and sesquioxides points to the absence, at the present time, of their migration through the soil profile. Among the absorbed bases, calcium shows a marked predominance. The reaction of the soils approaches neutrality or is slightly alkaline, which increases with depth.

The chernozems are characterised by a high humus content, which decreases to the south and south-east. A decrease of the humus content is also exhibited by the soils of slopes with a southern exposition, in connection with the local deficit of moisture. A characteristic feature of the humus of chernozems is that it contains a considerable amount of humic acids and possesses high solubility, reaching 0.02-0.05%. The humus contains 0.02-0.5% of nitrogen, whose ration to carbon is 1/10-1/11. The nitrogen balance of chernozems is, to a great extent, linked with the fixation of nitrogen from the air and with the nitrification processes which take place in them.

The chernozems possess a typical salty profile, with a constant accumulation of carbon salts of Ca and Mg, in the form of a carbonate illuvium lying below the humus horizon. The depth at which the carbonates are located is subjected to a certain periodic variation: it falls during wet periods and rises during dry ones. They migrate as bicarbonates $Ca(HCO_3)_2$. The amount of CO_2 in the illuvial horizon reaches 5-8% and more in the south of the zone. The thickness of the A horizon is greatest in leached and in high-in-humus chernozems (40-55 cm) and smallest in southern ones (15-25 cm). There is no effervescence with HCl in northern (podzoliser) chernozems; in leached chernozems, it arises in the C horizon, in deep chernozems, in the B_2 horizon, in ordinary ones, in the B_1 horizon and in southern ones, in the A+B horizon.

The chernozems possess a powerful and steady absorbing colloidal complex with a high base exchange capacity, saturated with Ca and Mg, and containing hardly any H and Na.

The amount of water-soluble salts in chernozems is not high, it usually seldom exceeds 0.1%. They are carried down to a depth of 4 m and deeper.

At a depth of 2-3 m, and in southern chernozems of 1.5-2.5 m, there is often some gypsum in the form of a sort of sulphate illuvium, above which, at a depth of approximately 1 m, lie the carbonates (lime).

The reaction of chernozems approaches neutrality and undergoes insignificant changes from season to season (pH—6.5-7.5). In the illuvial horizon, the pH has a value of 7.5-8.5. The northern chernozems may be slightly acid and the southern ones slightly alkaline. The mineral part of the chernozems changes relatively little through the profile. The silicate part remains relatively steady. The molecular $\frac{SiO_2}{R_2O}$ ratio is high (about 2).

The chernozems possess favourable water-air properties. Their specific gravity changes little through the profile: from 2.4 in the A horizon to 2.7 in the C horizon. The apparent density fluctuates accordingly from 1-1.2 to 1.61, and with it also the pore-space, from 55-60% in the A horizon to 45-55% in the C horizon.

Hygroscopic moisture fluctuates between 8 and 12%, maximum hygroscopic moisture reaching 12-14% in the A horizon, 7-10% in the B horizon and 6-7% in the C horizon. More often than not, the water balance is neutral, varying slightly from a positive one in the north to a negative one in the south of the zone.

The homogeneous crumble-granular structure of the chernozems conditions optimum water permeability, marked water retention (moisture capacity) and high aeration. A structure of this kind ensures the best possible combination of capillary with noncapillary porosity. From the point of view of their water-physical properties, the chernozems are particularly well suited to irrigation.

At a depth of about 4 m (deeper to the north and closer to the surface to the south of the zone) lies a so-called dead dry horizon. The depth at which it is found and its thickness vary considerably. Furthermore, it is not continuous, being absent in places.

The chernozems possess an abundant and varied microflora, which exhibits a sharp increase upon the application of organic manures. The number of bacteria reaches 3,000 million per gram of soil. In the autumn, there is an increase in the number of fungi (*Penicillium*, *Mucor*, etc.). The physico-chemical and biological properties vary appreciably from one subtype of chernozem to another.

Classification of Chernozems

The chernozemic zone is divided into three parts: northern (merging with forest-steppe), central and southern. To the chernozemic type of soil formation correspond therefore the following types of chernozems: northern (podzolised), leached, deep (high in humus), ordinary, southern.

All the chernozems in agricultural use are divided, according to the degree of taming, into weakly, medium and highly tame. In the zone of chernozemic steppes are also distributed soils of many other types: solonchaks and saline soils, solonetzes and solonetzic soils, solods and solodised soils, flood-plain, and even swamp and other soils.

Northern and leached chernozems. The northern chernozems are distributed in the northern part of the zone. They represent transitional soils between dark grey forest soils and leached chernozems. These soils exhibit signs of podzolisation, and for this reason they are referred to as podzolised.

The A horizon of northern chernozems has an average depth of 30-35 (40) cm, is of a black-grey colour, cloddy with a silicic powdering at the bottom, where traces of A_2 are still present and a horizontal lamination begins to show.

The B horizon has an average thickness of 40-50 cm, it is fairly dark brown, with humic tongues, cloddy-nutty with signs of a prismatic structure. A silicic powdering is present on the edges of the structural units. There are some indications of an illuvial horizon. The horizon is packed.

The C horizon has a thickness of 90-150 cm, it is brown with tongues of humus, prismatic. Towards the bottom the colour changes to pale yellow-brown. Not infrequently, krotowinas are found in the profile.

A characteristic feature of the northern chernozems is the absence of carbonates or the separation of the carbonate horizon from the humic one, with, between the two, a carbonate-free layer of appreciable thickness. These soils do not possess a horizon of readily soluble salts. They are slightly acid. A greyish tinge, grey patches and other indications are evidence of the fact that the northern chernozems are being leached. In their content of humus, they approach the dark grey forest soils or somewhat surpass them (4-8%).

Bordering the northern chernozems, to the south, lie the leached chernozems, which are distinguished from the former by the somewhat greater thickness of their soil profile and a higher humus content. They possess the following average characters:

A—35-45 (55) cm, dark grey, cloddy-granular with silicic powdering;

B—45-55 cm, brownish-grey, mottled, cloddy-nutty, with silicic powdering;

C—pale yellow-brown. Krotowinas are sometimes present. Effervesces from a depth of 120 cm.

These chernozems are subdivided into strongly and weakly leached. The former are found in the northernmost areas or on concave land forms, they contain 8-10% of humus or more.

The appreciable amount of humus (2.2%) present in leached chernozems can be traced down to a depth of 100 cm. The sesquioxides and silica are uniformly distributed throughout the depth of the soil profile. Only the upper part of the humic horizon exhibits an insignificant unsaturation with bases and there, the reaction is slightly acid, approaching neutrality (pH—6-6.5). Lower down, the reaction becomes slightly alkaline.

Deep (rich in humus) chernozems are mainly distributed in the central parts of the zone. Their average morphological characters are as follows:

A—40-55 cm, black-grey, granular, loose. Passes gradually into the B horizon;

B—55-65 cm, dark grey, brown with humus tongues at the bottom, cloddy-granular;

C—thick, pale yellow with carbonate pseudomycelium. Krotowinas are present. Considerable effervescence with acid. There are no salt efflorescences. The profile is characterised by a relative unifomity of structure.

Deep (rich in humus) chernozems contain 10-16% of humus. The reaction is neutral (pH-7, about 7.5 at the bottom). These soils are the best in the world, they possess the highest potential and effective fertility. They may be taken as a model when undertaking the accelerated taming of other soils. They have the deepest profile, which cannot have been inherited from the southern chernozems. These soils may have developed on the basis of deep leached chernozems. The distribution of the deep (rich in humus) chernozems as a subtype is still inadequate. Their areal is gradually widening. At the present time, the deep chernozems are scattered in the form of large massifs and growing pockets, penetrating in the north, into the subzone of leached chernozems and in the south, the subzone of ordinary chernozems.

Sometimes, this type of chernozems is differentiated into rich in humus (less deep) and deep chernozems proper.

Ordinary chernozems. The ordinary chernozems lie immediately to the south of the deep chernozems; in comparison with the latter, they are characterised by a somewhat lesser depth of the soil profile. The depth of the A+B horizons in deep chernozems reaches 120 cm, whereas the depth of the same horizons in ordinary chernozems is about 80 cm. Ordinary chernozems contain 6-9 (10)% of humus, i.e., appreciably less than the deep ones. With the decrease in humus content, the coloration gets lighter. The carbonate and accordingly also the sulphate horizons rise somewhat, reaching the lower limit of the humus horizon. Ordinary chernozems are more packed than deep chernozems. The other water-physical, chemical and biological characteristics change too, due to the positive and neutral water balance being replaced by a negative one. The reaction of the soil solution becomes alkaline.

The ordinary chernozems present the following morphological characters:

A—25-35 (40) cm, deep-dark grey, cloddy-granular, fairly loose. Passing gradually into the lower lying horizon.

B—35-40 cm, grey-brown with dark humus tongues, cloddy.

C—thick, pale yellow. Presence of carbonates in the form of lime nodules. Numerous krotowinas.

The ordinary chernozems are formed under a grassy steppe vegetation, the above ground mass of which reaches 1-1.5 tons of dry mass per hectare, the corresponding mass of the roots being 4-6 tons. The main mass of the roots reaches down to a depth of 45-60 cm, the proportion of roots in a layer of 30-40 cm reaching up to 80%. Accordingly, the maximum amount of humus is found in the top 30 cm, as well as the darkest coloration of the A horizon.

Ordinary chernozems possess favourable physical properties—good structure and aeration. They have a high noncapillary porosity, which decreases appreciably in the carbonate horizon, in connection with the soil getting more packed. As the soil becomes more packed with depth, there is a corresponding increase in its apparent density.

The humus-accumulative horizon and the horizon below it are characterised by a high water capacity and at the same time with considerable water permeability (nearly 30-40 mm in the first hour). Up to 500 mm of water, i.e., the whole of the atmospheric precipitations, can be retained in a layer of soil 1.5 m thick. And all the summer-autumn precipitations are indeed fully retained in the zone of mass distribution of the plant roots in the soil. In spring, the soil gets thoroughly wetted for a limited period down to a depth of 2-3 m.

The main reserve of water of deep and ordinary chernozems goes towards transpiration and synthesis of organic matter. The plants are constantly supplied with water and at the same time with nutrient elements in maximum required amount. The favourable physical properties of these chernozems and their high fertility develop in interdependence.

Ordinary chernozems possess high porosity (55-62%), water capacity and aeration, they are highly saturated with bases (up to 95% and more). With depth the neutral reaction of the top horizon passes into an alkaline one.

Tied with the accumulation of humus in soil is the accumulation of nitrogen, phosphorus and other important nutrient elements of plants. The absorbing capacity reaches 40-56 m-equiv. in the upper horizons, going down to 20-25 m-equiv. with depth.

In the present period, the climate of the southern part of the chernozemic zone is becoming drier and, consequently, the steppe is losing its former meadow character. As a consequence of the change of climate, the stable water regime of the meadow steppes is being replaced by an intermittent one. As the climate becomes drier, the native meadow steppe becomes gradually converted into a dry steppe. The grass stand, which formerly was continuous, gets sparser. Plants with a long vegetative period are eliminated, remain the xerophytes, ephemers and other early ripening grasses.

The soil dries out considerably, exhibiting deep cracks. The soil's humus undergoes intense decomposition, which leads to an increase in the content of readily soluble salts. The cloddy-granular structure collapses. The soil gets packed, becomes less water permeable and is weakly leached by water. The carbonate-content goes up and the soil becomes still more packed.

With the increase of readily soluble salts, Ca is partly replaced by Na and, as a result, the soil becomes alkaline. Under droughty conditions, the dissolved carbon dioxide comes out of solution, turning Ca bicarbonate to carbonate, which becomes precipitated in the soil and delimits the lower boundary of the layer subjected to thorough wetting. The carbonates, which accumulate from below upwards, may, with the decrease in leaching, become distributed up to the surface of the soil. The thick carbonate loess-like horizon appears as a consequence of the decrease in leaching of the soil.

Southern and alkaline chernozems—soils of the southern part of the chernozemic zone, passing from the chernozemic steppes to the dry steppes with the chestnut soils, where evaporation exceeds precipitations. Leaching being weak, the thickness of the A+B humus-accumulative horizon is lesser, not exceeding 50-60 cm.

In comparison with ordinary chernozems, the southern chernozems are less coloured by humus, whose content reaches 4-5% and more. The coloration is not uniform, the dominating colour being grey with a cinnamon tinge. The structure is less pronounced, more often than not it is cloddy and in the alkaline chernozems, it is stony-prismatic. The B horizon of these soils is strongly packed, fissured. At the bottom, appears a carbonate and sulphate horizon.

The generalised profile of a southern chernozem is as follows:

A—15-25 (30) cm thick. Grey, the structure is not pronounced, cloddy. At the top, the colour is somewhat lighter than at the bottom. The transition is gradual.

B—25-35 cm thick. Light grey with a brownish or cinnamon tinge. Packed, coarse-cloddy-prismatic, fissured Effervesces with HCl.

C—brownish-pale yellow, more packed than B horizon, with carbonate spots, beginning from a depth of 60-90 cm and lower, with concretions in the form of lime nodules. Below 120 cm, gypsum druses are sometimes found.

The absorbing capacity is lower than that of ordinary chernozems and has a value of 20-35 m-equiv. to 100 g of soil. The prevailing absorbed bases are Ca and Mg, the latter being relatively more abundant than the former. An insignificant amount of Na may occasionally be present among the absorbed bases with, as a consequence, a worsening of the properties of southern chernozems.

Southern chernozems are distributed jointly with alkaline ones. The latter are found as small patches disseminated among the southern nonalkaline chernozems. Alkaline chernozems are formed upon the entry of sodium into the absorbing complex of the soils, with displacement of the alkaline-earth bases. These chernozems are characterised by reduced thickness of the A+B horizon and low humus content. As a result of the entry of Na into the absorbing complex, the humus is broken down and some of it is translocated in solution from the A into the B horizon, as a consequence of which, the B horizon becomes packed and its dark coloration is intensified. The granular structure inherent to chernozems disappears, being replaced by lamination. The packed B horizon breaks up into large units.

Alkaline chernozems spread beyond the limits of the belt of distribution of southern chernozems, distributing themselves as small isolated areas and favouring the creation of a microrelief characterised by patelloid sink holes, which subsequently, here and there, condition the formation of leached soils of the solod type or of regraded solonetzic soils and solonetzes.

The morphological characters of chernozemic solonetzes are fairly stable but they always present features of the soils in the subzone of which they are distributed. Chernozemic soils exhibit pronounced changes in characteristics and properties when passing from alkaline soils to solonetzes.

Saturation of the colloids with sodium and peptisation with formation of sols conditions their translocation from the top layers of the soil into the lower ones where, meeting a higher concentration of salts, they coagulate and fall out in the form of gels. Tied with the latter occurrence is the formation of an illuvial horizon, which is fairly viscous when wet and dense, much fissured when dry. The vertical fissures give rise to column-like units and a particular column-like horizon. This horizon constitutes the characteristic morphological feature of solonetzes.

The upper part of the soil becomes enriched with SiO_2 due to the removal of salts and sesquioxides, which accumulate in the illuvial horizon. The actual reaction of the soil in the upper layers is almost neutral, in the lower ones it is distinctly alkaline.

Solodised chernozems, arising upon a slightly alkaline reaction with the formation of amorphous silicic acid in the A_2 horizon, are found on old alluvial plains and flat watersheds. These soils pass into solods with a thick whitish A_2 horizon, enriched with silicic acid. Apparently the typical solids should be distinguished from the soils of sink holes, arising under a forest vegetation, under birch-aspen groves.

The upper layers of solods are enriched with SiO_2 and the B horizon, with sesquioxides. The reaction of the upper part of the soil is acid and that of the C horizon is alkaline.

Certain chernozems of the southern half of the chernozemic zone are thoroughly dug up by burrowing animals, whereupon the carbonates become translocated from the horizon of mass accumulation into the upper part of the soil together with the soil-forming rock. Chernozems of this kind are referred to as dug up or krotowinas chernozems. Dug up chernozems are often found on the elevated parts of watersheds. Eroded chernozems are also distinguished, which are much eroded soils whose upper part has been washed way. On the surface of the earth the left-over lower part of the soil now appears, which is completely devoid of humus or which has retained fragments of the A+B horizon. Soils of this kind are distributed on slopes from watersheds to ravines and gullies.

Agromeliorative characteristics of the chernozems. The chernozems are characterised by high biological activity, intense ammonification and nitrification processes. The ammonia which forms in the soil oxidises to nitrates and provides plants with nitrogen. These soils contain large amounts of phosphorus, potassium, calcium and other ash nutrient elements of plants in available form. Virgin and tame chernozems possess a stable granular structure, ensuring optimum water-air and heat regime of the soil.

The most important point to bear in mind in connection with the improvement of chernozems is that they should be ploughed deep, attention should also be paid to the maintenance and improvement of their structure, the accumulation of ash nutrient elements, a supply of moisture and the setting up of other conditions favouring the development of plants.

The yields obtained on chernozemic soils, especially in the southern parts of the zone, are dependent upon the presence in them of moisture available to plants. The high fertility of these soils can be enhanced through the supply of complementary water in addition to that obtained from the atmosphere, against a background of high agrotechnical standards. In this connection, a number of measures should be adopted, starting with the proper system of soil cultivation (the spring jobs to be carried out with the shortest possible delay), manuring, sowing and ending with agromelioration (snow retention, levelling, etc.), forest improvement and irrigation.

The most important agricultural measure to carry out in the chernozem zone is the maintenance of structure. Poor husbandry leads to intense mineralisation of the humus, reduction of the biological accumulation of organic and ash substances and impoverishment of the soils. In well tamed soils, the amount of humus increases, its composition is improved on account of a higher ratio of humic acids to fulvic acids. This ratio goes up from 1.6-1.7 in untamed soils to 2.1 in tame ones. Under conditions of poor husbandry, the amount of water-resisting fine-cloddy structural units in A_p goes down in comparison with virgin soil from 60 to 40% and lower. Noncapillary porosity goes down from 20-25% to 5% and the water permeability from 20-30 to 6 mm in the first hour, which increases the loss of water on account of surface runoff. The structure is also subjected to destruction (is disaggregated) upon flooding irrigation, which does not take place upon correct, furrow or sprinkler, irrigation. High agrotechnical standards, lower watering rates, the maintenance of structure through suitable tillage and manuring, preclude disaggregation.

Intensive irrigation may lead to losses of nitrogen and ash plant nutrients. High as it is (0.2-0.5%), the nitrogen content of chernozems is not readily available. The conversion of nitrogen into an available form can be brought about by creating conditions favouring ammonification and nitrification, through manuring and the inoculation of bacterial cultures. Chernozems contain from 0.1 to 0.2% and more of phosphorus, but this is not in a readily available form either. All the chernozems of the northern part of the zone show good response to the application of phosphorous fertilisers (P_2O_5, up to 5 tons per hectare and more), whereupon their capacity to accumulate nitrates goes up.

It is indispensable to create and maintain the soddy soil-forming process in its cultural manifestation, through irrigation and high agrotechnical standards, putting into practice the most modern achievements of agricultural science.

In order to improve and maintain the qualities of the soils of the chernozem zone, the utmost attention should be paid to the fight against erosion (ablation, wash-out and deflation) of soils.

The effect of any meliorative measure is always more marked on chernozems than on other soils.

Chernozemic soils possess enormous latent possibilities and agricultural reserves whose realisation constitutes a task to be tackled in the nearest future.

Chapter XIV

SOILS OF DRY STEPPES, SEMIDESERTS AND DESERTS

Soils of Dry and Desertic Steppes

The chernozem-steppe zone borders, to the south, on the zone of dry and desertic steppes, which are distributed in a region characterised by soils with an inadequate moisture content and a high loss of water through surface evaporation.

Lying right next to the chernozems are the chestnut soils and to the south of them lies the belt of brown soils.

The zone of chestnut and brown soils is suited to irrigative meliorations.

The vegetation of the soils of the chestnut zone dies out, for the most part, in summer, during the droughty period. The dead organic matter is subjected to intense decomposition due to good aeration and the high temperature of air and soil. The soil being dry and the aeration favourable, aerobic decomposition is ensured not only of the above ground mass of plant remains but also of the roots to CO_2, H_2O and mineral compounds. Intense aerobic decomposition of organic masses is not, as a rule, accompanied by a progressive accumulation of humic substances. The amount of humus in the soils of the zone reaches 2-4% and seldom exceeds 4-5%. The thickness of the A+B horizon goes down to 50-30 cm, coinciding with the depth of the mass distribution of roots.

Chestnut and brown soils are saturated with bases, in which prevails Ca, to a lesser extent Mg and where the role of Na appreciably goes up. The carbonate content of chestnut soils increases and so does the amount of water-soluble salts. The reaction becomes alkaline in the upper horizons, reaching pH—7-7.5, increasing to 8-8.5 deeper down.

As the zone extends over vast areas in the longitudinal as well as latitudinal directions, its climate is characterised by lack of uniformity and continental features.

The vegetation of the zone of dry steppes and semideserts is characterised by a scanty range of species and the spareness of the grass stand. The mass of roots is considerably more abundant than the above ground part of the vegetation. The vegetation of the dark-chestnut soils of the northern part of the zone consists of sheep's fescue-feather grass associations.

On the chestnut soils predominate the wormwood-feather grass-sheep's fescue (Artemisia-Stipa-Festuca) and wormwood-sheep's fescue associations. On light chestnut and solonetzic chestnut soils, plants possessing a deep root system, such as feather grass and others, fall out in view of the accumulation of salts in the lower lying soil horizons. In the south of the zone predominate plants possessing a powerful root system but penetrating to a lesser depth. The plants growing here are wormwood, willow-bush, camphorosma. As the grass stand becomes sparser, lichens and blue-green algae develop. Brush is found, here and there, in sink-holes, and in flood plains grow broad-leaved forests.

The vegetation growing on chestnut soils yields approximately 0.8-1 ton of dry above-ground mass and 3-4 tons of dry organic root matter per hectare. The 0-25 cm layer of soil contains more than 50% of the roots of the grass vegetation and the 0-50 cm layer, nearly 80% of the whole amount of roots contained in a layer of soil one metre thick. The intense mineralisation of the plant remains, which is counteracted by the synthesis of organic matter, does not permit any appreciable accumulation of humus.

In the area of desertic steppes, where there is a formation of brown soils, the vegetation is still sparser and more stunted. Here, it consists of xerophytes, wormwood and steppe sward gramineous plants—sheep's fescue-wormwood and wormwood-Anabasis salsa associations, in which ephemers and ephemeroids* play a leading part. Saprophytic bacteria develop here which play a significant role in the transformation and intense mineralisation of organic remains. The amount of humus in brown soils goes sharply down, seldom exceeding 1-2%. Effervescence with HCl in brown soils comes nearer to the surface. The carbonates move upwards in the spring-summer period and become fixed at a depth of about 30 cm from the surface. The sulphates are located at a depth of about 50-60 cm and down to the depth of maximum wetting of the soil.

The chestnut-brown zone is characterised by a flat relief with a well-developed microrelief in the form of microsinks, constituting a labyrinth of microdepressions referred to as limans. This limanic microrelief is particularly noticeable in spring when the snow melts and water gathers in depressions. It conditions the unevenness of moistening and of the salt regime, exerting a marked influence on the distribution of vegetation and soils. As a result of suffosion and soil-forming processes, insignificant initial irregularities of the microrelief become progressively deeper. In places, the relief appears noticeably levelled out, due to alluviation. The vegetation growing in the depressions is relatively more moisture-loving and the soils forming there are of the leached type, whereas on the microrelief elevations, there is hardly any vegetation and the soils formed on them are saline. The microrelief, which develops in interdependence with soils, is the cause of their complex distribution.

The chestnut and brown soils were formed in the post-glacial period under a cover of steppe semidesert vegetation and, apparently, rapidly passed through the meadow steppe stage, owing to a change in the water regime of the country. The vegetation and the soils underwent radical changes. The soils became more packed, the porosity (noncapillary) decreased and so did the water permeability. The water regime of the soils became intermittent. The melted water now runs off from the surface almost entirely, the soil getting only slightly wetted. The meagre reserves of spring moisture are rapidly used up by the vegetation and the soil dries out to a great depth. Being provided with water only in spring, the vegetation is subjected to droughty conditions in summer and dries out, which happens every year. The plant remains are rapidly mineralised and the soil becomes poorer in humus and richer in salts.

---

* Ephemers are early ripening plants. Ephemeroids are perennial plants possessing bulbs or tubers; their above ground organs are short-lived.

In spring, the down-flowing water dissolves the sulphates, chlorides, nitrates and bicarbonates and translocates them downwards but very close to the surface, where they accumulate in the form of a carbonate and sulphate horizon some tens of centimetres thick. The readily soluble salts are translocated to a lower depth. The carbonates are located at a depth of 90-100 cm, where they cement the soil, packing it up. The bicarbonates become fixed at a depth of 100-150 cm and the sulphates at a depth of 150-200 cm, i.e., at the lower limit of the possible thorough wetting of the soil. At a depth of 250-300 cm a relict illuvial horizon is sometimes found, being evidence of a more humid period and greater intracontinental moisture circulation in the past.

As the climate became drier in the past, the vegetative period became shorter, and consequently there was a radical change in the vegetation, plants with a short vegetative period becoming dominant. The meadow steppes passed gradually into motley grass sheep's fescue-feather grass and sheep's fescue-feather grass-motley grass steppes in the northern part of the chestnut zone where are now formed dark chestnut soils, whose characteristics and properties are close to those of the present day southern chernozems.

The chestnut and brown soils acquired an alkaline reaction, which favours the fixation of atmospheric nitrogen and nitrification. That is why the soils of dry steppes are relatively rich in nitrogen.

The most developed in the chestnut zone are the solonetzic process of soil formation and the formation of chestnut solonetzic soils and solonetzes. These soils of the chestnut zone are distinguished from the solonetzic chernozems by a lower humus content, by the columnar horizon lying closer to the surface and by a smaller diameter of the columnar units. The solonetzes of dry and desertic steppes constitute a zonal phenomenon and are not a result of local conditions.

The solonetzic process, which grows in importance starting from the chernozem zone, reaches its maximum development in the south of the chestnut zone, whereas it is weak in the zone of desertic steppes and deserts, if we leave aside the takyrs and other solonetzic soils. The colloidal absorbing complex of the chestnut soils is saturated with calcium. The overall absorbing complex equals 25-30 m-equiv. The absorbing complex of the chestnut solonetzic soils and solonetzes contains sodium which reaches 5-20% of the absorbing complex. In the solonetzic varieties of the chestnut soils, the salt horizon lies closer to the surface.

Classification and general description of the soils. All the soils of the chestnut-brown zone can be grouped as follows: 1) dark chestnut; 2) chestnut; 3) light chestnut; 4) meadow chestnut; 5) brown; 6) solonchaks and solonchakous soils; 7) solonetzes and solonetzic soils; 8) solods and solodized soils; 9) sandy immatured soils.

The zone comprises subzones: a) dark chestnut and chestnut; b) light chestnut and c) brown soils.

The chestnut soils are characterised by the following general morphological features:

A—approximately 20-25 cm thick, dark-cinnamon (dark chestnut), cloddy, fairly loose, penetrated by the roots of the grassy vegetation;

B_1—about 25-35 cm thick, cinnamonish (chestnut), coarse-cloddy, carbonate, packed, fissured;

B_2—about 30-50 cm thick, cinnamon-pale yellow with darkish-brown stripes of humic streaks (marbly), with carbonate spots in the form of lime nodules, packed, indistinctly cloddy.

B_3—about 50 cm thick, pale yellow to yellow, loessy, uniform, packed, carbonate. At the bottom, lies a belt of soluble salts.

These morphological features vary considerably with changes in the zonal conditions of soil formation. In the northern regions of the zone are formed dark chestnut soils with a deeper profile, darker coloration, a more pronounced structure. In a southern direction, the depth decreases and the coloration of the humic horizon gets lighter, but the coloration of the B horizon often gets more intense, the soil gets more packed, effervescence with HCl reaches the humic horizon and is present throughout the whole of the section; the salt belt does not leave the upper one-metre-thick layer of soil. The signs of alkalinity of clayey chestnut soils increase in the southern and eastern directions, from which it can be deducted that the properties of the soils get somewhat worse in the eastern and southern directions.

The brown soils are much less distinctly differentiated into genetical horizons.

Here are the morphological features of these soils:

A—0-10 cm. Pale yellow-grey (brown), indistinctly cloddy, sometimes layer-laminated in the upper part.

B_1—10-40 (70) cm. Light cinnamon (brown) with, not infrequently, a reddish tinge, structureless, packed, porous, sometimes fissured.

B_2—40-80 cm. Yellowish with a cinnamon tinge, not infrequently light straw-coloured or whitish due to the presence of carbonates, structureless, packed. Intense effervescence with HCl. Passes gradually into the lower horizon.

C—80-110 (150) cm. Yellowish with cinnamon tinge, structureless, packed, effervesces with HCl. Sulphate efflorescences are met with.

The brown soils are distributed in a complex with the solonetzic ones. This complexness is more pronounced on clayey deposits and less pronounced on sandy and sandy-loamy formations.

In light chestnut and brown soils, the differentiation between the genetical horizons is insufficiently distinct, in contradistinction to the dark chestnut soils, which, in this respect, are closer to the southern chernozems.

The dark chestnut soils which were formed under a continuous gramineous (sheep's fescue-feather grass) stand of grass, are distinguished by a humus content of up to 4-5% in the upper layer, the thickness of the humus horizon being approximately 40-50 cm.

The chestnut soils contain up to 3-4% of humus. The humus content goes down to 2-3% in light chestnut soils and the thickness of the humus horizon decreases down to 30 cm. In addition, the degree of alkalinity in the latter is more pronounced; in strongly solonetzic soils, this entails denser packing in the lower part of the humus horizon. In the upper part of the humus horizon, the structure loses its water-resistance, the horizon acquiring a dusty character. The decrease of humus in light chestnut and brown soils is tied with the fact that the soil receives but a small amount of plant remains, due to the grass stand being sparse and the litter-fall meagre and also in view of the fact that the humus formed, which contains a certain amount of mobile humates, possesses less stability. Chestnut soils are characterised by the possession of three layers of salt accumulation in their profile: carbonates of alkaline earths bases in the upper part, carbonates in the middle, sulphates (gypsum) and chlorides of alkaline earths and alkalis at the bottom.

The sulphates are distributed below the limit of wetting, down to 180-190 cm. Water-soluble salts in the form of illuvium are found in chestnut soils at a depth of 2 m and in light chestnut ones, at a depth of 1 m. The amount of water-soluble salts is about 0.05-0.1% (0.4%), a proportion which does not inhibit the development of plants. In the presence of such an amount of water-soluble salts, with correct irrigation, secondary salinisation of the soils does not occur.

Chestnut soils contain about 0.06-0.15% of phosphoric acid and 0.1-0.2% of nitrogen. Like in chernozems, the ratio of C to N in the humus is equal to 10-12. In brown soils, this ratio goes down to 7-9. Chestnut and brown soils contain a high amount of nitrates, whose quantity changes in accordance with the agrotechnical standard. With depth, the amount of nitrates goes down; at a depth of 100 cm, they are present in the proportion of 1-13 mg to 1 kg of soil.

Chestnut soils are characterised by a high content of Ca and the predominance of the mineral part of the soil over the organic one. The base exchange capacity and the total of exchangeable cations are lower in chestnut soils than in chernozems, higher in the A and lower in the B horizon. The total of absorbed bases amounts to 30-35 (40) m-equiv. to 100 g of soil, of which: 25-28 m-equiv.

of Ca (70-75% of the total), 6-9m-equiv. of Mg (20-25% of the total). The amount of absorbed sodium in chestnut nonsolonetzic soils does not, as a rule, exceed 5% of the capacity.

Chestnut soils are characterised by the following water-physical indices: apparent density 1.2-1.5, specific weight 2.6-2.7, pore-space 40-45%. These soils are not easily wetted. Atmospheric precipitations do not penetrate beyond 100 cm, more often down to 50 cm. At a depth of 2 m is found a so-called dead—physiologically dry horizon. But even this contains maximum hygroscopic moisture amounting to 8% and more. The reserve of productive moisture in chestnut soils is inadequate. The soils of the Volgograd region, for example, in a layer one metre thick, contain about 60-80 mm of moisture in spring, about 20 mm in summer and about 50 mm in autumn.

In morphological features and properties, the dark chestnut soils are close to the southern chernozems, whereas the light chestnut soils are close to the brown ones.

Typical of the chestnut and brown soils is the complexness of their distribution, which is expressed by the presence of solonetzic soils and even solonetzes, appearing as isolated areas in the form of patches against a background of one or the other type of soil. In the northern part of the zone, the solonetzes are distributed over pronounced microdepressions whereas in the southern part, over microdepressions are often formed solonchakous soils and solonchaks, which, not infrequently, border with water bodies. In the northern part of the zone, in depressions, known under the name of limans, develop meadow dark-coloured limanic soils.

The regular association of solonchakous soils with slight depressions, that of solonetzic soils, solonetzes and solods with more pronounced depressions and that of dark-coloured limanic soils with still more pronounced ones, points to the sequence of their development. As the soils become solothised, the depressions become more pronounced and this leads to an increase in moistening. Pronounced moistening promotes a better growth of the meadow vegetation and higher yields, leading to an increase of the amount of humus, a darkening of the coloration and more pronounced structurisation of limanic soils. The soils of such depressions contain up to 4-6% of humus and more. The dark-coloured humus-accumulative horizon reaches a thickness of 50-60 cm. Not infrequently, the meadow dark-coloured limanic soils of depressions are bordered with solonchakous and solonetzic soils. Such a distribution of soils gives an even clearer indication regarding their origin.

Changes in the microrelief are accompanied by changes in the salt regime of soils and also in the composition of the vegetation. The proportion of halophytes increases from the centre to the periphery and their presence leads, in turn, to biological salt ac-

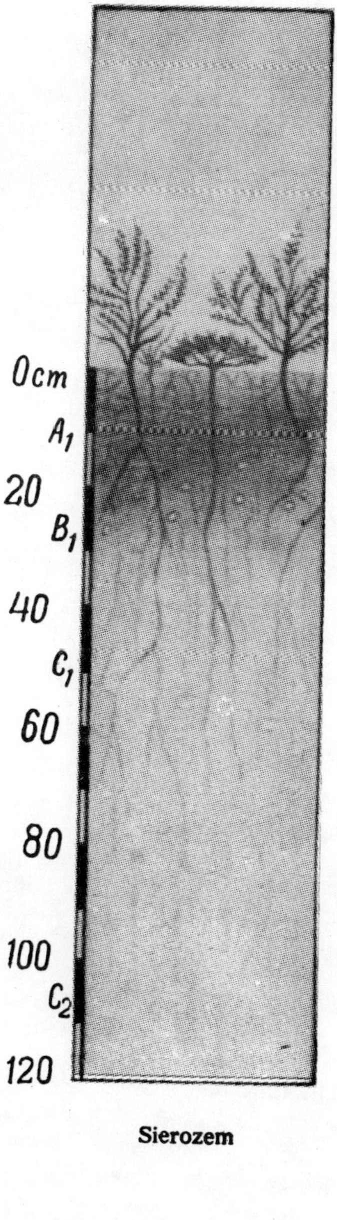

Sierozem

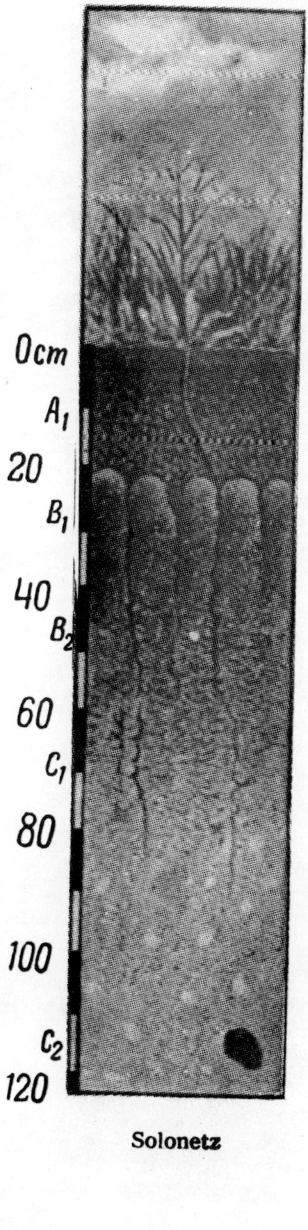

Solonetz

cumulation and the development of solonchakous soils. There is an increase in the proportion of normal gramineous vegetation from the periphery towards the depression, which promotes the enrichment of the soil in calcium and the displacement of exchangeable sodium. This leads to the improvement of the physico-chemical properties of the soils and their regradation—passage to structural soils of the meadow-chestnut type.

Many other soils are also distributed in the chestnut-brown zone: saline, swamped and bog soils, soddy-alluvial soils, etc.

Agromeliorative characteristics. The soils of the chestnut zone are potentially rich soils. The main obstacle hindering the obtention of high yields is the lack of moisture, as well as the low structurisation of the soils.

The productive capacity of these soils can be greatly enhanced through the appropriate agrotechnical, agromeliorative and hydromeliorative measures. These measures include, in the first place, the creation of a thick structural A_p horizon, possessing high non-capillary porosity and water capacity. It is capable of containing and retaining the whole of the melted water and atmospheric precipitations. It loses little moisture through evaporation from the surface and, at the same time, supplies the plants with adequate moisture. However, due to the complexness of the soils, the creation of a deep structural A_p horizon constitutes an exceptionally difficult task requiring considerable capital and labour outlays.

A most important measure in this respect is the improvement of solonetzic soils and solonetzes; this can be achieved through gypsuming, which, in view of the complexness of the soils mentioned earlier, is difficult to carry out. The patches of solonetzic soils are not clearly outlined, forming small disseminated areas, so that it is difficult to carry out discriminating gypsuming. In a number of cases, it will be found necessary to effect deep levelling ploughing and the entire gypsuming of large areas. The meliorative measures and technique to adopt should suit each particular set of complex conditions.

In addition, all the soils of the zone are in need of an improved water regime through the adoption of correct irrigation, using the local water resources as well as large hydrotechnical installations.

So far, the great possibilities for watering the soils afforded by catchwork irrigation, snow retention and regulation of surface runoff, have not been fully utilised. Water reserves in the soil are ensured here by the creation of ponds, reservoirs and afforestation. Catchwork irrigation is quite promising, all the more so since it is easy to carry out and profitable. When in some districts, this system of irrigation fails to give adequate results, due to an irregular supply of moisture, the difficulty can be remedied by disposing an amphitheatre of small limans.

Catchwork irrigation works satisfactorily not only in the case of level areas, but also in the case of gentle slopes, where the water can be retained by erecting horizontal banks along the surface.

Brown as well as chestnut soils play an active part in agriculture. With irrigation and progressive agricultural methods, they give high yields. The latter lead to the improvement (structurisation) and enrichment of the soils with humus. A well-organised system of irrigation leads, in the long run, to a rise of the effective and potential fertility.

Soils of Desertic Steppes and Deserts. Sands

The zone of desertic steppes and deserts constitutes a relatively level area; it is subjected to heating by the sun and deflation by winds, the role of water erosion being reduced to a minimum. There are no deep erosional valleys here, with a well-defined amphitheatre of river terraces. But there is a complex system of old channels (uzbois, wadies), left by erratic streams, which flowed during more humid periods in the past. Nowadays, these channels are periodically followed by flows forming after cloud-bursts and by mud flows developing on mountain slopes, which form layers of peculiar proluvial-alluvial deposits.

To the south, the desertic steppes border on desert, and along river valleys and low plains they are wedged into it. The desertic part of the zone represents a weakly drained flat plain.

The climate of the zone is a peculiar variety of the continental type; it is rather arid.

The vegetation of the desertic steppes is represented by a sparse stand of grass with a predominance of wormwood and willow-bush (*Kochia prostrata*—a subshrub) with Russian thistle, camel-thorn, numerous bulbaceous, spring ephemers. It develops in the early part of spring during 1-2 months, after which it dries out, with a new growth in the latter part of autumn, so that in the course of the year there occur two short vegetative periods, as it were, or one sharp seasonal change of vegetation. On sands grow calligonum and saksaul which reach a height of up to 3-3.5 m and a diameter of up to 35 cm. The vegetation of desertic steppes develops a powerful root system, which is several times more abundant than the above ground part of the plants. In light sierozems, the dry above ground mass reaches 1 ton to one hectare, the corresponding mass of the roots reaching 15 tons; in dark sierozems, the corresponding amounts are 4.5 and 30 tons. A considerable proportion of the roots belongs to perennial plants which do not die out every year, and does not therefore contribute to the replenishment of the lost humus.

The vegetation of the desert is still sparser and consists of Russian thistle-shrub (*Salsola arbuscula,* kerauk, tetyr, *Anabasis salsa,* etc.) with ephemers (sedge, meadow grass) and wormwood (grey wormwood, southern wormwood). It is poorer in rocky desert and richer in sandy desert, where it is represented by shrub and grasses with a powerful root system. On sandy hillocks grows a species of *Calligonum,* which is devoid of leaves, *Aristida pennata* among the gramineae, marram grass or giant ryegrass (*Elymus sabulosus*). Among the shrubs grows sand acacia (*Ammodendron Conollyi*) which reaches a height of 7-8 m. At the foot of sandy hillocks, in depressions grows sedge (*Carex physodes*), locally called "ilyak".

At the bottom of interrow depressions, saksaul (*Haloxylon aphyllum*) attains a good development, reaching a height of 5-6 m. Its ash contains a considerable amount of sodium and potassium salts. Under *Haloxylon,* a surface crust 10 cm thick and what are known as haloxylon salined soils are formed.

In clayey deserts, in early spring, grow fine-leaved sedges (*Carex Hostii*), bulbiferous meadow grass (*Poa bulbosa vivipara*) and others. Here predominate the so-called ephemeral and saliniferous plant formations.

The saliniferous formations are associated with salined soils. In turn, the saliniferous vegetation exerts an influence on the salt regime of soils. The ash of the plants growing on the saline soils of desertic steppes is characterised by a high content of sulphates and even chlorides. Soil formation is not concerned with the various roles played by representatives of the animal kingdom, but we must note the role played by grubs in the formation of passages—foraminated pores, which permeate the soil. Ants and termites, which build underground nests in the soil, increase filtration.

In the zone of desertic steppes, the soils are formed on fairly varied soil-forming rocks.

Among the soil-forming rocks, the loesses and loess-like clays figure in good place.

In the zone of desertic steppes, prevails the sierozemic type of soil formation. The characteristic features of the sierozemic type of soil formation are as follows: seasonal fluctuation of the biological processes (spring, autumn), alternation of ascending and descending water-salt regime, development of intrasoil weathering, manifestation of an illuvial process with reference to the readily soluble salts and extremely weak translocation through the profile of the silty fraction, abrupt replacement of the process of humus accumulation by that of its destruction, leaching and ascent of the solutions, etc. Structure formation proceeds fairly weakly, with rapid breaking up of macroaggregates upon high dropping-liquid moistening; at the same time, a marked formation of stable microaggregates takes place in the soil. These particularities of the sierozemic type of soil formation are manifested upon a relatively low humus content, a relatively considerable thickness of the humic horizon, high and stable carbonate content, full saturation of the absorbing complex chiefly with alkaline earths bases, slightly alkaline reaction of the solution, high capillary porosity, fairly high water-raising capacity and water permeability of the soils and, finally, upon the existence of a contrast between the water and heat regimes of the soils. During cold periods, the soil is very slightly leached and, on the contrary, dries out considerably in summer.

The main particularity of the sierozemic type of soil formation is the rapid destruction of the humus. The high capillarity conditions the movement of a water flow to the evaporating surface of the soil and as a consequence, the soil dries out to a condition when only the dead reserve of water remains, incapable of capillary movement.

Sierozemic soils are formed under the conditions of an extremely arid climate with a prolonged warm period during the year, there being two brief vegetative periods, one in spring, the other in autumn, separated by a droughty summer.

Being tied with such dynamics of bioclimatic conditions, the soil formation proceeds in a most irregular fashion. Active soil formation lasts only 2 to 3 months. The rest of the time, the changes occurring in the soil are quite insignificant and soil formation proceeds at a very slow pace. At the same time, there is a marked biological accumulation of salts in the soil. Not infrequently this zone is referred to as the sierozem-solonchakous zone. In the main, the presence of carbonates in the soil-forming rocks under sierozems, is a consequence of a sierozemic type of soil formation. The absence of steady leaching and the enrichment with calcium salts strongly depress, here, the development of a solonetzic type of soil formation. The chlorides and sulphates reach appreciable concentrations at a depth of 0.5=1 m and lower down a sulphate horizon (accumulation of gypsum) is formed.

As a result of prolonged uninterrupted soil formation and biological accumulation, in the sierozemic zone, the ground water and soils are markedly saline. The soils and soil-forming rocks of the sierozemic zone possess a high capillary capacity for raising the ground water, which leads to salinisation of the soils. Here and there, salinisation causes a reduction of the area of agricultural land.

Every year, the organic matter manufactured by the vegetation is completely destroyed by microorganisms, which influences the properties of the soil, the vegetation and the microorganisms.

The sierozems are saturated with absorbed calcium, which conditions the microaggregation and uniformity of texture of the soil.

Classification and physico-chemical characteristics of the sierozems. Associated with desertic steppes are the following types of soils: northern sierozems (brown sierozems); dark sierozems; ordinary (typical) sierozems; light sierozems; irrigational sierozems; meadow-sierozemic soils; takyr-like sierozems; takyrs; meadow and meadow-boggy soils; solonchaks; solonetzes; immatured sandy sierozems.

All the above-mentioned sierozems are differentiated according to their salinisation (saline, alkaline), and according to the degree of taming (weakly, medium, highly tame).

All the sierozems are characterised by the uniformity of their profile. The A horizon is light grey, slightly coloured with humus. The B horizon is structureless, slightly packed. The C horizon is straw-coloured-yellow, loess-like.

Under the effect of irrigation, forms a so-called irrigational sierozem, which acquires a uniform dirty grey coloration down to a depth of 1-1.5 m. Its water-soluble salts get leached down to a depth of 1-2 m. This type is found everywhere in the zone of irrigation.

On plains situated at the foot of mountains are formed northern (brown) sierozems. They are characterised by low-carbonate-content in the upper part of the soil profile. The A+B horizon reaches a thickness of 30-60 cm in the plain and 50-80 cm at the foot of the mountains.

In the regions situated at the foot of mountains and in regions of low mountains, where the amount of atmospheric precipitations somewhat exceeds that of plains, are formed dark grey submontane sierozems with a deeper soil profile and more deeply coloured with humus. The depth of A+B reaches 100-130 cm and the amount of humus in them goes up to 2-4%. Submontane sierozems exhibit marked differentiation, depending on their geographical situation.

Above the dark grey sierozems lie leached sierozems of the northern sierozems type with an amount of humus reaching 4.5%. At great heights, they are replaced by brown forest soils. On submontane plains and hills, ordinary sierozems are formed, which contain up to 1.5-2.5% of humus and more. They possess the following morphological features (see Fig. 42*b*):

A—16-18 cm thick. Light grey with a pale yellow tinge, flaky-slaty at the top, weakly cloddy at the bottom, somewhat darker on the general light background;

B_1—30-40 cm thick. Light grey, fine-cloddy, whitish with a slight yellowish-brown tinge at the bottom, due to the presence of carbonates. Has sometimes a foraminated (vesicular) structure raised by the passages dug by insect grubs;

B_2—40-50 cm deep. Greyish-pale yellow, fairly loose, with lime concretions;

C—pale yellow, loess-like with rare and small lime concretions and gypsum crystals from a depth of about 2 m.

The light sierozems are distributed in the southern part of the zone on plains and river terraces. Their A+B horizon has a depth of 40-70 cm and they contain 1-1.5 (2)% of humus in the A horizon.

The sierozems are characterised by a low content of nitrogen, whose amount does not, as a rule, exceed 0.2%. But they possess a considerable reserve of phosphorus, up to 0.2-0.3%. The high content of water-soluble salts conditions an alkaline reaction (pH—7-8.5). The absorbing capacity amounts to 10-14 m-equiv. to 100 g of soil. The sum of absorbed Ca and Mg reaches 92-98%, and that of Na and K, 2-8% of the absorbing capacity. The sierozems are characterised by a higher content of microelements in comparison with other soils, which is tied, apparently, with a more prolonged biological accumulation of ash elements.

Due to an inadequate content of humus, the structurisation of sierozems is not high. The proportion of macroaggregates is less than 15%. The composition of the alumosilicate portion is uniform in all sierozems.

The sierozems possess the following water-physical indices: their apparent density is 1.2-1.3, their specific gravity 2.7-2.75. They have a relatively high pore-space [50-55 (60)%], which decreases somewhat with depth. Maximum hygroscopic moisture reaches 5% in light and up to 7% in dark sierozems. The wilting percentage is about 7.5 % in light grey sierozems, 9.7% in typical ones and 11.4% in dark ones. The absorption and filtration coefficient in the loess-like soil-forming rocks of sierozems amounts to about 0.0005 m/sec. The water permeability of these rocks amounts to 0.4 mm/min for light sierozems, 1.3 mm/min for typical ones and 1.8 mm/min for dark ones. The high porosity of the sierozems ensures a fairly high water permeability.

Nonirrigated sierozem is characterised by drying out in summer and relatively shallow spring wetting: down to a depth of 1 m for light and of 1.5 m for typical sierozems. The reserve of moisture is small and ephemeral. Sierozems contain their maximum amount of moisture in March, in connection with atmospheric precipitations. The moisture reaches limit field water capacity, attaining a depth of 1 m and more.

The soil absorbing complex of the sierozems is inadequate, due to the low content of colloids. Furthermore, the mineral colloids (SiO_2 and R_2O_3) prevail over the organic ones.

The absorbing capacity in milligram equivalents to 100 g of soil is:

in light	sierozems,	in the A horizon	10.4,
"	"	in the B horizon	6.4;
in typical	sierozems,	in the A horizon	14.1,
"	"	in the B horizon	9.3.

The sum of absorbed Ca and Mg equals 10-12 m-equiv., Na—2-4 m-equiv., Ca representing more than 80% of the sum of absorbed bases.

The mechanical and gross composition of the mineral part of the sierozems can be expressed diagrammatically (Fig. 43). The diagram reveals the relatively uniform distribution of SiO_2 through the soil profile. The amount of sesquioxides increases somewhat

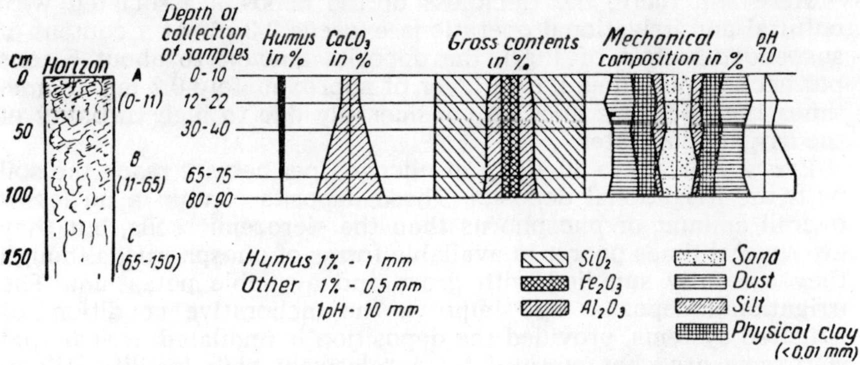

Fig. 43. Diagram illustrating the composition of sierozem

with depth, which is seen in dark grey sierozems. In the upper and medium parts of the soil profile, the amount of silt is 1.5 times higher than in the soil-forming rock, which indicates that the sierozems are undergoing a process of clayisation.

Owing to the low humus content, the sierozems are poor in nitrogen. The gross content of potassium reaches 1.5-3% and only an insignificant part of it is in the available form. Certain sierozems are characterised by a considerable content of water-soluble salts, which is in accordance with their biological accumulation and more advanced stage of soil formation. In the upper part of the section prevail the carbonates of alkaline earths elements, in the medium part prevail the bicarbonates and at the bottom, the sulphates and chlorides of alkalis and alkaline earths. The packed carbonate horizon of sierozems contains up to 6-7% of CO_2. The sulphates are located immediately below the limit of wetting.

Distributed in the sierozemic zone are the so-called chalno-meadow, tugaic-forest flood-plain soils and meadow-sierozemic soils, formed when the ground water table lies close to the surface. They are characterised by a high content of humus and by gleisation. Chalno-meadow soils are also characterised by alkalinity, they have a humic horizon reaching 30 cm and more, coloured in dark grey with a dove-coloured tinge. The B horizon is packed. Below are diffuse carbonate spots. These soils usually possess a heavy mechanical composition. Every year, soddy-alluvial forest soils with chloride-sulphate salinisation are renovated by silt.

As a result of prolonged (millennial) precipitation of suspended deposits, which get on to the fields with irrigation water, so-called irrigational sierozems are formed. They are characterised by a deep profile (several metres) with a relatively high amount of organic matter down to a depth of 1.3 m and more.

Here and there, the thickness of the deposits associated with cultural and irrigational operations exceeds 2-3 m. At a content of suspended matter of 1 g/l, the deposits amount to about 5 tons per hectare per annum, or a layer of approximately 0.2 mm. Sometimes the thickness goes up considerably due to high turbidity of the irrigational water.

Every year, up to 30-40 kg of nitrogen per hectare reach the soil with the irrigational deposits. These deposits contain a no lesser overall amount of phosphorus than the sierozemic soils, but they are several times poorer in available forms of phosphorus, although they are fully supplied with gross and available potassium. The irrigational deposits may improve the meliorative condition of irrigated systems, provided the deposition is regulated. Irrigational sierozems are characterised by a relatively high fertility. When cultivated, their fertility steadily rises.

Agromeliorative characteristics of the sierozems. Sierozems are rich in ash nutrient elements but poor in humus and nitrogen. They possess favourable water-physical properties, viz., fairly high pore-space, uniform carbonate distribution, a relatively homogeneous from the mechanical viewpoint vertical profile. This ensures a favourable water-air and salt regime of the soils when they are irrigated.

The only crops that can be grown in the sierozem zone without irrigation are early ripening crops, even so, they can only be cultivated in low areas or on the lower terraces of river valleys, where the ground water lies close to the surface. Farming without irrigation (dry farming) is conducted on the dark sierozems of submontane regions. This type of farming rests on the maximum utilisation of autumn-winter-spring precipitations and earlier sowing.

The sierozem-desertic zone possesses great latent possibilities. The prolonged frost-free period, lasting on the average up to 200 days, and the sum of temperatures during the vegetative period amounting to more than 4,000 degrees favour the culture of many valuable southern crops.

The sierozem zone receives abundant warmth and sunshine but extremely little moisture. The fight for water has been going on here for thousands of years.

Well-planned irrigation, avoiding leaching of the soils, exerts a favourable influence on the development of soil formation and

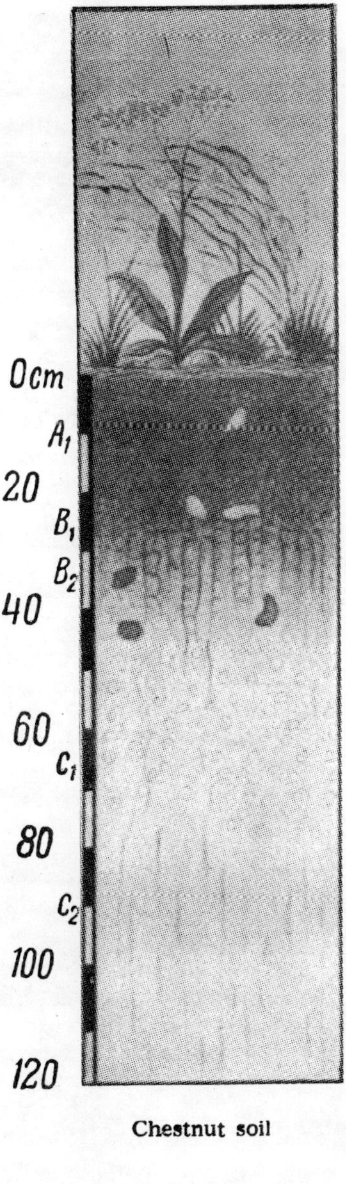

Chestnut soil

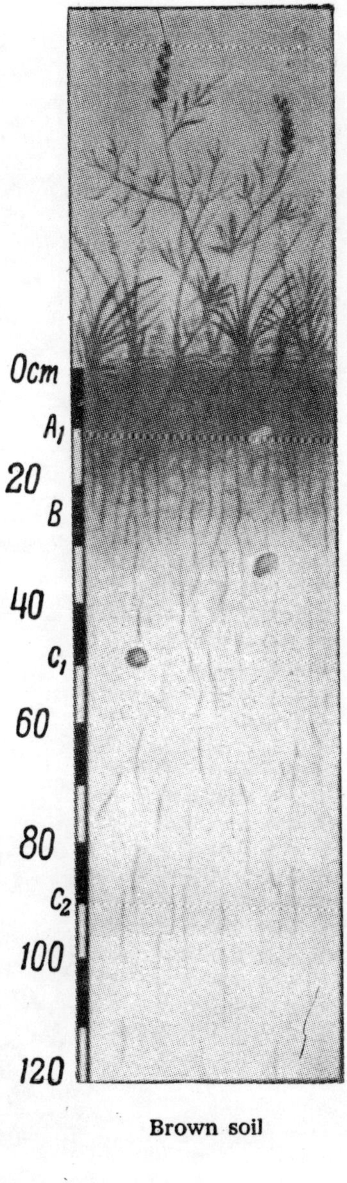

Brown soil

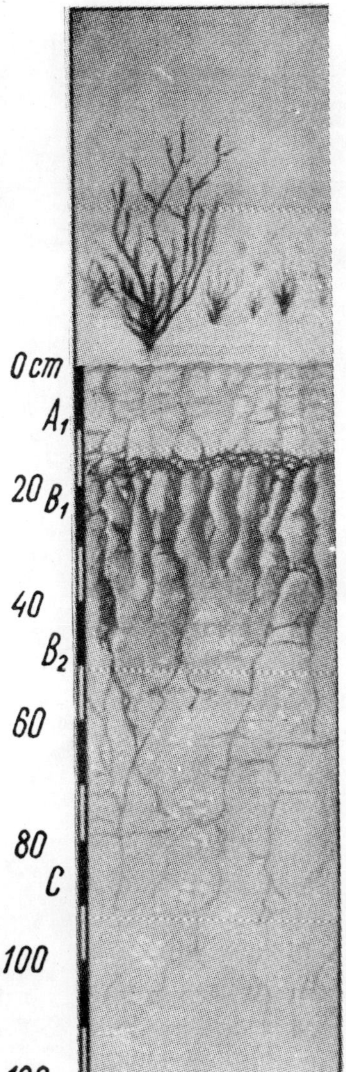

Solonchak

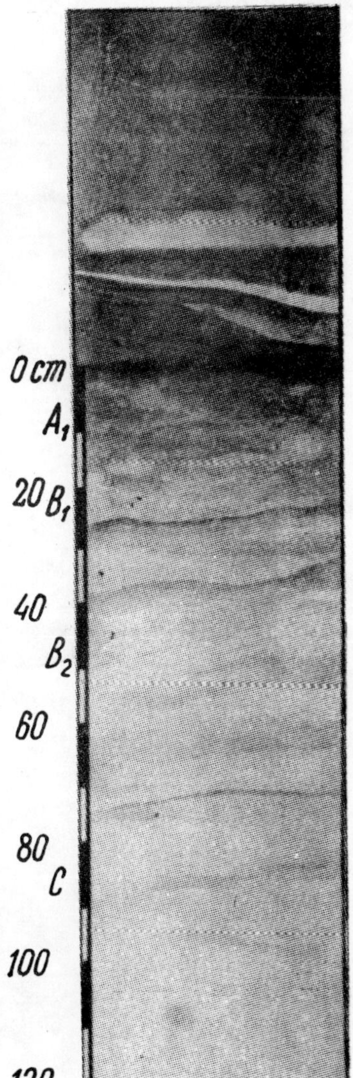

Alluvial-sod soil

creates the necessary water regime. Irrigation promotes the development of azotobacter and other bacteria. 1 g of irrigated sierozem contains up to 2,000 million microorganisms, which is far above the number contained in nonirrigated sierozems. Irrigational type of farming favours cotton-lucerne rotations. The irrigation of sierozems accompanied by the application of organic and mineral fertilisers, green manuring, etc., makes it possible to obtain high yields of the most diverse crops, in particular of cotton, which is of great economic importance. The production of cotton goes up every year thanks to the increase of the area devoted to that crop and higher yields. Apart from that most important crop, the sierozems are devoted to mulberry growing which is needed for silkworm breeding, rice, rubber plants, plants used in the textile industry, etc. The areas of orchards, vineyards and plantations are on the increase.

Among the most important measures to adopt for the reclamation and improvement of the lands of the zone are: the better use of old and new irrigation systems; the installation of new irrigational systems; the radical improvement of land (taming of soils); the reclamation of virgin and long-fallow land.

The accelerated taming of the sierozems will permit the realisation of enormous new agricultural reserves. Of great importance in this respect is the creation under irrigational conditions of a thick A_p horizon, with a depth of 30 cm and more.

Like the other soils, the sierozems need to be enriched with organic matter, which can be achieved through the application of organic manures, such as farmyard manure, compost, green manures, and the cultivation of lucerne. When taken after lucerne, cotton and other crops give high yields. Upon the application of nitrogenous and phosphorous fertilisers to sierozems, the yields of cotton rise to 10 tons to the hectare and more, i.e., they go up several times.

The most important measure to adopt in the zone is to fight and prevent salinisation. It is indispensable to slow down the intensive decomposition of the organic matter and the accumulation of salts in the soil. The sierozemic soil-forming process should be directed towards the formation of dark sierozems. In the plain, this can be achieved through a series of measures and in the first place through irrigation. By creating optimum moisture conditions and slowing down the processes of decomposition and mineralisation of the organic matter, the amount of humus in the soil can be increased several times.

In order to transform the sierozems and raise their fertility, an abundant supply of sweet water is necessary for irrigation purposes. An abundance of water will allow us to take the offensive against sandy deserts, colmating sandy soils and reclaiming them using reliable methods. The local water reserves of deserts are absolutely inadequate. It is indispensable to look for new sources of water using ground and artesian water and installing underground water reserves.

Attention should be devoted to the fight against secondary salinisation. Secondary salinisation, like swamping, excessive ploughing, loss of structure, overgrowing with weeds, erodibility, deflation, is a consequence of low agricultural standards. The "salt balance of salinisation" or process of salt accumulation, should, under productive conditions, be replaced by a "salt balance of desalinisation"—the process of translocation and removal of harmful salts. In order to achieve this, recourse should be had to a complex of measures aimed at lowering the ground water table and desalinising salined areas. The fight against salinisation can be carried out through the installation of a network of drains and collectors and by resorting to biomelioration.

Soils of deserts. The most fully developed zonal soils of deserts are the grey-brown soils, which form under a sparse Russian thistle-wormwood vegetation, under the conditions of a contrasting hydroterminal regime, i.e., light and shallow moistening in spring, rapid subsequent drying out of the surface and formation of a crust under conditions of pronounced overheating.

The characteristic particularities of soil formation in the deserts of the U.S.S.R. are: a) the formation of a porous crust and under it, of a clayised and weakly ferruginised foliated horizon; b) weak accumulation of humus; c) small depth of the soil profile as a consequence of inadequate and shallow wetting, salinisation and alkalinity of the soils, due to the absence of drainage on that territory and the high mineralisation of plant remains; d) accumulation of secondary biogenic carbonates in the upper horizons of the soil, gypsum at a depth of about 50 cm; e) the appearance of a yellow coloration, ferruginisation in the soils and sands; f) weak weathering of the soil-forming rocks (enrichment with primary minerals).

The formation in grey-brown soils of a packed horizon of a brown coloration with a reddish-ferruginous tinge is due to the oxidation, dehydration and subsequent fixation on the spot of mobile forms of iron. In connection with the over-heating of the soil and the inadequacy of newly-forming humus, the structure of the soil deteriorates. The high carbonatisation of the soils in the upper horizons is tied with the character of the decomposition of the vegetation in deserts and the distribution of the breakdown products. There is a considerable amount of magnesium in the carbonates present in these soils.

The absorption capacity of grey-brown soils reaches 30-46 m-equiv., more often 18-26 m-equiv. These soils contain only 0.3-0.5% of humus. The humus is associated with fulvic acid and has a narrow ratio of C to N (=4-5) as a consequence of the intense mineralisation of the plant remains. The readily-soluble salts are located close to the surface of the soil, which often leads to their salinisation and solonetrisation.

Takyrs. The flat clayey depressions of deserts, in which the surface melted water or that of downpours collects, are called takyrs. In spring they are impassable due to the swelling and overmoistening of the silt, whereas in summer upon its drying out, the silty-clayey rock is transformed into a dense mass distinctly fissured, testaceous on the surface. The fraction of particles of mechanical composition smaller than 0.01 mm in diameter reaches 95%. The heavy mechanical composition of the takyrs is mainly a consequence of their takyrisation (silting and solonetrisation), but it may also be the cause of an acceleration of this process. Not infrequently, the takyrisation process affects neglected arable lands which become flooded with a thin layer of water containing alkaline salts in solution. The formation of takyrs precedes and accompanies colmatage (silting) of the surface with silt, brought from elsewhere and settling on the bottom of temporary shallow water bodies. In this fashion, the area of the takyrs gradually increases, as a result of a widening of the surface subjected to colmatage and an increase in area of the temporary water bodies. Colmatage may be assisted by the soil being subjected to a brief spell of frost and a stagnant water regime. The silt, which results from the elutriation of deposits, may arise from the aeolian abrasion of quartz grains. The size of the areas covered by takyrs varies from small patches surrounded by sands to large tracts spread amid proluvial and alluvio-deltaic plains.

The vegetation of takyrs consists of algae and lichens. The deposits reveal traces of boggy vegetation, which is evidence of the biogenic and hydrogenic nature of the takyrs. On the surface of takyrs forms a dense porous crust, which subsequently acquires a porous vesicular structure (Fig. 44). This crust hinders the development of plants. The annual deposition of sediments and the formation of crusts conditions the formation of a lamellar horizon, which reaches a thickness of several tens of centimetres. Takyr-like soils and takyrs possess the features of grey-brown soils of deserts. They constitute hydromorphous desertic soils. The profile of takyrs is approximately as follows:

0-3 cm—dense crust, porous where fractured;
3-12 cm—bright brown (chocolate) clay, markedly scaly (shelly);
12-50 cm—dense cloddy-lumpy clay.
Below, lie loams or sand.

The extent to which the genetical horizons are marked and their thickness give direct indications concerning the age of the takyrs.

By nature, the takyrs are saline-alkaline soils. The degree of the salinity varies greatly. At a depth of 10-20 cm from the surface, lies a sulphate (gypsum) horizon, which is sometimes used

Fig. 44. The surface of a takyr

for the self-melioration of solonetzic takyrs. A distinctive feature of the salt profile of the takyrs is the pronounced sweetness of the uppermost horizon.

The takyrs are characterised by the following water-physical and chemical indices: apparent density of surface crust (0-2 cm)—1.45; apparent density of underlying horizon—1.6-1.7; pore-space—47-49%, in the solonetzic horizon it is 42%, the capillary porosity reaching 74-93% of the total pore-space, whereas the noncapillary porosity does not usually exceed 25% of the pore-space. This accounts for the fairly low filtration capacity of the takyrs. The depth to which they are wetted does not exceed 20-25 cm.

The takyrs do not acquire a steady reserve of water. The overall absorbing capacity of the takyrs amounts to 7-10 m-equiv. Nearly 20% of the capacity is often captured by sodium (up to 2 m-equiv.), which conditions the high degree of dispersiveness and mobility of the colloids upon wetting.

In spring and autumn, the upper part of the profile of takyrs is leached down to a depth of 10 cm, where the dense residue, amounting to 0.2-0.6%, consists chiefly of sodium and partly of magnesium bicarbonates. In summer occurs seasonal salinisation, whereupon the dense residue goes up to 1-2%, the chlorides in it reaching 90%, and, as a result of exchange reactions, sodium displaces calcium and magnesium and gains entry into the absorbing complex.

Like the solonetzes and the solods, the takyrs are ephemeral soils. They arise relatively rapidly, "mature" and regrade. A plano-concave relief develops in interdependence with takyr formation (like solothisation). Here and there, the takyrisation process passes relatively rapidly through its stages of development and gets rid of itself as it were. Takyrisation of an originally concave surface is, to begin with, accompanied by its very slight subsidence, after which it becomes stabilised and then levelled out, due to silting, which goes as far as the elimination of its drainless character. In this connection, the takyrs regrade from hydromorphous soils to soils of a zonal type. The removal of the hydromorphism through artificial means (drainage, levelling) may accelerate this process. Under natural conditions, the fertility of the takyrs is fairly low. During the process of transformation of takyrs into deserts, the crust is partly loosened and ephemers make their appearance, whereupon the takyrs acquire the features of grey-brown immature soils. The low fertility of the takyrs is due to their residual alkalinity and salinity, most unsatisfactory physical properties, weak biogenic activity and extremely low organic matter content. The amount of humus does not exceed 1%. In summer, the surface horizons of the soil are heated to a temperature of 65-70%, which precludes vegetative life.

Like all the other soils which are unsatisfactory under natural conditions, the takyrs are susceptible to radical improvement and transformation. They should be thoroughly loosened, ploughed over, subjected to leaching operations, sometimes to sanding, and to self-melioration, using the gypsum horizon of the takyrs themselves, and fertilisers.

Very good results are obtained through plantage ploughing to a depth of 40-45 cm and irrigation. Good results are obtained through the application of farmyard manure and other organic manures, which bring about a radical improvement of the physical properties of takyrs.

Sands. Care should be taken to distinguish between sands proper (free-flowing, bare or weakly fixed) and sandy soils. We shall limit ourselves to certain general data, mainly regarding sands. Sands, which constitute a loose geological formation, are a product of disintegration (weathering) of rocks. They consist of mineral particles (chiefly SiO_2) of a diameter of from 3 to 0.05 mm. The region of sands constitutes a new, almost untouched, enormous agricultural reserve. The sands are divided into eluvial, deluvial, glacial, alluvial, lacustrine, marine and aeolian sands, of various granulometric compositions (Fig. 45).

The structure of sands depends upon the sizes of the particles, their shapes, reciprocal arrangement and the character of their surface. Sands may be rounded, semirounded and angular. As the sands get coarser, the number of rounded and semirounded grains goes up and that of angular particles goes down. The surface of the sand grains is sometimes covered with a coating (cover) of other substances, chiefly on account of the adsorption of colloids, particularly sesquioxides. The water properties of sands and sandy soils, viz., filtration, absorbing power, formation of quicksands, etc., depend to a great extent upon these coatings. The coatings may be clayey, ferruginous and mixed. Clayey coatings communicate to sand the capacity to adsorb cations. The coatings become fixed by exchangeable calcium. Displacement of the calcium from the coating causes its destruction. Coloured sand looses its colora-

Fig. 45. Sandy desert

tion, giving coloured suspensions. Mixed coatings are usually double, the inner one being ferruginous, the outer one clayey. Clayey coatings also contain coarser particles. The reciprocal arrangement of the particles can be loose or packed. According to size, there are coarse, medium, fine and very fine sands. Sands are also differentiated according to the arrangement of the grains (loose and packed), to the character of the surface (even, smooth, dull, polished). The specific gravity of sands is 2.64-2.71, the apparent density being 1.62-1.76. Depending on the size of the grains, the sands possess varying pore-space; that of coarse sands is 35-39%, of medium sand—40-41, of fine sand—42-45 and of dusty sand—47-55%. In heterogeneous sands, the pore-space may be lower, due to the spaces between the coarse particles becoming filled with smaller particles. An admixture of clay raises the pore-space. The pore-space may also go up on account of the angularity of the grains. It depends upon the degree of packing, which decreases with time, as a result of pressure from above.

The surface of sands is, in nature, subjected to constant renovation, which hinders the formation of soils. But here and there, occurs a natural fixation of the sands. As the sands become fixed, there is a gradual formation of fine earths and their coherence goes up. Sands and sandy soils possess a number of adverse and favourable properties. Sands and sandy soils are poor in colloids and humus, possess low absorbing capacity and low coherence.

Owing to their high water permeability, they lose their water-soluble salts including the ash elements of plant nutrition. On the other hand, they possess relatively stable physical properties, do not get covered with a crust, water on them does not stagnate, seedlings do not suffer from excess of moisture and do not decay. Sands are much easier to work, possess good heat conductivity, the vegetation on them begins to develop earlier and crops mature more rapidly.

The shortage of ash nutrient elements in sands is compensated by dressings of organic manures (farmyard manure, peat), claying, marling, etc., which brings about a rise of the absorbing capacity of sands and sandy soils. The excessive water permeability and inadequate water capacity of these soils are counteracted in the same manner, as well as through colmatage, the application of gluey substances, the creation of artificial water impermeable or semipermeable screens at the necessary depth from the soil surface. On irrigated fields, sands are converted into highly fertile soils.

Soil formation strongly alters sands. Not only do they become more ferruginous and coherent, they also appreciably change in composition. A soil absorbing complex gradually makes its appearance, in the shape, to begin with, of colloidal coatings surrounding the separate sand grains and small clay crusts, and then of organic matter itself, which enriches the sands.

The ferruginisation of the sands of deserts, the accumulation in them of carbonates, gypsum and water-soluble salts is evidence of a process of mineralisation and accumulation of products of decay of plant remains. It can thus be seen that the leading biological factor of plant formation does not lose its significance even in hot sandy deserts. Semidesertic and desertic zones constitute very promising territories, where, through watering, irrigation and colmatage, useful soils enriched with humus and possessing relatively high fertility can be created. Supplying them with water in addition to that received from the atmosphere, is tantamount to displacing them, so to speak, towards more humid subtropics, in other words, it brings their properties and characteristics somewhat closer to that of the latter zone.

Chapter XV

SOILS OF HUMID SUBTROPICS, TROPICS AND MOUNTAIN REGIONS

Soils of Humid Subtropics and Tropics

The soils of the subtropics and tropics cover an area of 28 million sq km, or 19% of the earth's surface. Among them, the krasnozems of the subtropics occupy 3%, the red-brown soils 7%, the krasnozems of the tropics 7% and the lateritic soils 2% (according to L. I. Prasolov). These soils possess many clearly marked particularities, conditioned by a soil formation which goes on uninterruptedly under a cover of evergreen vegetation.

The annual temperature regime is characterised by a narrow average monthly amplitude. The average temperature of all the months is above 18°C, the average annual temperature being 25-30°. The total of the annual atmospheric precipitations reaches 1,500-2,500 mm, and, here and there, it even goes up to 3,000 mm and higher. Such narrow seasonal variations in temperature and moisture are only found in the subtropics. Winter is characterised by abundant rainfall and a certain lowering of the temperature. Summer is relatively drier and warmer. The difference between the mean temperatures from season to season reaches 8°. The dry period is typical of the region of monsoon forests. During one half of the year, the winter monsoons blow from the ocean, bringing rain, and during the other half, the dry summer monsoons blow from the inland.

The vegetation growing in this humid tropical climate consists of evergreen, continuous forests, which produce an enormous amount of above-ground vegetative mass, reaching 100-200 t/ha. Tropical forests consist of palms, lianas and arborescent ferns with multicoloured leaves. The grass cover is relatively uniform and consists of club-mosses and ferns. Undergrowth and grass are absent, here and there, in very dense forests. But multiple-storey forests comprising undergrowth and epiphytes often form almost impassable thickets. Monsoon forests are less varied. During the dry period (June-October), the predominant species, viz., teak, sheds its leaves. But also present here, are numerous representatives of evergreen vegetation. Monsoon forests are lighter than humid (rain) forests. They contain fewer lianas and epiphytes. The production of the above ground mass goes down to 50-60 t/ha. In connection with the increase in the duration of the dry period and the general decrease in annual rainfall down to 1,200-1,000 mm and lower, monsoon forests are replaced by savannahs with a tropical grassy vegetation. The above ground mass goes down to 30 t/ha of organic matter. The woody vegetation is represented here by isolated trees or clusters of trees.

The soil-forming rocks of the zone of tropics and subtropics are particularly varied. The soils are formed on the products of weathering of dense crystalline rocks, on deep eluvium and deluvium. In wide river valleys, in flood plains and deltas, the soils are formed on old and recent alluvium. Here and there, the parent rocks consist of deposits of stagnant water bodies of the lacustrine type, marine deposits, moraine formations and even volcanic deposits.

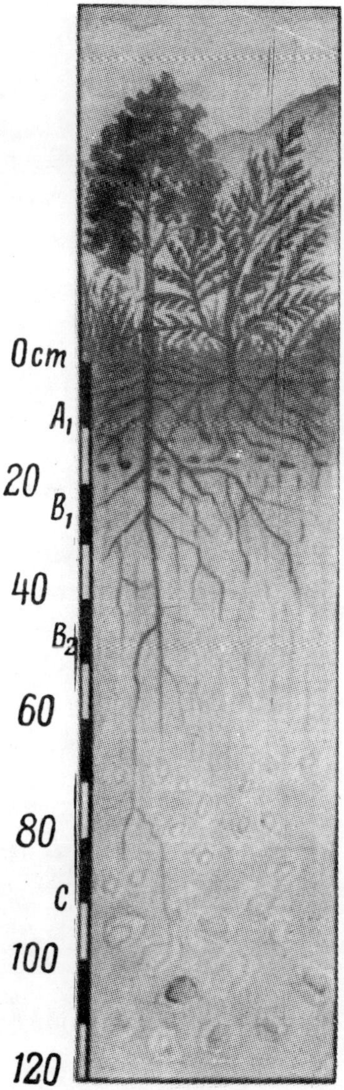

Krasnozem

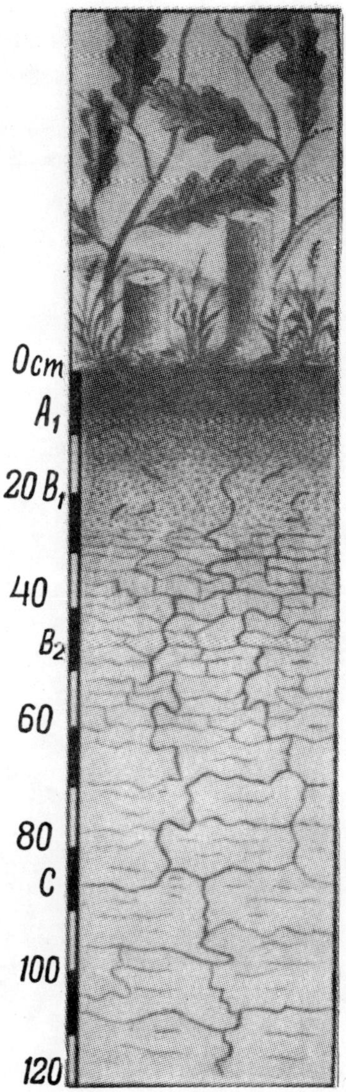

Yellow soil

In view of the high temperatures and heavy rainfall of the zone of tropics and subtropics, chemical weathering is intensive. The water, here, is subjected to high electrolytic dissociation and plays the role of an active chemical reagent and powerful factor of weathering and soil formation. The water of tropical regions contains CO_2, HNO_3, H_2SO_4 and much silicic acid, which is removed from the soil as hydrates. The disintegration of unstable rocks is accompanied by the formation of hydroxides of aluminium and iron in the shape of sols of low stability, which are precipitated by electrolytes.

The silica is washed down, playing a protecting role in the lithogenesis and causing the silicification of limestones and sandstones. As a result of secular intensive mineral weathering, there forms a uniform mantle of clayey eluvium of the lateritic type, 5-25 (50) m thick, in the upper part, with numerous reddish concretions of iron hydroxide and white-pink alumina. The products of weathering are coloured in red of various tinges, owing to their enrichment with hydrates of the sesquioxides of iron and aluminium (Table 29).

Table 29

Gross Composition of a Laterite from Etacot (India)

Layer	Percentage composition					
	SiO_2	Al_2O_3	Fe_2O_3	SiO	CaO	MgO
1. Laterite (enrichment) . .	14.1	36.9	29.8	2.3	0.7	0.3
2. Allitic siallite* (decomposition)	52.8	28.4	7.2	0.5	0.3	0.2
3. Siallite	58.6	25.2	5.9	0.8	0.7	0.2
4. Gneiss	64.8	18.4	3.8	0.6	5.2	0.9

* *Note.* The word "allite" is composed of the initial letters of aluminium and lithos, which in Greek means stone. Siallite is composed of the initial letters of silicon and aluminium.

The soils and soil-forming rocks are very poor in silicic acid and bases. Under the brightly coloured upper horizons of the crust of weathering with a high R_2O_3 content, lie the bleached kaolinised products of weathering of rocks.

Under the conditions of the uninterrupted vegetation and soil formation dating back to very old geological times, there proceeds a biological cycle of changes, viz., an intensive accretion of organic matter, which goes on all the year round, and a no less intensive decompostion or mineralisation of organic remains. Furthermore, the products of decomposition of the organic matter are removed, hence the brown and even dark coloration of the water in the tropical regions. For this reason, the soils of the subtropics and tropics do not possess a high humus content, with humus not easily noticed in any case, since its colour is not very dark.

Soil formation in this zone is accompanied by an intense disintegration of the minerals of the soil-forming rocks, removal of the soluble products and the formation of horizon enriched with iron and aluminium compounds, especially in the upper layers.

As a result of erosion, accumulation and repeated redeposition of the products of weathering which have been going on for centuries, there forms a deep red crust of weathering. At the present time, it constitutes the new soil-forming rock of the more recent soils.

The soils formed under the so-called rain-forests are peculiar tropical podzols and soddy-podzolic soils. The layer of duff forming under the cover of such forests is shallow, due to the fact that in a warm and humid climate, the litter-fall is subjected to an energetic decomposition. The soils get thoroughly leached, with attending removal of the bases, this being so pronounced that not enough is left for neutralising the humic acids. In the latter case, the iron and aluminium sesquioxides are also washed out from the soil together with the bases. Immediately below the forest floor, forms the podzolised horizon (A_2) of tropical podzols, arising under rain forests devoid of grassy cover. In relatively light forests, where a grass cover is present, tropical soddy-podzolic soils develop, in which, overlying the A_2 horizon and at its expense an A_1 horizon containing up to 10% and more of humus is formed. Under monsoon forests with a grassy vegetative cover a soddy type of soil formation proceeds. Here arise soils enriched with soft humus, of the tropical brown soils type. The soddy type of soil formation leads, under savannahs, to a noticeable accumulation of humus, nitrogen and ash plant nutrient elements. The soils forming here are the red-brown soils of tropical savannahs. During dry periods, ascending flows of water arise, which lead to the formation, on the surface of the soil, of a crust enriched by Fe_2O_3 (iron crust) which the plants are unable to penetrate. Underneath the iron crust a layer enriched with aluminium is formed. In the tropical and subtropical zone forest and grass bogs are also distributed. Under papyrus brush and, in tropical Asia, under juncaceous bogs, here and there, arises a thick layer of grassy peat of up to 10 m and more. In forest bogs, under a 1 m thick peat horizon, lies a thick horizon of gleisation.

The soils of the zone of tropics and subtropics bear the mark of man's direct and indirect intervention in no lesser degree than those of the other zones. Annihilation of the forests aggravates the droughty conditions; as a consequence, the savannahs infringe upon the monsoon forests and the latter encroach on the rain-forests. Desert advances upon the savannahs. On the whole, the intracontinental moisture cycle decreases. At the same time, water and wind erosion are intensified.

In the humid subtropics, yellow and red soils are formed.

The soil-forming rock is the red-coloured crust of weathering on tuffs, igneous and sedimentary rocks, which reaches a thickness of 15-20 m. Most of it is relic eluvium. Here and there, the structure of the weathering crystalline rocks is well preserved. The eluvium constitutes a clayey mass of a bright red or dark yellow colour. It contains up to 60-80% of physical clay. The rock contains 40-60% of R_2O_3, little Ca and K and no readily soluble salts.

The vegetation of the humid subtropics consists of broad-leaved forests.

Some part in the formation of rèd earths is played by the biochemical processes of disintegration of the alumosilicates occurring under the influence of the dissolved CO_2 given off upon the decomposition of organic matter and as a result of the life activity of microorganisms. The formation of red earths is accompanied by losses of Ca, to a lesser extent of Mg and still lesser of SiO_2, and the accumulation of sesquioxides. As a result of biogenic accumulation, the soil's humic horizon retains Ca, P_2O_5, SO_4, and other elements, which are indispensable to the following crops. The process of weathering of the soil-forming rock is so thoroughgoing in the red earths that it leads to the pronounced disintegration of a number of primary minerals, with the formation of secondary minerals. The silty fraction of the red earths consists of a mixture of the kaolinite group (kaolinite and halloysite) and of minerals of sesquioxides (goethite, hydrargillite).

Typical red earths have a brick-red coloration and the differentiation of their soil profile is weak. The distribution of the particules of mechanical composition through the profile of these soils is relatively uniform. It is only in the podzolised red earths that an illuvial horizon begins to show at a depth of 30-50 cm. However, the presence of secondary clays is evidence not of a podzolic but of an eluvial process in the soils of the red earths.

The morphological features of the profile of red earths soils are as follows:

A_0—forest floor, which may reach a depth of 10 cm. The lower layers sometimes possess a peaty character and are black;

A_1—0-15 (25) cm—humus-accumulative horizon, dark cinnamon or brown-red cloddy-granular (granular-nutty);

B—15-60 (85) cm—red or orange, uniform, of a curdy structure. Somewhat packed;

C—red, with pink, more seldom yellow, spots.

In mechanical composition, it is a clayey loam or even a clay.

The red earths contain 5-8 (10%) of humus. The predominance in the composition of the humus of fulvic acids, brings it close to the humus of podzolic soils. The humic substances of the red earths are bound with R_2O_3. The total of the absorbed bases amounts to about 19-25 m-equiv. to 100 g of dry soil. There is a marked predominance in the composition of the exchangeable cations, of absorbed aluminium; hydrogen may also be present, but in smaller amount. The amount of absorbed calcium and magnesium is exceptionally low, being only 1-4% of the exchange capacity. This is fairly typical for the red earths. The reaction of the soil solution is acid, the pH of the water suspension being 4.6-4.9, that of the salt suspension 3.9-4.3. The removal of the bases from the red soils, the accumulation of R_2O_3 of the goethite and hydrargillite type, bring the red soils close to the laterites.

The yellow soils are found in association with the red soils. With age, they may turn into the latter. The conversion of red soils to yellow soils is also possible when there is a marked change in the conditions of soil formation. Yellow soils are formed under broad-leaved forests, under the conditions of a humid subtropical climate, with a mean annual temperature of approximately 13-15° and 1,500-2,000 mm of rainfall. Yellow soils arise on clayey deposits, shales and other sedimentary rocks, whereas red soils arise on igneous basic rocks (trachytes, andesito-basaltic porphyrites) and sedimentary striped clays. The differences between the soil-forming rocks tell on the mineralogical composition and on the developing secondary clayey minerals.

The difference between red soils and yellow-podzolic soils is still more pronounced. Yellow-podzolic soils have a bright or pale yellow coloration and the genetical horizons of their soil profile are fairly well differentiated. On clayey rocks, yellow-podzolic soils exhibit marked signs of podzolisation. Here and there, the thickness of the A_2 podzolised horizon reaches 45-60 cm and more. Numerous large ortstein grains become fixed in the B horizon which underlies the A_2 horizon and has a thickness of 1 m and more. The morphological features of the yellow soils are as follows:

A_1—shallow soddy straw-coloured-yellow horizon;

A_2—greyish-pale-yellow horizon, with an unstable cloddy structure;

B—yellow or reddish-yellow illuvial horizon, with iron-manganese concretions.

The upper part of the profile is characterised by an increase of the SiO_2 content, and the illuvial horizon by an increase in sesquioxides. The upper horizon of the soil lying on sandy loamy soil-forming rocks, contains 1.6-2% of humus, the corresponding amount on shales being 4.5%. The absorbing capacity of these soils is insignificant. The lowest total of absorbed bases is exhibited by the yellow soils which form on sandy loamy deposits. Yellow-podzolic soils have an acid reaction, the pH of the water suspension being 5-5.6, that of the salt suspension 3.8-4.9. The exchangeable acidity of the yellow soils is governed, in the main, by the exchangeable aluminium, but to a lesser degree than in the red soils. It amounts to 0.2-4.5 m-equiv. to 100 g of soil. Apart from aluminium, the yellow soils also contain hydrogen in an absorbed condition and during the periods of possible anaerobiosis, it even predominates. The exchangeable acidity is somewhat higher in the red soils, reaching 7-10 m-equiv. to 100 g of soil. With depth, the exchangeable acidity goes up in both types of soils. The red soils contain twice as much mobile aluminium as the yellow-podzolic soils. The yellow-podzolic soils contain considerably more SiO_2 (70-85%) than the red soils (40%). The amount of sesquioxides is higher, on the contrary, in the red soils (55%) than in the yellow-podzolic ones (12-20%).

A distinction should be drawn between the yellow and red soils of the subtropics and the red-coloured soils (*terra rosa*) arising in the region of semideserts on the eluvium of limestones and on other carbonate rocks with a high content of iron hydroxides, kaolinite and other minerals.

The fertility status of the red and yellow soils is quite satisfactory. The warm and humid climate favours the culture of highly valuable crops (citrus, grapes, tobacco, tea, mulberry, southern fruit trees and vegetables). However, due to the acid reaction and the abundance of R_2O_3, the phosphoric acid is in a strongly-bound unavailable form, viz., iron and aluminium phosphates; the application of large amounts of phosphorous fertilisers is therefore required, preferably as phosphorite meal and basic slag, to raise the productivity of these soils. Liming does not always give a positive effect, all the more so since a number of cultures, such as tea, require a slightly acid reaction. Good results are obtained from the combined application of nitrogen and P_2O_5, organic manures and green manuring (soya, white lupins and others). The region of development of red soils is not free from the risk of wind and water erosion, hence the necessity to resort to the terracing of slopes, ridging and trellis planting of tea.

In tropical regions are formed laterites, which cover enormous areas. These soils are formed under conditions of pronounced atmospheric humidity and high mean annual temperature. They are characterised by a marked enrichment of the upper horizons with free hydrates of iron and aluminium oxides and extreme depletion with bases. The brick-red colour of these soils is conditioned by a high content of iron hydroxide of a low degree of hydration. In bauxitic laterite, which is of a less bright coloration, predominates Al_2O_3. The red colour is a characteristic feature of all soils located in warm regions of the earth.

Under the laterites, there forms a thick layer of slag-like red rock. The upper horizons of these soils are brightly coloured. Lower down, the colour turns paler and still lower, it passes into a mottled and bleached zone, characterised by the presence of koalin. Such kaolino-lateritic and bauxito-lateritic soils are widely distributed in the fossil condition.

Laterites are formed as a result of the intense decomposition (hydrolysis) of the alumosilicates, under conditions characterised by high humidity of the soil and high temperatures of the air.

Of all the soils of the earth, the laterites are the oldest, in absolute as well as relative age. They were formed as a result of a soil formation which began in very remote geological times and has been going on uninterruptedly all the year round for millions of years. Furthermore, the intensity of the soil formation is more pronounced here than in the other zones.

All biological, physico-chemical and especially chemical processes proceed several times faster in a warm climate than in a temperate one. Whereas in a cold and temperate climate, the decomposition of a mineral demands thousands of years or centuries, in a hot and humid climate, it requires but decades and even shorter periods. A fairly exact indication regarding age is afforded by the secondary aluminous and ferruginous minerals.

Crystalline alumina is present in laterites and red soils in the form of hydrargillite. Hydrargillite is considered as the most characteristic mineral of the true laterites.

The lateritic soil formation of the tropics bears some resemblance to the podzol formation of cold and temperate regions with permacidous leaching.

Lateritic weathering and soil formation present the following particularities:

a) intensive thorough decomposition of the alumosilicate core. The rocks are decomposed down to kaolin, with the liberation in the free state, with a high degree of hydration of SiO_2, Al_2O_3, Fe_2O_3, etc.;

b) outwash and removal of a large amount of SiO_2 and bases (K, Na, Ca, Mg) and removal to a lesser degree of R_2O_3 hydrates, their fixation and concentration in the layer of weathering rock;

c) change in the composition of the weathering rock, without removal of mechanical elements, with retention of the previous structure.

The essence and sequence of the process of laterisation consists in the following:

1st stage. In the process of soil formation, the primary mineral substances pass from the crystalloidal condition into the relatively mobile colloidal condition.

2nd stage. Formation of SiO_2 and R_2O_3 gels, which form the basis for the subsequent formation of hydrargillite ($Al_2O_3 \cdot 3H_2O$), limonite ($2Fe_2O_3 \cdot 3H_2O$) and other minerals.

3rd stage. Formation of hydrates of a low degree of hydration and anhydrous R_2O_3.

The soil gradually loses bases and SiO_2. The content of aluminium and iron goes gradually up. The iron is only partly lost in the form of soluble silicates and carbonates.

Finally, as a result of prolonged laterite formation, are formed secondary clayey minerals—hydrargillite ($Al_2O_3 \cdot 3H_2O$), bauxite ($Al_2O_3 \cdot 2H_2O$), diaspore ($Al_2O_3 \cdot H_2O$). The soil presents a peculiar kind of mottling and figure banding (zebra-like) resulting from the formation of so-called Liesegang rings.

Lateritic soils are, like podzolic ones, characterised by an acid reaction of the medium and are base-unsaturated. The humate and alumosilicate part of their colloidal complex is energetically dissociated. The absorbing colloidal complex of lateritic soils contains aluminium. Desiliconisation and ferruginisation in lateritic soils are clearly marked throughout the soil profile. An abundance of electrolytes and an energetic mineral weathering cause the aluminium and iron hydroxides in laterites to be precipitated on the spot as soon as they become liberated from the alumosilicate complex, without any appreciable translocation. Under the influence of high temperatures, they age and, losing water, become irreversible. The lateritic process leads therefore to the accumulation of hydrates of R_2O_3 throughout the whole of the profile.

Laterites do not contain carbonates, but carbonate soil-forming rocks may give rise to humus-calcareous soils of a special kind.

The products of lateritic weathering and soil formation give rise to fairly varied multicoloured soils of secondary origin, including even tropical bogs and gleyed soils.

Under native conditions, lateritic soils are not very productive. This is apparently due to the fact that they are very thoroughly leached. The effective and potential fertility of these soils may be considerably raised by bringing about the saturation in bases of their absorbing complex and creating a soddy type of soil formation.

Soils of Mountain Regions

Mountain regions are characterised by complex native phenomena, so that the conditions of soil formation in the mountains are considerably more varied than in the plain. In the plain, the soil-forming material consists mainly of sedimentary rocks and the products of their redeposition, whereas in the mountains, the soil-forming material often consists of massive-crystalline igneous rocks and the products of their disintegration on the spot. Due to the relief conditions and as a result of denudation, there is a wide distribution in mountain regions of soil-forming rocks consisting of debris and gravel, and skeletal soils.

In the mountains, due to the complex character there of the land forms, the microclimatic and together with them, the corresponding biological conditions of soil formation show considerable variation. The vegetation found in the mountains differs considerably in composition and development from that of plains.

In mountain regions, all the factors of soil formation stand out particularly clearly. Without dwelling on the fluctuations in climatic conditions connected with changes in altitude, we may mention the marked role played in the mountains by the various elements of the relief and the outcrops of the diverse soil-forming rocks, which govern the development and location of the mountain vegetation and soils. Mountainous regions present particularities of relief and climate (abundant insolation) which are not met with in plains. That is why the soils which arise here, have no analogues in the precincts of level zones. Such are the mountain-meadow soils under alpine meadows, or the brown soils—mountain-forest soils—which occupy special zones on mountain slopes.

At the foot of mountains, are found soils which correspond to the horizontal soil zones in the precincts of which the mountains are distributed.

Thus, at the equator, at the foot of the snow-capped Kilimanjaro and Kenya mounts, develop soils of the tropical zone, whereas at the southern foot of the Caucasian mountains, we find sierozems. Going up along the slopes of the Caucasian mountains, one can observe how the sierozems are being replaced by a belt of mountain-steppe, chestnut and chernozemic soils. Higher up, they are followed by mountain-forest soils, arising under broad leaved oak and beech forests. Beyond them, lies a belt of mountain-forest podzolic soils, arising under coniferous forests. At high elevations, the trees disappear; the vegetation there consists of subalpine meadows with mountain meadow soils. Still higher, we find bare rock, which, higher up, is covered with perpetual snow and ice. Some resemblance to this disposition in belts can be traced from the foot of the Caucasus to the shores of the northern seas, which is direct evidence of the influence of climate on the development of vegetation and soils and their geographical distribution over plains and mountains.

Vertical zones are more clearly marked in the submontane and lower parts of mountains. Higher up, the picture becomes more complicated, due to the fact that above a certain absolute height in the mountains, the fall in temperature and the increase in humidity do not follow a direct arithmetic progression but depend on the local mountain conditions. Of great importance is the geographical position of the mountainous region, as well as the geological and geomorphological structure. On the slopes of low mountains, there is a clear sequence in the change of climatic conditions and soils as the absolute height goes up. No such sequence is observed in medium and especially high mountains of the alpine type, where one should take into account a certain inversion in soil formation due to specific conditions. The soils of high mountains are more skeletal, their profile is less deep and they are not well-formed. The genetic horizons are not clearly marked. The soil cover of high mountains is characterised by the ruptured and mosaic-like character of the soil contours. Middle-height relief is characterised by the presence of skeletal soils in the upper parts of slopes and small-grained soils in the lower parts. The soils of low mountains and highlands are characterised by the relative continuity of the soil cover and full formation of the soil profile.

The distribution of soils in mountainous regions is characterised by the fact that above the podzolic soils, higher in the mountains, lies a zone not of tundra but of mountainous (Alpine) meadows and meadow soils. Lower down on the slopes, the podzolic zone may pass directly into the chernozemic or even the chestnut zone, the forest-steppe zone falling out.

Of great importance in the mountains are the insolation, the inflow of water from the mountains and surface leaching.

Divergence (inversion) of the soils from the theoretical types of soil formation in the mountains is also due to the local topography and the aspect of the slope. Inversion is more pronounced in mountain valleys with slopes of varying steepness and various surface forms.

Even although in places, mountain soils may be disposed as in horizontal zones, yet they differ considerably from the soils of plains. Thus, mountain podzols are relatively weakly podzolised and show signs of immaturity, they are insufficiently differentiated into genetic horizons and usually devoid of illuvium. They constitute young underdeveloped soils. Mountain chernozems do not possess a deep profile, do not effervesce with HCl, are characterised by a loose structure and a brownish colour due to the browning of the rootlets which thoroughly permeate the soil and do not become fully humificated.

In connection with the occurrence of sheet and linear erosion and accumulation on slopes, the processes of humification and leaching are reduced. Podzolisation, as it occurs in the plain, does not go far in the mountains and is, at the same time, accompanied by the accumulation of peat-like masses. The chestnut and brown soils of mountains differ also from those of plains by the possession of a more friable texture, a humic horizon of lesser thickness and a sharp delimitation between the humic horizon and the lower lying horizons devoid of humus. The depth at which the soil effervesces with acids depends upon the local conditions of the relief, infiltration and evaporation of moisture from the soil. In mountainous regions, according to V. V. Dokuchayev, "the relief rules over the destinies of soils". Of decisive importance are the absolute height of the place, the angle of inclination, the aspect and the geographical position.

The soils of mountainous regions have been in agricultural use for a long time. The taming and meliorative methods adopted for these soils differ sharply from those in use on level land. In the first place, it is absolutely indispensable to protect the soils from destruction and the native vegetation from annihilation. Apart from leading to the erosion and ablation of soils, the felling, burning and stubbing of forests sharply reduce the alimentation of mountain rivers, leading, in turn, to the drying out of valleys and wide mountain areas. Most important with regard to the protection of mountain soils is the careful management of the grazing and crop husbandry, taking into careful account the climatic and relief particularities of the spot. All the agromeliorative, forest improvement and hydrotechnical measures should be adapted to mountain conditions.

Owing to the abundance of moisture, their rich soils, the deposits of useful minerals, including agricultural ores (phosphorite, gypsum, lime, microelements), mountain regions are very promising from the agricultural point of view.

Thus, from the polar circle to the equator, from the snowy peaks of mountains to their foot, wherever the temperature of the earth's surface and of the air is above zero and plants and microorganisms can live, everywhere on the surface of the earth develops a process of soil formation and we get the formation of soils. Everywhere the soils arise from soil-forming rocks. As a result of soil formation, arise rocks of a special kind such as loess and loess-like formations, as gley, ortstein, bauxito-laterites, peat, sapropel, brown coal, etc.

Chapter XVI

FLOOD PLAIN SOILS

Flood Plains and Their Elements

Flood plain soils are formed in the flood plains and deltas of rivers. A flood plain is the depressed part of a river valley, annually or periodically subjected to flooding. It constitutes a band of newest land, forming as a result of the migration of a water flow and the accumulation of alluvium. Also regarded as flood plains are continental deltas—"oveflows" and flooded depressions, forming part of the meander belt.

Flood plains are formed by interacting turbulent fluvial and flood plain flows, carrying suspended and dissolved substances. The movement of these flows is always curvilinear and gives rise, consequently, to a transversal circulation, which conditions the peculiarity of the land forms and deposits in bed and flood plain.

A flood plain forms as a result of the development of an alluvial process consisting of two continuously alternating stages: in the bed and in the flood plain. The first stage spreads over a longer period and manifests itself only within the boundaries of the normal water level banks, the second one occurring during floods lasting from several days to several weeks, takes place in the whole of the flood plain, including the river-bed.

As a consequence of transversal circulation, the water flow in the bed accumulates alluvium and erodes its bottom and the banks. River-bed erosion here is combined with lateral erosion, which manifest themselves in the form of the bed and the character of the deposits. Sand spits are formed in the bed, orientated at a certain angle in relation to the main current. The subsequent shifting of the sand spits lower downstream and the deposition of sediments one on top of the other give rise to the typical cross bedding of fluvial formations. Furthermore, fluvial deposits usually lie in tessellated fashion, in the form of adjoining inclined lenses, the general inclination being directed towards the steady shift of the migrating bed. From the bedding and inclination of the limiting planes of fluvial deposits, one determines the direction of the migration of the river flow in the past (Fig. 46).

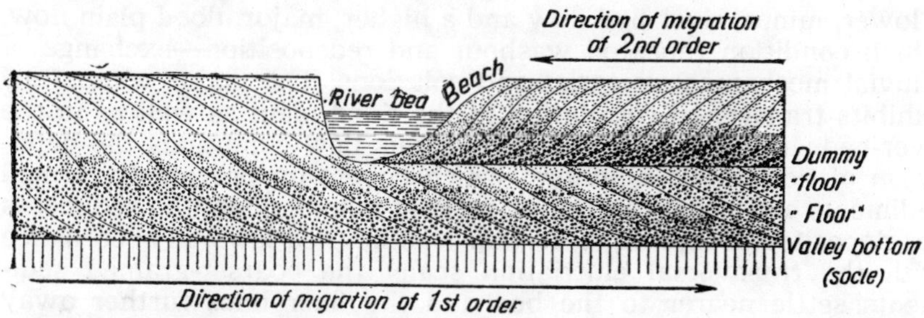

Fig. 46. River-bed obliquely-laminated alluvial deposits of a two-phase flood plain

The river-bed erosion conditions the formation of the complex surface of the flood plain bottom—socle or floor. This floor is sometimes framed from above with placers of useful minerals, including gold and diamondiferous placers. A transversal section through a flood plain, along the floor, gives a curve which can quite legitimately be called a live natural diagram, illustrating the combination in time of river-bed and lateral erosion. Such a section of the floor is sometimes expressed in the form of a fairly complex curve, representing the migration of the flow under the conditions of an unstable position of the base level of erosion.

A rise of the base level of erosion causes the floor to assume the shape of a complex surface, its transverse section presenting a zigzag-like curve. Over the old, main floor, arise so-called false or dummy floors. They can be seen in the section of wells, numbering sometimes 2-3 and more, and accordingly, there occurs the same number of horizons of useful minerals. Under migration conditions, with a lowering of the base level of erosion, the form of the surface of the floor corresponds to the last "paces" of the migration, due to the fact that the traces of earlier planations are annihilated by erosion. Fragments of old floors are only conserved in places of wide amplitudes of former migrations, at the periphery of the flood plain.

In river-beds in which the water flow is not subjected to great fluctuations of level, are only formed river-bed deposits of a sandy-clayey composition. This can be observed in the estuarine parts of rivers or, for instance, in rivers flowing out of lakes, in which forms a flood plain of a special kind, devoid of flood plain deposits. The same thing occurs in the valleys of mountain rivers, where the violent water flow spends its energy on river-bed erosion and repeated redeposition of sandy-pebbly alluvium.

Two flows interact in a flood plain during the high water period: a lower, minor, river-bed flow and a higher, major, flood plain flow, which condition repeated washout and redeposition—exchange of alluvial mechanical material and solutions. The major water flow exhibits transversal circulation in the flood plain as well as in the river-bed. In the flood spate period, this flow conditions the deposition of flood plain alluvium, which covers up previously formed sediments and levels out the surface. In a flood plain, the sediments are deposited according to certain well-marked laws, in accordance with the transversal elutriation scale: the coarser-grained sediments settle nearer to the bed and the finer ones further away from it. The mechanical composition of flood plain deposits changes from the bed to the periphery in a fairly sequential fashion. The sands of the areas of the flood plain closer to the bed pass imperceptibly into the loams and clays of the peripheral parts of the flood plain. But this scale of deposition in a horizontal direction is not infrequently disturbed by the transverse circulation of the secondary water flows which arise in the flood plain; tied with this is the formation of irregularities on the surface of the flood plain. The vegetation consisting of forests, brush and grasses of various heights conditions the uneven surface of the flood plain and exerts a corresponding influence on the high water regime.

The water flow in the river-bed possesses a general spiral movement and, accordingly, it migrates in three planes. The migration of the flow on the plan is represented by meanders, traces of old beds, from which one can reconstitute the position of the river on the plan in the past. But migration (planation of the river-bed) does not always occur from one of the original banks right to the other one; more often than not, it is limited within narrower belts, corresponding to separate phases of the flood plain. The horizontal migration of river-beds is of great importance. It exerts a strong influence on changes in the whole of the flood plain conditions, including the ground water regime. The horizontal displacement of the water flow in one direction or another is necessarily accompanied by an alteration of the hydrological regime of the flood plain. As the river-bed moves away from the flood plain mass, the ground water table necessarily rises, in which connection the water and salt regimes of the soils change, which may go as far as swamping and salinisation of the soils. Such a horizontal displacement of the river-bed in the flood plain is assimilated to the vertical displacement of the base level of erosion, in other words: the ground water regime is influenced not only by a deepening of the flow channel but also by its horizontal migration. The vertical migration is represented by a longitudinal curve along the thalweg of the floor, as well as by its transverse section. The vertical migration of the water flow is limited, here and there, by outcrops of fairly dense rocks, which form a kind of bulkheads, where erosion is very much reduced. Such bulkheads in the flood plains of mountain rivers condition the peculiar character and step-like disposition of the mountain alluvium.

Any mature flood plain is divided along the stream into a series of large sections or links of flood plain, characterised by the possession of common external features and conditions of formation. A widened, depressed, more levelled, clayey part of the flood plain may pass into a loamy or sandy loamy, less wide, but more elevated section with a crested relief, etc. The length, relative width and height of such sections or links of the flood plain vary within a wide range. The depressed links are granular whereas the elevated ones are stratified, according to the character of the prevailing soils. Granular and stratified flood plains do not exclude each other but form simultaneously in different links, or the stratified flood plain forms in the fore-part (closer to the bed) of the flood plain link, whereas the granular one forms in the rear, peripheral part, closer to the terrace. In a number of cases, the granular and stratified flood plains overlap with their deposits, repeatedly overlying one another. The latter occurrence can be observed in the section of polyphase flood plains. Granular stratified flood plains are formed mostly in accordance with the gradient, speed and direction of the current, the active power of the water flow and transverse circulation and with a number of other factors, including the presence or absence of forests in the drainage area.

Each link is further subdivided into separate smaller flood plain sections, which are, in turn, divided into segments. In a flood plain where the water flow constantly shifts to one side and undermines one bank, there forms only one washed-in flood plain segment which gradually increases in area (Fig. 47).

Such a flood plain area, whose formation follows the shifting river-bed, represents the so-called progressive segment of a mon-

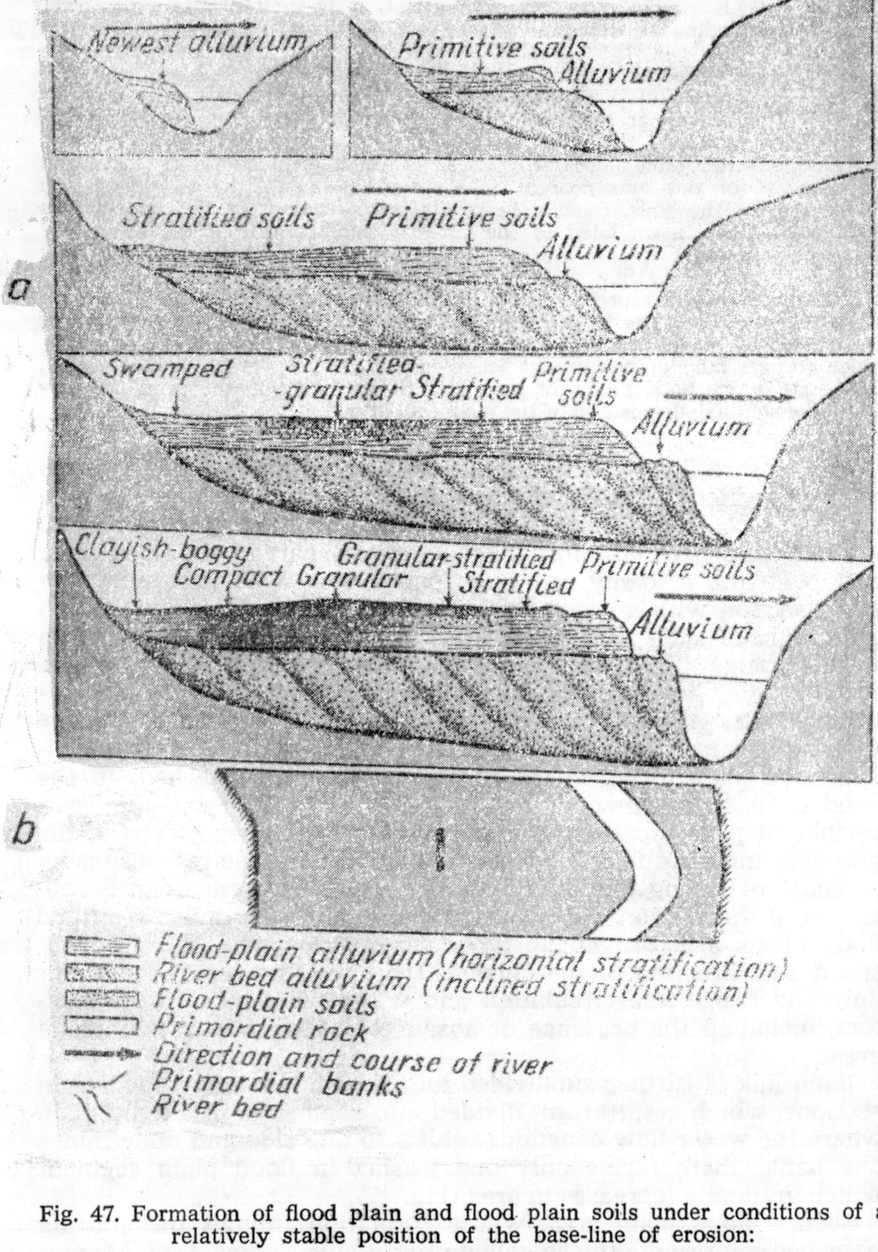

Fig. 47. Formation of flood plain and flood plain soils under conditions of a relatively stable position of the base-line of erosion:
a—formation of a progressive segment and of soils of a monophase flood plain (in section); *b*—progressive segment of a monophase flood plain (in plan I)

ophase flood plain. If the water flow should afterwards be displaced in the opposite direction, then the previously formed progressive segment will necessarily be undermined and transformed into an eroded (regressive) segment and following this (second) displacement of the bed, there will be formed a new progressive segment (Fig. 48).

Fig. 48. Flood plain. Undermined bank

In that case, the flood plain enters upon its second phase of development, i.e., there arises a two-phase-bilateral flood plain. Once the water flow has, for the third time, changed the direction of the migration into the opposite side, there invariably again forms a newest progressive segment of the third phase, on account of the scouring of the former progressive segment of the second flood plain phase. Thus arise the segments of the first, second and third stages and is formed a three-phase flood plain. The formation of the polyphase flood plains occurring in nature takes place in the same manner. A section of a link of bilateral-two-phase flood plain is made up of two segments of different ages (one young, one old), disposed on either side of the migrating bed. The young, gradually growing segment is bordered with the convex bank of the shifting bed. On the other side, the bed is bordered with the scoured (regressive) old segment, bounded by the concave undermined bank of the river-bed advancing onto the flood plain.

The mechanical composition of the flood plain alluvium of a progressive (convex) segment is characterised by the fact that it becomes progressively heavier from the bottom to the top, whereas in a regressive, concave segment, on the contrary, the heavier deposits are covered up by lighter ones, being rejuvenated as it were. This is due, in the first case, to the withdrawal of the bed from the flood plain area, and, in the second case, to its approach towards it. The unsteady (pulsating) position of the migrating water flow is accompanied by the deposition of the markedly stratified flood plain alluvium of the segments of the right and left banks, characterised by the alternation

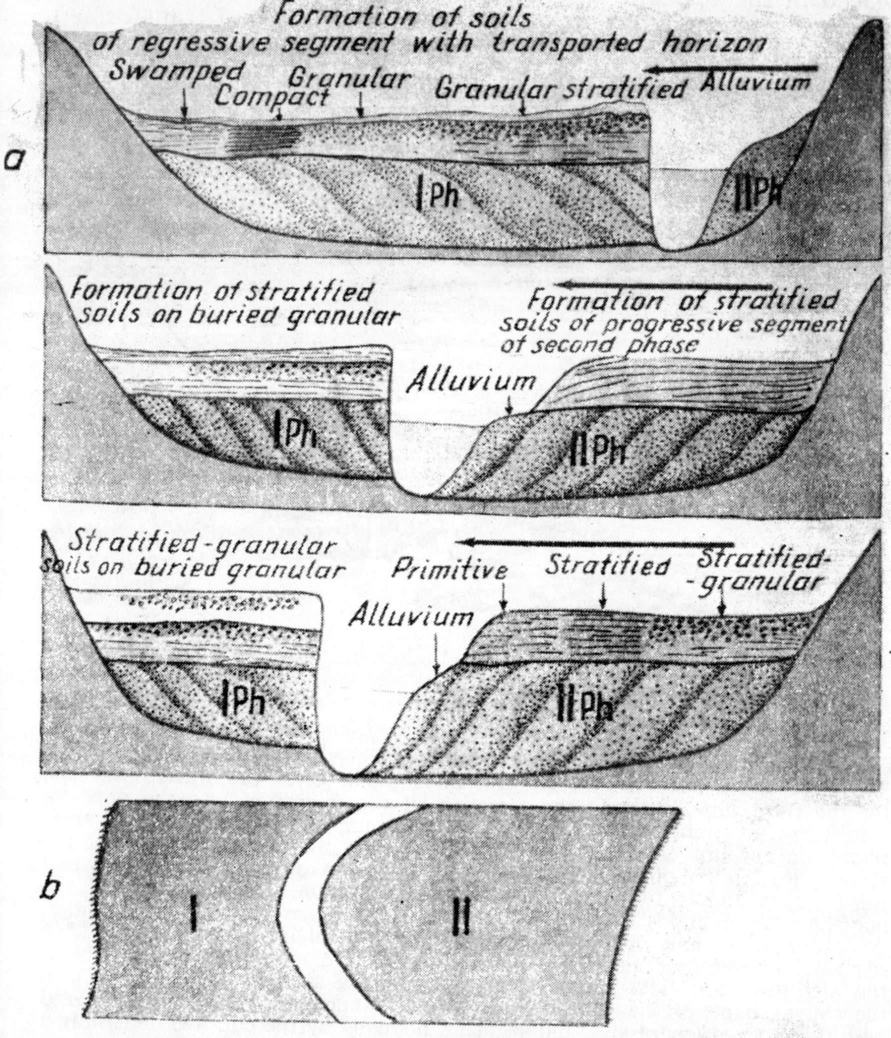

Fig. 49. Formation of a two-phase flood plain and flood plain soils:
a—in section; *b*—in plan. I—regressive segment of first phase; *II*—progressive segment of second phase of flood plain

of layers of different, mechanical composition. Straightened segments, which from above appear similar, are often clearly differentiated at the base, where they retain the features of the geological structure of preceding convex and concave segments. As a rule, all the flood plain segments constitute complex geological formations.

In contradistinction to a regressive one, a progressive flood plain segment is outlined in the plan by a "fan" of crescent-shaped ridges situated near the bed, reflecting the shift of the modern bed, whereas on the old (regressive) segment, such ridges are hardly visible, seen through, as it were, having been buried under top-set deposits. In both segments of the flood plain, the upstream parts situated near the bed, composed on the surface of relatively lighter flood plain alluvium, are more elevated.

Simultaneously with the migration of the bed, all the flood plain conditions undergo changes and with them also soil formation, the ground water regime and their mineralisation. The latter increases in the flood plains of the zone of dry steppes, in connection with the recession of the bed and a worsening of the drainage conditions. Mineralisation may, on the contrary, decrease somewhat in a regressive segment, in connection with more pronounced drainage of the flood plain, caused by the advance of the bed.

The stages in age, corresponding to the part situated close to the bed, the central part and that which is situated on the terrace side, are usually clearly seen in a progressive segment of a monophase flood plain. A young regressive flood plain segment, in the initial stage of scouring is likewise subdivided into these parts. But it is no longer possible to divide an old regressive flood plain segment into the above named parts, especially when the greater part of the segment has been destroyed by scouring.

In the segments of a monophase flood plain, the flood plain alluvium, deposited in a tessellated fashion in the form of curved lenses, is arranged strictly according to age. The lenses-tiles of a younger flood plain alluvium continue to be laid down, coming to rest obliquely on top of the lenses of the older alluvium, situated relatively further away from the retreating bed. Flood plain alluvium proper is deposited in a normal fashion in a vertical direction.

The picture is considerably more complicated with regard to the arrangement of alluvium and soils in bilateral, two-phase and polyphase flood plains (Fig. 49).

Flood plains of this type are formed when there is a decrease in the amplitude of migration of the bed, where the voungest part of the flood plain, with new soils and newest alluvium, is inserted into an older flood plain, which usually corresponds to their low and high stages. The flood plain phases are not always expressed in the relief. But they are easily recognisable from the vegetation, soils and geological structure. The flood plain phases are also easily delimited on photomaps, where the junctions between the phases stand out clearly along the lines of demarcation between adjacent unconformable forms of the mesorelief, sometimes as if showing through from under the top-set deposits (Fig. 50). The formation of a deltaic flood plain is somewhat different; there occurs, here, a migration of the branches, which form their own flood plains, inserted into the area of previously deposited alluvium, their surface being sometimes larger than that of the zone of migration of neighbouring branches. Here, the flood plain elements, and with them also the soils, are arranged in overlapping fashion. A similar thing occurs also in flood plains of the insular and lacustrine types. The elements of a polyphase flood plain are also arranged in overlapping fashion, even though they may not always be clearly expressed, especially where the modern flood plain regime has effaced the traces of the preceding phases. In this case, all preceding flood plain phases may be found to be buried under thick top-set flood plain deposits and the modern elements of the flood plain are formed independently of the fossil ones.

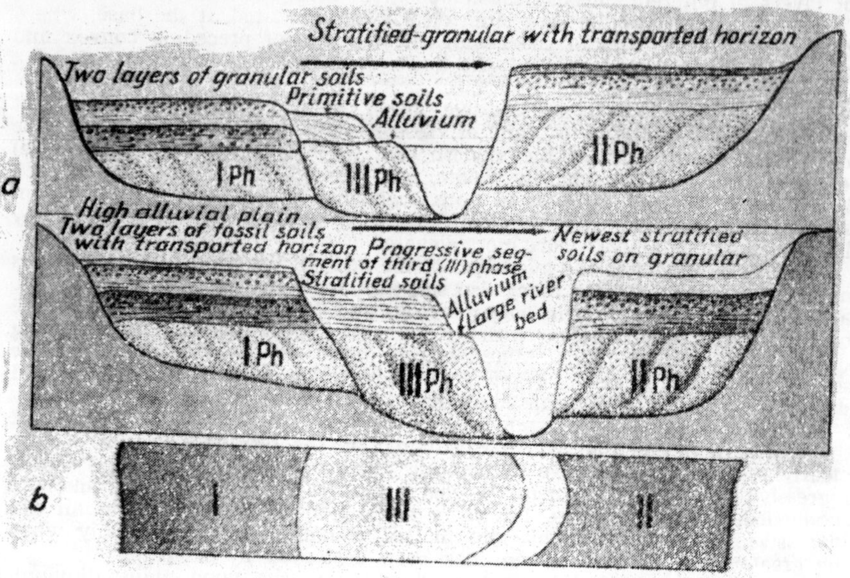

Fig. 50. Formation of a three-phase flood plain and flood plain soils: *a*—in section; *b*—in plan. *I*—unscoured part of flood plain of first phase; *II*—part of regressive segment of second phase of flood plain remaining after scouring; *III*—progressive segment of third phase of flood plain

When, due to maximum filling of the flood plain depression with alluvium, the bed becomes widened and overdeepened, a newest flood plain is invariably formed at a lower level. This newest flood plain, arising on the basis of the bed thus modified, afterwards develops to the full, with all its elements; as for the old flood plain, it becomes a terrace above the flood plain (Fig. 51).

It is found that the relative height of the upper terrace is usually twice that of the lower one. Terraces may, like flood plains, be monophase or polyphase. Different parts of the terrace are not synchronous either. The differences in age are determined by the intermittent character of the migration and the irregularity of the scouring during the flood plain period of their development.

Flood Plain Soil Formation

Flood plain soils differ markedly from those arising under other conditions of soil formation, due to the fact that they are formed under the influence of a powerful factor of soil formation, viz., the activity of a pulsating water flow. Every time the river is in spate, which happens every year, the soils become supplied with water up to their limit water capacity, the plants receive a dressing of nutrient elements, there occurs a useful warming up in the north and an appreciable cooling down in the south. All this favours the development of a rich flood plain vegetation, which, in the flood plain, is also a leading factor of soil formation. Being retained in drainless depressions, the flood water is partly responsible for a reduction in nitrification and aggravates swamping of the soils. Swamping does not, however, affect the whole of the flood plain, in view of the fact that the flood water is enriched with oxygen and does not favour reduction processes. Although flood plain soils are widely distributed from north to south, yet their pH fluctuates but within narrow limits, nearing 7. The infiltration of flood water conditions the leaching of flood plain soils. But in regions where the evaporation from the surface is intense, here and there occurs a considerable accumulation of salts in the soil, which reaches up to 1-2 t to the hectare and more.

In contradistinction to other soils, flood plain soils are formed in conjunction with the soil-forming rocks. The flood plain alluvium does not require a prolonged preparatory weathering stage prior to the formation of soil. It already contains plant nutrient

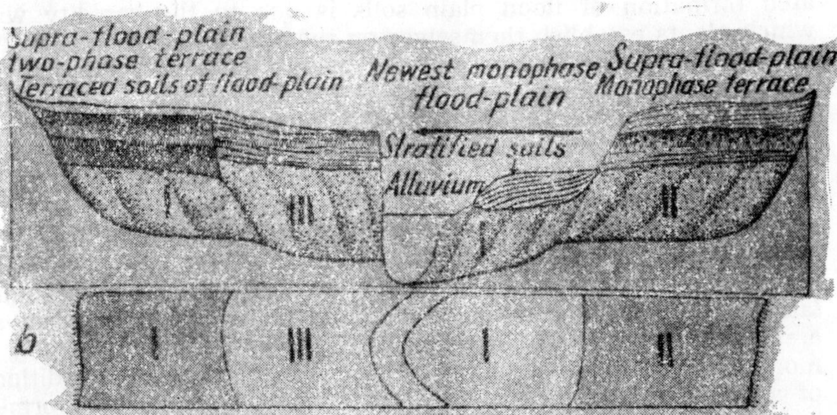

Fig. 51. Formation of terraces with fossil soils above the flood plain:
a—in section; *b*—in plan; *I-III*—terrace above the flood plain consisting of remaining parts of first and third phases of former flood plain; *II*—terrace above the flood plain consisting of remaining part of second phase of former flood plain; *I*—progressive segment of newest flood plain formed on the basis of the large bed of the former flood plain

elements in available form and posseses fertility. Chemical analyses of flood plain deposits indicate a considerable content of organic matter, nitrogen and ash plant nutrient elements. The newest flood plain deposits may therefore be regarded as primary flood plain soils.

In flood plain deposits, the silty fraction (0.05-0.001 mm) predominates. In irrigational deposits, the silty fraction reaches 50-80%.

In flood plain conditions, soil formation begins already on the bottom of small water bodies, in the riparian shallow zone of overgrown beds and rivers and in their estuarine parts, where there is enough light, warmth, plant ash and nitrogenous nutrient elements as well as oxygen, brought in by the flow. Here, water plants, which take root on the bottom, reach a good development. It should be noted that in the vicinity of river beds, in the zone situated along the bank, are formed only immature soils, whose soil-forming material consists of fluvial alluvial deposits. As for flood plain soils, their soil-forming material consists, as a rule, of flood plain alluvial deposits.

In flood plain conditions, on the newest deposits, under meadow grassy vegetation, the soils develop comparatively rapidly, reaching relatively high development in but a few decades. The accelerated formation of flood plain soils is due to the rapidity with which plants establish themselves on the alluvium and to their accelerated development in the flood plain, where they find all the necessary conditions.

Usually, in flood plains, there is a simultaneous formation of soils of the meadow and boggy types of soil formation, which may pass one into the other. In other flood plains, develop forest soils. In any flood plain the influence of the corresponding zonal conditions of soil formation is always manifested and all the more strongly as the influence of the flood conditions is weaker, i.e., the shorter the flood period and the weaker the erosive-accumulative activity of the flows. In the flood plains of small rivers, as a rule, the influence of the zonal conditions of soil formation is more pronounced. It should also be noted that under conditions of artificial flooding of the nature of limans, zonal soils formed outside flood plains undergo changes and come closer to flood plain soils. There arise peculiar flood plain-limanic meadow soils. Also formed here is a peculiar fine-lamellar limanic alluvium.

The differences between the flood plain soils of various native zones are far less pronounced than the differences between the soils of plains, formed outside flood plains, of the same zones. The common characters exhibited by flood plain soils of different zones is due to the hydrogenous nature of the soils or the levelling role of the fluvial factor of soil formation, favouring the development of soils of the soddy-meadow type. This accounts for the wide distribution in flood plains of the soddy type of soil formation, i.e., for the fact that it is found far beyond the boundaries of the soddy-podzolic and chernozemic zones. To the south, it penetrates into the region of semideserts, which is tied with additional moistening and lowering of the alkalinity, the flood plain soils being subjected to annual leaching. The displacement of soddy soil formation to the south along rivers, is conditioned by the moderating thermal and acidifying effect of flood waters, which promote the development of a meadow vegetation. As for the advance of soddy soil formation along river valleys far to the north, it is due to the warming up effect of flood waters, which, in addition, transport electrolytes. The latter coagulate the colloids of the flood water suspensions and of the flood plain deposits and soils, saturating them with bases. The carbonates brought by the flood water mitigate the reaction of the soil solution. All this favours the biological processes. Owing to the fact that flood plain deposits are rich in nutrient elements, there develops a rich meadow flood plain grassy vegetation in fluvial plains, which ensures the abundant accumulation in soil of organic matter, this being insufficiently mineralised due to relatively excessive moisture. Flood plain soils contain considerably more humus than any other soils except chernozems.

In the flood plains of the podzolic zone, the process of podzolisation is slowed down owing to the absorption by the soil of calcium and magnesium, brought along by the flood water, as well as liberated when the meadow vegetation dies out. In the podzolic zone, flood plain bog soils spread onto relatively elevated elements of the relief of flood plains. In the enlarged parts of the flood plains of this zone, here and there are even found peat bogs. In the steppe zone, on the contrary, soils of the bog type are, on the greatest part of the territory, replaced by soddy-meadow chernozem-like soils with a thick A+B horizon, which sometimes reaches 100 cm. Here, peat bogs have but a limited distribution and, in the zone of semideserts and deserts, they are extremely rare, being replaced by so-called tugaic, meadow-boggy flood plain soils. In the flood plain soils of these zones, here and there, occurs an accumulation of readily soluble salts, which is accompanied by the formation of meadow-solonchakous soils and flood plain takyrs.

Flood plain soils, which are bound, in one way or another, to find themselves outside the area subjected to flooding, undergo a transformation and are converted into the soils of the zonal type and all the faster as is more pronounced the contrast between the level land zonal and flood plain conditions of soil formation. The characteristics of flood plain soils are retained for a more prolonged period in zones characterised by a positive water balance of the soils. In the flood plains of small rivers, where the area subjected to flooding gradually diminishes, the additional moisture due to overflowing at high water maintains the soddy process, without weakening thereupon the manifestation of the zonal conditions of soil formation and even, somehow, reinforcing them.

Soils of Flood Plain Segments

The successive stages in the formation of the flood plain soils of a progressive segment may be represented by the following soil formation series:

Type I. Flood plain meadow soils; 1st stage. Weakly developed meadow sandy-loamy, light loamy, more seldom loamy, soils; usually formed on elevations situated near the bed.

2nd stage. a) Soddy-meadow laminated light loamy and loamy soils; b) soddy-meadow granular-laminated loamy soils. The soils of this stage are usually formed on a gently inclined plain, in the transition belt from the elevations situated near the bed to the median (central) plain.

3rd stage. a) Soddy-meadow granular loamy, heavy loamy and clayey soils, formed in the flat depressed areas of flood plains; b) meadow-boggy, clayey and heavy clayey soils; formed in the concave parts of the depressed flood plain.

Type II. Flood plain boggy soils; formed in the depressions of the flood plain.

Type III. Flood plain salined soils; formed on all the elements of the flood plain.

Type IV. Flood plain forest soils; formed on all the elements of the flood plain.

The soils of these types and stages may pass one into the other. The sequential character of the transition in stages is usually maintained in all the flood plain links, but is expressed in different ways. Some flood plain soils pass into others when the bed and flood plain water flows interact. The passages proceed in diametrically opposed directions in the progressive and regressive segments.

In a progressive segment, one can trace a progressive passage from young, weakly-formed meadow soils to lamellar ones, to begin with, and then to granular, soddy-meadow and further to bog soils (see Fig. 49). Weakly-formed meadow soils are usually formed in the immediate vicinity of the bed on the newest, light, abundantly accumulating flood plain deposits. As the bed retreats and the area of the progressive segment of the flood plain increases, the weakly-formed meadow soils serve as the basis for the progressive formation of lamellar, soddy-meadow soils of a heavier mechanical composition. This increased heaviness is due to silt deposition and partial suffossion of fine suspensions. In connection with the continuing retreat of the bed and the accumulation of silty deposits, the soddy-meadow light loamy lamellar soils serve as the basis for the formation of, to begin with, loamy granular-lamellar soils and then also of clayey granular soddy-meadow soils. In the course of its relatively prolonged migration, here and there the bed manages to withdraw to a considerable distance from the seat of the initial formation of weakly-formed soils and where, after having passed through the stage of lamellar soils, now develop clayey soddy-meadow granular soils. With each new stage, new deposits accumulate on the surface. However here, earlier formed soils do not become covered up with thick deposits. As the bed retreats, to each successive stage of the soils correspond gradually decreasing in thickness and increasing in heaviness new horizons of flood plain deposits, which succeed to the previous ones. There occurs therefore not a simple superposition of deposits, nor a vertical juxtaposition of soils one on top of the other, but concurrently with the deposition of flood plain alluvium and suffossion, older soils are formed on the basis of previous ones. On the plan, soils which correspond to these stages are also distributed successively in the form of bands, forming behind the retreating bed. The relative age of the soils of these bands will be all the greater as they are located further away from the bed. As for the absolute age of the soils of the same stage, it will vary appreciably from one part or link of the flood plain to another, in connection with the fact that their formation proceeded at unequal paces.

The passage of lamellar flood plain soils to granular ones is tied with the alluvial process and the fluvial factor of soil formation. The prolonged but moderate annual accumulation of flood plain deposits amounting to a few tons per hectare (a layer of 1 mm and more) does not interfere with the process of formation of soils on the basis of preceding stages. The silty elements of the deposits partially penetrate downwards with the infiltrating water, thus also rendering heavier the mechanical composition of preceding lower lying deposits and soils, and, at the same time, attenuating their laminated character.

But in spite of this, the illuvial horizon in the profile of flood plain soils is not clearly marked, even in well-formed soddy-meadow granular soil. The inwash horizon does not exhibit clearly outlined lower and upper limits in a flood plain, owing to the fact that the illuvial process is masked by suffossion, which is renewed every year at the expense of the newest deposits. The formation of illuvium is not favoured by the soddy horizon of meadow flood plain soils, which moderate infiltration; as for the flood water electrolytes, they distinctly hamper the migration of colloids, especially in the presence of water impermeable layers. In the flood plain soils of the zone of semideserts and deserts, the saline illuvium, appearing in the form of sulphate and carbonate horizons, is not pronounced either. A horizon characterised by a high content of R_2O_3 is a rare occurrence in flood plains, and furthermore, it is only found in highly developed soils passing into zonal ones.

In a regressive segment of a flood plain, the formation of soils proceeds not from meadow weakly developed soils to boggy ones but, on the contrary, from boggy soils to soddy-meadow and weakly formed meadow renovated flood plain soils (Fig. 50*a*). A formation of soils of this kind occurs only in a concave segment of a flood plain area undergoing a reduction and subjected to destruction, as a result of the erosive action of the river bed as it advances on to the flood plain. In front of the advancing bed, on the surface of the flood plain area undergoing erosion, there occurs an intensive accumulation of newest flood plain deposits, which renovate the old soils, even to the extent of taking their place and completely covering them up. The heavy soils of preceding stages of soil formation become buried under newest deposits of a lighter mechanical composition, on which, usually, young soils of a new, upper, layer of flood plain alluvium and of a new series of flood plain soils of varied modifications have the time to form. The sequential character of the displacement of the bed conditions the corresponding order in the passage of the soils of certain stages into others, concurrently with the increase in thickness of the newest deposits. This transition proceeds at a faster pace in the part of the flood plain situated near the bed and is less pronounced to the periphery of the segment, away from the advancing bed. The soils of regressive segments have achieved maximum development in time. Here, as a rule, are encountered fossil soils, sometimes arranged in several layers, whereas no fossil soils are found on progressive young segments, if we overlook the unimportant burials conditioned by the local transversal pulsations of the flood plain flow. The presence of fossil soils in old segments is a natural phenomenon, tied with the constant overlaying of alluvium, resulting from the migration of the flows and transverse circulation. The age of fossil soils, like that of all the other soils and elements of the flood plain, may be determined with a very close approximation. To this end, one reconstitutes the historical march of the migration of the flow in the bed, using aerophotographs of the flood plain and geological transverse profiles. The fossil soils met with in flood plains are of all ages, from very young to quite old. That is why their formation cannot be linked with changes in the general physico-geographical conditions of prolonged periods of time, such as the Glacial and Interglacial epochs.

The number of layers of fossil soils is usually inferior or equal to the number of phases gone through by the flood plain.

With the passage of the flood plain first into a high flood plain, and then, into a low terrace situated above the flood plain, the soils of the flood plain gradually evolve towards the zonal ones. With age, such transitional soils are subjected to intense degradation; they become podzolised under forests in the north or, being packed, lose their humus and undergo salinisation in the south.

On the high terraces of river valleys there are formed zonal terrace soils, whose past association with a flood plain is revealed by some slight traces. The difference between them and typical level land zonal soils is governed by the modern geomorphological situation. A common feature of terrace soils is that they are all formed on old flood plain, usually loamy loess-like, more seldom clayey, alluvial deposits. Terrace soils are characterised by an extended profile and a relatively thick humus horizon.

Classification and Description of Flood Plain Soils

A classification of flood plain soils will be found in Table 30.

Flood plain soils possess well-marked typical characteristics and distinguishing features. They are characterised by their relatively extended profile devoid of a clear differentiation between the genetic soil horizons. Only flood plain soddy-meadow granular soils possess clearly marked genetic horizons.

In contradistinction to soils of a level land soil formation, all types and subtypes of flood plain soils are characterised by an inherent mechanical composition and their arrangement in the transverse section of the flood plain in strict accordance with the elutriation process and the transverse circulation of the flows.

Meadow weakly-formed sandy-loamy flood plain soils are formed under the conditions of a prevalence of the hydrologic factor (intense alluviation) over the biological soil-forming role of the vegetation. In other words, the intense deposition of sediments hampers the normal development of a distinct humus accumulative horizon. The alternation of humic layers with new deposits devoid of humus which occurs here, is responsible for the coarsely "zebra-like" (striped) character of the whole soil profile. Soils of this subtype contain a relatively small amount of plant nutrient elements and humus and are characterised by a relatively low productive capacity. But they are usually utilised in no lesser degree than the other soils due to their accessibility, their location in the immediate vicinity of river beds and the fact that they are drained.

Soddy-meadow lamellar soils, arising on the basis of preceding weakly-formed soils, acquire a more definite aspect. They show a certain differentiation into genetic horizons. But the unsteady hydrologic regime of these soils in the transitional zone of

Table 30

Classification of Flood Plain Soils

Soil-geographical zone	Flood plain conditions of soil formation	Soils		
		type (series)	genus (stage)	species and variety
Tundra, taiga-forest, forest-steppe, steppe, desert-steppe, subtropical	Type of flood plain (mountain, flat land, deltaic)	Flood plain meadow	Weakly formed Meadow lamellar	
	Phase (monophase, two-phase, polyphase)		Soddy-meadow lamellar-granular	
	Link—high (sandy), low (clayey)		Soddy-meadow granular	
	Segment (progressive, regressive)		Boggy-meadow (compact)	
	Part of segment (situated near the bed, central, situated near the terrace)	Flood plain forest	Meadow-forest (granular) associated with oak-groves Forest podzolised Swampy-forest (sogras, mangroves, tropical forests)	Differentiated according to mechanical composition, moisture content, podzolisation, gleisation, degree and character of salinisation, amount of humus, structure, degree of taming, etc.
		Flood plain boggy	Meadow-boggy Uliginous-boggy (tugaic, of plavnis) Humus-boggy (peaty-humus) Peaty-gley Peat	
		Flood plain salined	Solonchaks Solonetzes	

interaction of bed and flood plain flows, conditions the marked lamination of their profile.

In contradistinction to weakly formed meadow soils, soddy-meadow lamellar soils possess a thinly lamellar (striped), up to 50-60 cm and more thick humus horizon, somewhat packed in flood plains situated in the southern regions of the U.S.S.R. In the zone of forest-steppes, these soils acquire a more or less pronounced cloddy-granular structure.

Flood plain soddy-meadow granular soils are the most mature and best formed soils. They possess the deepest profile, clearly differentiated into genetic horizons, the granular humus-accumulative horizon reaching a thickness of 70-80 cm and more. These soils are rather rich in humus, the proportion reaching 8-9% in the zone of forest-steppes. They are also characterised by a high content of phosphorus in connection with the accumulation here of flood plain deposits, rich in plant nutrient elements.

Under forests and fluvial plains, forest flood plain soils are formed. Under thin broad-leaved forests, in flood plains, develop structural forest granular soils with a high humus content.

In flood plains situated in the zone of arid steppes, forests create their own microclimate in the surface layer of air and their own climate in the soils themselves, which favour the development of a dense stand of grass under which arise structural forest soils rich in humus. Under coniferous forests, in the soddy-podzolic zone, flood plain forest podzolised and so-called "sogras"—flood plain boggy-forest gleised soils are formed.

Also distributed in flood plains are meadow-boggy clay-uliginous structureless packed (compact) pseudosolonetzic soils with a monotonous deep profile. Even in the zone of semideserts, the amount of absorbed sodium present in the soil absorbing colloidal complex of these soils is not sufficient to cause the alkalinity of the soils. The pseudosolonetzic character of these soils is tied with their heavy mechanical composition and the inwash of suspensions. Such compact meadow-boggy soils are met with here and there in flood plain areas located outside the meanders, in isolated depressions, which get flooded at high water, where there occurs a deposition of fine mud (Fig. 52). After the flood water has receded, some of the water remains stagnant, giving rise to anaerobic processes and gleisation.

In alluvial plains bog soils are distributed, ranging from mineral uliginous-boggy soils and plavni soils to peaty, peaty-humus, peaty-gley soils and turf-peats.

To the south of the zone of dry steppes, in fluvial plains, are distributed salined soils. The prevailing salts are usually sulphates, which do not harm the vegetation. Chlorides and other readily soluble salts are evacuated from the soil and conditions favourable for their accumulation in significant amounts obtain only,

Fig. 52. Silt deposition in a terrace flood plain

here and there, in the flood plains of semideserts and deserts. Salined flood plain soils are often associated with the older peripheral parts of flood plains. These soils are notable for their highly dynamic character with regard to the degree as well as the nature of the salinisation. The presence in the aforesaid zones of flood plain solonchakous soils and solonchaks is by no means a rare occurrence, but the same cannot be said of the presence of solonetzes in fluvial plains. Fluvial plains do not possess the conditions necessary for the formation of solonetzes, due to the abundance in the water and soils of Ca and Mg ions, which hamper the entry of Na into the soil absorbing complex.

Flood plain soils are characterised by their relatively low apparent density and specific gravity, which change according to the mechanical composition of the flood plain soils. Owing to their high pore-space and structure, they have a high water capacity (3,000-5,000 m^3/ha), possess a considerable reserve of water in the upper one-metre layer of soil (1,000-1,500 m^3/ha and more), which is at the disposal of plants during the vegetative period.

Flood plain soils are rather rich in nitrogen, phosphorus and potassium, yet they show a good response to the application of compound organic and mineral fertilisers and in particular of microelements. The relative abundance of plant nutrient elements in flood plain soils, including phosphorus, is due to their century-old accumulation in flood plains, this being due, in turn, to their particular position on the continent, viz., their proximity to the denudation bases and base-lines of erosion, and also to the selecting capacity of plants and microorganisms, which, in flood plains, achieve a vigorous growth.

Agricultural Value and Melioration of Flood Plains

The latent, so far little utilised, agricultural possibilities of flood plain soils are enormous. Their high productivity is well-known. When tamed, flood plain soils invariably give increased yields.

With the erection of dams and the regulation of rivers, flood plain land gets flooded in the head-water area or, on the contrary, loses its flood plain nature altogether in the tail-water area and becomes subjected to degradation (podzolisation in the north or steppe formation in the south). The risk of soil degradation in the tail-water area should be prevented by artificially maintaining flood plain conditions through the erection of simple hydrotechnical installations. The restoration, so to speak, of their former flood plain nature to the soils of extensive terraces, using the water reserves stored in reservoirs and through the installation of the necessary irrigation systems, appears as an entirely practical proposition. Quite promising in this connection are the irrigation and water supply of the catchwork type, using a network of water-pipes, for watering crops.

Possibilities exist in flood plains with regard to the regulation of high water, particularly with a view to controlling the accumulation and formation (sedimentation) of flood plain deposits, but these possibilities have not yet been realised. The regime of high waters and the deposition of sediments can be regulated through the attenuation of transverse pulsation and eddies in the flood plain flow, the creation in the flood plain of special strips of trees and brushwood aimed at directing the current where it is wanted, the alignment of the corners of flood plain forest areas, the increase or decrease of the uneven character of the surface of the flood plain, etc. Special installations are needed here and there, such as embankments, to regulate the interaction of flood plain and bed flows and prevent the access of suspended sandy material from the bed into the flood plain. Other special installations are needed in flood plains to bring about the conditions necessary for a moderate elutriation of the aluminous material, which would exert an indirect influence on the formation of soddy-meadow granular loamy soils. It is highly advisable to try and bring about a moderate mineralisation of turf-peats and to intensify their enrichment with silt from flood water, which will convert them into meadow-boggy soils of the best type. The covering up (burial) of thick turf-peats with deposits prior to the carting out of most of the peat is to be avoided. The flood plain peat should be transported onto the podzolised and podzolic soils of watersheds and slopes, thus compensating the losses of substances suffered by them in the course of century-old leaching.

When proceeding to the reclamation of flood plain-boggy soils, provisions should be made for the bilateral regulation of the water and with it of the air, heat and even salt (nutrient) regime of the soils, using, for example, mole drains and simple hydrotechnical installations, viz., water gates and pumps, combining drainage of the soil with its moistening. Meliorative measures in flood plains should be directed in the first place towards controlling the soddy-meadow soil-forming process.

An improvement of the drainage conditions alone, simply by digging trenches through the banks bordering the bed and the dikes in the flood plain may radically improve mineral swamp soils.

The presence of a series of reservoirs along large rivers leads, apart from their extremely beneficial influence, to certain losses due to submersion, underflooding, dampening and icing in winter; part of the flood plain areas situated below the dams are also lost. Plant nutrient elements will inevitably find their way into the reservoirs with suspensions, whereupon they will, in addition, jeopardise the maintenance of their full capacity and good wear. Here and there, together with very rich flood plain soils, deposits of peat, sapropel, building materials and other useful minerals may become submerged. Considering the present high level of machinery at our disposal, it appears advisable, here and there, to remove from the flood plain destined to become submerged not only peat and sapropel, but also the thick, richest in humus, semipeaty-soddy horizons of meadow flood plain soils and to use them for taming podzolised and podzolic sandy soils situated in the immediate vicinity of flood plains. It can thus be seen that the flooding of flood plains necessitated by the building of dams raises most important questions:

1) the preclusion of losses of nitrogenous substances, plant nutrient elements and suspensions from watersheds, and the simultaneous protection of reservoirs from silting up;

2) the compensation of losses of flood plain land caused by their submergence, through the radical improvement of soils located outside flood plains.

These tasks can be carried to a successful conclusion through reducing surface runoff to a minimum, causing it to flow inwards into the soil, through fighting soil erosion in every possible way and creating artificial flood plains of the catchwork type on watersheds and on the surface of river terraces, as well as through the regulation of the flood plain regime and the reclamation of the small flood plains of the tributaries of large regulated rivers. The losses of flood plain soils through their submergence should be compensated by creating new soils of the hydrogenous type, intercepting water, suspended and dissolved substances on their way to the nearest reservoirs.

The study of flood plain soil formation may serve as the foundation for working out hydromeliorative methods for the improvement and transformation of soils situated outside flood plains, by creating the necessary conditions for soddy soil formation on irrigated and drained land.

Chapter XVII

BOG SOILS

Bog and swamped soils are characterised by constant or prolonged overmoistening, attended, as a consequence, by insufficient aeration, by gleisation and, as a rule, by peat formation. These soils carry a specific boggy vegetation, adapted to conditions of overmoistening and shortage of oxygen.

In zones where atmospheric precipitations exceed evaporation of water from the surface of the earth and where, as a consequence, the upper horizons of the soil are constantly wet, the boggy soil-forming process is the most widespread and bears a zonal character. In zones with an irregular and inadequate moistening, the boggy areas are only found in concave land forms, where overmoistening of the soil arises as a result of an additional inflow of surface and ground water. The widespread distribution of bogs in the north is also due to the lack of warmth and the accumulation of water-retaining organic matter in the soil, in connection with the fact that its decomposition (mineralisation) proceeds at a much slower pace than the accretion of new vegetative mass. But in a cold climate, the yearly accretion of the vegetative mass is quite small. Further south, the increase of the average annual temperature and of the temperature during the vegetative period leads to an increase in the accretion of the vegetative mass, but also, at the same time, to an acceleration of the mineralisation. That is why the maximum amount of organic remains in the form of a peaty mass, accumulates in a moderately cold climate and not in a cold or warm one.

Bog soils are mostly distributed in the tundra and taiga zone, especially in the northern part of the latter. They have but a secondary distribution in the south of the country.

Bog and swamped soils are characterised by the occurrence in them of reduction processes—oxides being converted into protoxides, available plant nutrients into unavailable ones.

A bog is a tract of land or element of the geographical landscape characterised by a cover of specific moisture-loving vegetation, a special hydrologic regime and bog soils.

There is a certain lag between the setting up of conditions of overmoistening and the establishment on soils subjected to bogging up of a specific boggy type of vegetation. When the thickness of the peaty horizon does not exceed 70 cm, the root system of the main plants (edificators) penetrates into the mineral part of the soil. With an increase in the thickness of the peaty horizon, the root system does not get beyond the organic horizon of the soil and fails to reach the mineral horizon. The roots of trees growing on bogs sometimes manage to penetrate through a layer of peat 1 m thick and more. The main mass of the roots remains suspended in the peaty layer, as it were, and does not come into contact with the mineral one.

In more southern regions, bogs are considered to include wet tracts of land with uliginous-bog and meadow-bog soils, enriched with organic matter.

Mineral overmoistened soils constitute bogged up land, which may be in various stages of bogginess.

It is customary to classify bogs into simple and complex boggy areas. The former are usually characterised by the existence of small separate areas and individual development. A complex boggy area presents a higher stage of development and represents a grouping into one system of several simple boggy areas formerly developing in isolation.

Reasons for the Formation of Bogs and Origin of Bog Soils

The origin and development of bog and swamp soils is linked, in the first place, with overmoistening, which is mostly due to an excess of the atmospheric precipitations over evaporation, or else is the result of a discharge of soil-ground water.

Overmoistening of the soil is favoured by the presence of flat and concave land forms, as well as the water impermeable character and high water capacity of the soils. One of the reasons for the excess of moisture is the lowering of its evaporation from the soil as a consequence of inadequate thermal energy and reduced transpiration. In the tundra and taiga, the factor responsible for the bogging up of soils is what is referred to as permafrost, with its low temperatures and waterproof properties. A most important cause of the overmoistening of soils is the intensive accumulation of water-retaining, organic, weakly mineralising plant remains. The excess of moisture in the soil is a consequence of the high moisture capacity of these substances. Ample evidence of this is the frequent presence of bogs even on convex sandy watersheds.

The presence of an excessive amount of moisture in the soil

leads to oxygen starvation, losses of heat due to evaporation, reduction processes, shortage of plant nutrient elements and higher acidity. All these factors exert a depressing influence on the biological activity going on in the soil and lead to the accumulation of raw organic matter and gleisation—the main features of a boggy process of soil formation.

The main causes for the formation of bogs can be grouped as follows:

1. *Biologo-hydrologic*: a) high water capacity of the organic matter and shortage of ash plant nutrient elements in the soil;

b) replacement of forests by meadows and swamps, natural or as a result of burning or felling of forests;

c) overgrowing of shoaled water bodies (lakes, cut-off lakes, limans, detached branches of deltas and gulfs) with water-bog vegetation.

2. *Hydrologic*: a) excess of atmospheric precipitations due to relatively weak evaporation of soil moisture and low air and soil temperatures;

b) excessive inflow of water on account of surface runoff and irrigation of the land;

c) excessive ground water discharge;

d) stagnation of water in the upper soil horizons resulting from the formation of a dense illuvial horizon, or the low water permeability of soil-forming rocks, or the presence in the soil of a horizon of permanent or prolonged frost.

The formation of bogs and bog soils can be of three types: a) swamping of dry land, b) overgrowing of water bodies, c) overmoistening due to local discharge of ground water.

Swamping of dry land refers to swamping of soils of forests, meadows, brushwood, layland, pastures and fields. The soils of forest areas are more often swamped under conifers as well as in places where the forest was felled or burnt. Coniferous forests, which, as they develop, condition the formation of podzolic soils with a strongly leached upper eluvial and a dense waterproof illuvial horizons, bring about conditions favouring the formation of bogs. In the taiga zone, bogging up of forests begins with the establishment of grasses under the cover of a thin stand of trees. The first plants to appear in the forest are the gramineae (reed grass and others), followed by sedges, irises, etc. The grasses are succeeded by mosses, which promote the retention of moisture and overmoistening of the soil's surface. The first plant to appear is haircap, which possesses a water capacity of up to 600-700%, then hypnum moss, followed by sphagnum. The water capacity of the latter reaches 1,500-3,000%. Sphagnum leads to pronounced bogging up of the forest and its subsequent inevitable ruin. On the seat of the former forest arises a bog, which retains the remains of the buried forest in the form of very slowly decaying trunks, stumps, branches, bark, pollen and seeds of forest vegetation. The prolonged period during which the buried plant remains (pollen, seeds) are conserved in the bog is bound up with the acidity of the sphagnum. Swamping of forests is usually slow, due to the fact that the forests, transpiring abundantly, put up a long fight against the onset of boggy vegetation. The soils of felled or burnt forests submit to swamping considerably faster, owing to a deterioration of the physical properties of the soil (packing of the surface) and rise of the ground water table, which is no longer reduced through transpiration. Following the felling or burning of trees, there appears a weedy vegetation composed of various plants: willow-herbs (*Epilobium*), golden-rod (*Solidago virgaurea L.*) and others, followed by hair grass (*Deschampsia caespitosa*), sow thistle (*Sonchus*), rush (*Juncus effusus*), after which come mosses, first in depressions, from where, widening their areal of distribution, they gradually invade large areas. In the upper portion of the soil a water-retaining semipeaty sod arises, which later is converted into a turfy and peaty horizon of the soil. The site of the annihilated or receding forest becomes the seat of a relatively brief meadow stage of soil development, which constitutes a sort of intermediate stage between forest podzolic and peat-boggy soils.

Dry meadows and pastures undergoing swamping go through the stages of development of rhizomatous grasses, thin shrub, dense shrub, sedges and mosses. These stages present fairly marked differences in accordance with the differences in physico-geographic conditions (climatic, relief, hydrologic, geologic and biologic).

In the course of the progressive formation of a boggy area the ground wetting of flood plain marshes changes to the mixed atmospheric-ground wetting of transitional swamps and the atmospheric wetting of high bogs. As the role of the atmospheric precipitations in the alimentation in water of the bog grows in importance, the hardness (mineralisation) of bog waters goes down, and this leads to a corresponding change in the specific composition of the vegetation.

The ash content of each successive higher lying peat horizon decreases, in view of the fact that the roots of the boggy vegetation, which, to begin with, become less and less associated with the mineral part of the soil, finally leave it altogether and distribute themselves in the upper parts of the growing peaty layer, i.e., they are, so to speak, suspended in them and fail to obtain any mineral nutrients from the lower horizons. As a consequence, the content of plant ash nutrient elements gradually falls, the exacting vegetation dies out (grasses), which leads to the predominance of green mosses—plants devoid of roots, capable of assimilating food with all their cells. The development of green mosses is attended by an accretion of the peat. At the same time the degree of its decomposition goes down, in connection with the intensification of anaerobiosis, acidity and the attenuation in the mineralisation of the organic matter. The process of the impoverishment of the upper horizons in ash substances progresses. The solution of such boggy soils contains hardly any ash elements and when traces of them are present, they are of atmospheric origin. On bogs, there may be a periodical appearance of unexacting woody vegetation—long-leaved pine, subshrub: heather, ledum, dwarf arctic birch, willow, cloudberry, cranberry and other plants with a powerful root system, ensuring their nourishment. The uninterrupted accretion of the peat conditions the eventual death of these plants and leads to the complete predominance of sphagnum in the central parts of the bog, where the best conditions obtain for its growth. The intensive development of sphagnum and the accelerated accumulation of peat in the central parts of the bog condition the appearance of its convex surface. It is due to this feature that such bogs are referred to as upland or high bogs, as distinct from transitional, flat and concave, lowland bogs.

Here and there on the surface of old high moors, there occurs peat undergoing weathering, in the form of dark bare spots, surrounded by live mosses. These spots widen and eventually give rise to shallow water bodies—peaty lakes with unstable margins and a variable water regime. In this fashion, the high bog disappears, being converted into a geological formation, viz., turf-peat. Such is the scheme of the evolution of soils from podzols to high,

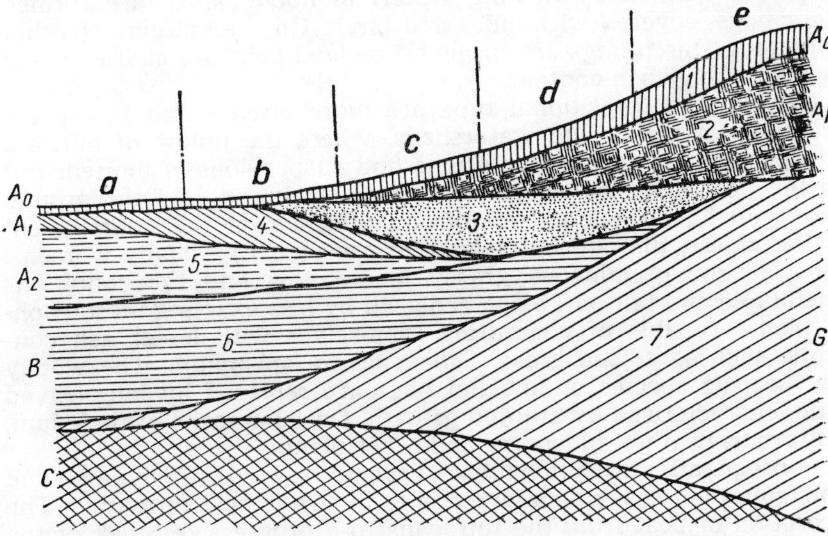

Fig. 53. Diagram illustrating the transition from podzolic to peat-boggy soils: *a*—podzolic soils; b—soddy-podzolic; c—peaty-gley; *d*—peat-gley; *e*—turf-peats. Soil horizons: *1*—rough-matted layer; *2*—peat; *3*—semiturfy; *4*—humus; *5*—podzolic; *6*—ore; *7*—gley

prominent mossy bogs and turf-peats (Fig. 53). This scheme is often complicated by particularities of local physico-geographic conditions. Certain stages of soil formation are lengthened whereas others, on the contrary, are reduced to a minimum and even eliminated.

Lowland bogs—meadow grassy bogs—are formed in depressions, where the ground water, which lies close to the surface, contains a certain amount of dissolved substances, and where develops an automorphous, varied, bog vegetation, exacting with regard to nutrient elements. Here grow gramineae, sedges, mosses, and among the woody species, the deciduous: willow, birch, alder. The situation of the surface of lowland bogs below the surrounding territory conditions the inflow of solutions and suspensions to the bog, where grow woody and grassy vegetations and develop potentially rich soils. Such are back-water or terrace swamps, as well as bogs alimented from the ground water on slopes and hollows where the ground water table comes to the surface. Here are gathered and conserved plant nutrient elements in the form of accumulating organic remains.

Lowland bogs are usually found in river valleys, on the margins of lakes, around issuing springs and in depressions wetted by ground and surface waters. Comprised in the lowland bogs are the grassy, green-mossy and forest bogs. Among the grassy bogs are distinguished sedge, rush, reed, horsetail bogs, etc. Green-mossy bogs are covered with hypnum mosses. Of the forest bogs, in places washed by flowing water, in flood plains, are formed alder bogs, covered with alder and birch. On the margins of other bogs sedge birch bogs are formed. Lowland bogs are characterised by peat of high ash-content.

Bogs of the transitional type are more often found in the depressions of plains and watersheds, where the inflow of nutrient elements in the form of solutions and suspensions is limited. But as the peat layer of soils increases in thickness, and the ground water no longer exerts any direct influence on the nourishment of the plants, the nutrient regime of the plants gradually worsens. Most of the gramineous grassy plants now find the conditions unsuitable and fall out, being replaced by less exacting mesotrophic plants capable of contenting themselves with lesser ash contents, such as cotton grass, *Scheuchzeria*, sphagnum. The woody leaf-bearing species of lowland bogs are replaced by long-leaved pine, characterised by stunted growth. On hummocks grow ledum, butterbur, heather, blueberry and other plants.

Subsequently, transitional bogs become covered with hypnum and sphagnum mosses, which give rise to intensive peat formation. The accretion of peat from the top leads to a progressive break of the upper part of the bog from the mineralised ground water.

High mossy bogs constitute the final stage of the boggy process of soil formation and must therefore in some or other way pass through the preceding lowland and transitional bog stages. In narrow depressions such as hollows, the lowland grassy bog stage may develop during an indefinitely long period. In wide depressions, the lowland bog stage lasts longer at the periphery of the bog whereas in the central part, in connection with a certain break from the margins and limited inflow of plant nutrient elements, it passes more rapidly into the transitional grassy-mossy and then the mossy high bog stages. In extensive plains, including watersheds, after a grassy lowland stage of short duration and a somewhat longer grassy-mossy transitional bog stage, follows the high bog stage which is of long duration. High bogs are mostly found in the forest zone.

The designation of bogs under the names of lowland, transitional and high bogs corresponds, in the main, to their geomorphological scheme of distribution. However, these names have a genetical meaning. In nature, the named types of bogs are often found on the same relief forms and elevations. It would be more correct to designate lowland, transitional and high bogs under the name of grassy, grassy-mossy and mossy bogs respectively.

Here and there, bogs are formed in places where the ground water table lies right next to the surface of the ground. These bogs usually pass through the same stages of development, from the grassy to the mossy ones, being dependent upon the regime and mineralisation of the ground water. Also comprised in bogs are the swampy parts of slopes in places of discharge of ground, usually mineralised, water, where grow sedges, rushes, alders and green mosses, but not sphagnum. Here are formed spring suspended bogs, where sometimes arise boggy tufas.

The formation of bogs arising as a result of the overgrowing of open water bodies—lakes, lagoons, limans, blind channels, pools, etc.—proceeds differently. The process of overgrowing starts from the shores. As the water-boggy vegetation gradually invades the water spaces, it brings about an accumulation of organic matter on the bottom of the water bodies and all around them, and eventually takes over the whole of the space previously occupied by the lake, lagoon, bed, etc. In an overgrown water body, going from the shores to the centre, the following belts are found: 1) a sedge belt, which occupied the shallow waters down to a depth of 0.5 m, more seldom 1 m; 2) a rush belt, distributed at a depth of 2 (3) m; 3) pond-lily belt, reaching a depth of 4 m; 4) a broad-leaved pond-weed belt, extending to a depth of 5 m; 5) a belt of macrophytes, consisting chiefly of flowering and partly of sporophytic (submerged), occupied mostly by charophytic algae banks; 6) a belt of microphytes, occupied by sporophytic, mainly (floating) blue-green algae, certain green algae *(Vaucheria, Cladophora* and others) and diatoms.

Overgrowing of water bodies begins with their shoaling, caused by the deposition of mineral and organic ooze—sapropel, consisting of plankton and benthos remains, as well as mineral and organic suspensions from the shores. In the riparian part of the basin settle deep water, submerged plants (narrow-leaved pondweeds, hornwort, stoneworts, etc.). As shoaling becomes more pronounced, they are replaced by plants with floating leaves (waterlilies, yellow water-lilies, followed by rushes, reeds, horsetails) (Fig. 54). The first plants to settle in a water body subjected to overgrowing are floating plants with vegetative organs, keeping afloat on the surface of the water or submerged [duckweed (*Lemna*), water soldiers (*Stratiotes aloides*), pondweed (*Potamogeton natans*) and others].

Shoaling often leads to an increase of the water body area together with its translocation away from the prevailing wind, sometimes reaching full displacement. Irrespective of the position of the water body itself, the above-mentioned belts undergo translocation and, in addition, there occurs a superposition of successive deposits, which leads to still more pronounced shoaling of the water body, followed by its complete overgrowing and transformation into a bog. The lacustrine-boggy ooze, together with the plant remains, assume the character of peat, which, afterwards, is taken over by grasses. There forms a sedge bog with grassy peat. But long before that, the surface becomes covered by a shifty mat consisting of mosses and the rhizomes of flowering plants [marsh cinquefoil (*Comarum palustre*), buck or bogbean (*Menyanthes trifoliata*), marsh-marigold (*Calta palustris*)]. Their long stalks trailing on the water form a felt (floating mat) on which water horsetail (*Equisetum limosum*) settles, the green mat becoming subsequently invaded by sedges (*Carex rostrata, C. limosa*) and mosses (*Calliergon, Drepanocladus, Sphagnum*), which form a peaty mat of increasing thickness. In this fashion are formed grassy-mossy and mossy bogs. Prior to the water body becoming completely covered up, some tracts of water remain open, which afterwards also become overgrown. From the moment the water body subjected to overgrowing becomes filled up with an organic mass, the bog which arises in its stead passes, just as upon bogging up of dry land, through the same stages of grassy, grassy-mossy, mossy and finally sphagnum bogs. In the central part of the boggy area, the accretion of sphagnum leads also to the formation of a prominent surface, which brings about a break of the surface soft water fed from the atmosphere from the slightly hard ground water of the water body. In the south, the overgrowing of water bodies does not end in peat formation, as in the north, but in silting up. The silt of such bogs, which is enriched with organic matter, only occasionally acquires a semipeaty character.

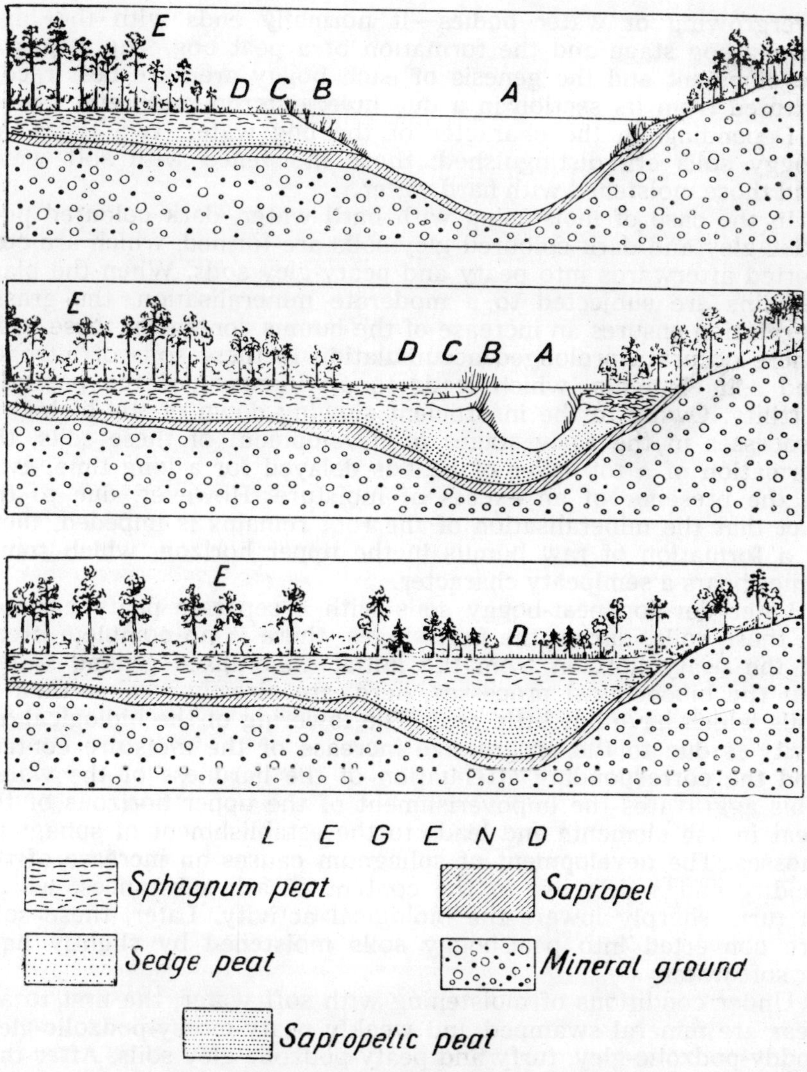

Fig. 54. Stages in swamping of water bodies:
A—open water space; *B*—riparian-water vegetation; *C,D*—sedge lowland bog; *E*—pine forest on sphagnum bog

The overgrowing of the water bodies of flood plains (lakes, cut-off lakes, limans, plavnis) under conditions of accumulation of flood plain deposits ends in the formation of mineral bogs with uliginous-boggy soils. In the taiga zone, regardless of how the bogging up process began—whether from swamping of dry land or overgrowing of water bodies—it normally ends with the high, mossy bog stage and the formation of a peat bog. The stages of development and the genesis of each boggy area are best reconstituted from its section in a dug hole, outcrop, well, etc.

Depending on the character of the bog water, two types of boggy soils are distinguished: those moistened with soft water and those moistened with hard water.

In the case of moistening with hard water, dark coloured podzolic-gley and dark coloured gley soils are formed, which are converted afterwards into peaty and peaty-gley soils. When the plant remains are subjected to a moderate mineralisation, the grassy vegetation ensures an increase of the humus content of these soils. There occurs a prolonged accumulation in these soils of nitrogen and ash elements, which leads to an increase of the potential fertility. Owing to the insufficient slowing down of the biological processes in the early stages of development of these soils, the formation of a top layer of peat is delayed for a long time, even in the presence of an excess of moisture. However, due to the fact that the mineralisation of the root remains is impeded, there is a formation of raw humus in the upper horizon, which sometimes bears a semipeaty character.

Later develop peat-boggy soils with a separate peaty horizon. In these soils, during the first stages, there is no complete break of the peaty horizon from the initial mineralised ground water. But the biochemical processes, here, are already weaker than in soils which have but little peat. The lowering of the biological activity is due to the progressive increase of the moisture content and the corresponding attenuation of the hardness of the water. This aggravates the impoverishment of the upper horizons of the peat in ash elements and leads to the establishment of sphagnum mosses. The development of sphagnum causes an increase of the acidity and a decrease of the content of ash substances, which, in turn, sharply lowers the biological activity. Later, these soils are converted into peat-boggy soils moistened by slightly hard or soft water.

Under conditions of moistening with soft water, the first to appear are mineral swamped and weakly peaty-mossy-podzolic-gley, soddy-podzolic-gley, turfy and peaty-podzolic-gley soils. After that shallow-peat soils are formed, which serve as the basis for the subsequent formation of peat-boggy soils. They are all characterised by an insignificant content of alkaline-earth bases and ash substances, high acidity and weak biogeny. This leads to a slow humification of the plant remains and the accumulation of peat.

Peat-boggy soils with soft water moistening become highly acid and their ash content goes down to a very low level. As moistening goes up and the inflow of ash substances weakens, the biological processes proceed at a slower pace. Maximum weakening of the biochemical processes occurs in peat-boggy soils with sphagnum peat overlaid with a thick rough matted layer representing the final link in the evolution of boggy soils.

The formation of peat-boggy soils is a zonal phenomenon, even though this term is used in a wider sense than in the case of the other soil zones, since they are distributed from the tundra to forest-steppes. Furthermore, the southern boundary of peat-boggy soils does not quite coincide with the boundaries of the other native zones. The zonality of boggy soils presents meridional and latitudinal variations; it is notably complicated by physico-geographical conditions in general and geological conditions in particular. Standing apart are maritime bogs, referred to as marshes, laidas, tidal marshes, as well as river plavnis, arising as a result of the filling in of estuaries which occurs upon deltaic formation. These bogs constitute overmoistened mineral soils, in connection with the rise of the ground water table right up to the surface and with flooding by the sea at high tide or overflowing with flood water of the parts of rivers situated near their mouths (Fig. 55).

The swamping of a number of flood plains is tied with the fact that the rivers have reached their maximum profile and are aging, owing to a rise of the base-line of erosion. The development of bog soils in fluvial plains proceeds in two diametrically opposed

Fig. 55. Juncaceous microlandscape (plavnis)

directions, one tending to cause bogging up, the other, the reverse (regradation). In both cases, the soils develop progressively from low to high forms.

Peat-boggy soils consist of two distinct horizons: a) an organic—peaty horizon and b) a mineral—gley horizon at the bottom. There is a clearly marked separation between them. They arise in the process of formation of the bog.

Gleisation

As a rule, under the layer of peat, lies a horizon of gleisation, or gley (a Ukrainian folk term), clay-like, viscous when wet, of a grey colour with a bluish tinge. Gleisation is a biochemical reducing process. Gley formation is accompanied by the conversion of ferric to ferrous oxides, these being subjected to translocation, mostly through leaching and partly through a rise along the capillaries, where they become oxidised and accumulate in the form of spots, streaks, nodules, concretions. Above the gley is sometimes found a bright ochre mottled horizon containing bright ochre streaks and small packed ferruginous pipes along the plant roots, where there occurs a process of oxidation of ferrous iron. As for the reduction of iron, it is attended by the formation of a number of minerals, viz., salts of ferrous iron, such as vivianite [$Fe_3 (PO_4)_2 \cdot 8H_2O$)] and siderite ($FeCO_3$). Serobacteria often take part in the reducing processes, forming FeS_2 (iron disulphide), which may be converted to ferrous sulphate, with the liberation of sulphuric acid according to the equation: $FeS_2+H_2O+7O= =FeSO_4+H_2SO_4$. The oxidation of ferrous iron is caused by the life activity of iron bacteria. This biochemical process is accompanied by the formation of brown iron ore ($Fe_2O_3 \cdot nH_2O$), settling sometimes in the shape of concretions forming seams of bog ore. The first stages of bogging up are indicated by greyish-bluish mottlings of gleisation along cracks and on the surfaces of large soil elements and abundant ochre efflorescences in the upper part of the gley horizon, in connection with the inflow there of a certain amount of oxygen. Subsequently, at the bottom, the bluish-grey spots widen and merge into a continuous, gley, dove-coloured-grey with a blue-tinged horizon. In the process of gleisation, the soil loses ferrous iron and becomes enriched with silicic acid and aluminium. Gleisation is accompanied by an increase of the soil's dispersiveness and of the mobility of the organic matter, and by a collapse of the structure. At the same time, the absorption capacity of these soils goes down somewhat, but then so does also the degree of base-unsaturation. Gleisation is attended by a sharp decrease of the porosity (down to 25-30%) and an increase in compaction of the soil-ground.

Peat Formation, Composition and Properties of Peat

To begin with, peat formation, just like humus formation, proceeds on the whole under aerobic conditions, followed by decay. This is brought about by microorganisms (bacteria and fungi), which are most active in the upper "peatogenic" layer of dying out mosses, whose thickness reaches 0.5 m and more. The number of microorganisms in high peat bogs reaches 0.7-0.8 thousand million in 1 g of peat in a naturally wet condition. Peat formation depends upon the degree and duration of the periods during which the soil is aerated as well as upon the supply of nitrogen and ash elements necessary to ensure the normal life activity of the microorganisms. As for the preservation (conservation) of peat, it is due to the prolonged anaerobic conditions obtaining in boggy soils. As the bogging up process develops, the aeration of the soils is lowered and so is the activity of the microorganisms, which leads to an acceleration of the accumulation of undecayed organic remains in the soil. As for the degree of decay, it is determined from the relative content of the products of decomposition of the tissues which have lost their cellular structure. The layer of peat usually consists of separate subhorizons of various degrees of decomposition, which points to changes in the conditions of peat formation. From the thickness of the subhorizons one can determine the rate of peat formation which, on the average, approaches 0.7 mm annually, this corresponding to 1 ton to the hectare of mossy peat. The fact should be taken into account that the decomposition of mosses proceeds at a less intensive pace than that of the remains of a herbaceous vegetation. Compared with the forest zone of a temperate climate, the annual accretion of organic remains and the rate of peat formation in a cold climate are not high. As was pointed out before, the accumulation of peat also decreases to the south, due to the increase, as we move southwards, of the mineralisation of the organic matter. Far north, in the zone of cold tundras, the thickness of the peat hardly reaches 25-30 cm. In the north of the taiga zone, here and there, the thickness of the peat reaches up to 3-4 m. In the central part of the forest belt, the thickness of the peat exceeds 8-9 m. To the south, the average thickness of the peat decreases. No peat formation takes place in subtropical zones. In a tropical climate, the annual mineralisation of the organic remains is not compensated by the accretion of new organic matter.

Many peat-forming plants can be found in the peat in a semi-disintegrated condition, which often still show the characteristic features of their anatomical structure (grasses, shrubs, woody species, mosses).

Peat-forming plants often consist of grasses, mostly monocotyledonous ones and in particular various species of the sedge family

(*Cyperaceae*). Of the mosses, those which give rise to peat are the white mosses (*Sphagnales*) and hypnum green mosses (*Bryales, Hypnum*) possessing rhizoids—multicellular hair-like filaments which replace roots. The pioneer among the mosses is haircap moss (*Polytrichum pagonatum*), which promotes the bogging up of soils.

In composition, peat represents an accumulation of incompletely decomposed plant remains, forming under conditions of boggy soil formation. As a rock, peat belongs to the group of so-called caustobioliths. The following peats are distinguished: of herbaceous, forest and mossy lowland, of transitional and high bogs.

The ash content of the peat corresponds to that of the plants of which it was formed and depends on the type of the bog.

The ash content in peat decreases from lowland bog to high bog and increases as its decomposition becomes more pronounced.

As for the content of organic matter, it increases, on the contrary, from lowland bogs to high bogs: from 50-80% in lowland bog, 92-97% in transitional bog, to 95-99% of the weight of absolutely dry matter of the peat, in high bog. The amount of nitrogen in peat sometimes reaches 4%, whereas in chernozem it does not exceed 0.5%. In peaty soils with hard water moistening, the nitrogen content amounts to 1.5-3.5%, whereas in peat deposits moistened with soft water, it amounts to 1-2.25% (0.5-0.9% in the rough matted layer horizons). The content of phosphoric acid and potassium is low in the peat of high and transitional bogs (P_2O_5—0.2-0.3%, K_2O—0.05-0.01%). But in the peats of sedge bogs, containing vivianite [$Fe_3(PO_4)_2 \cdot 8H_2O$)], the content of phosphoric acid may reach 7.5% and more. They contain up to 0.3-0.9% of potassium. The peats of high and transitional bogs contain 0.1-0.7% and 0.5-1.7% of calcium respectively, whereas the content of Ca in the peat of lowland bogs may be appreciable (1.5-7.5-15%).

In physical properties, peat is characterised by the following particularities. Peat exhibits various degrees of decomposition: 5-25% (weak), 25-45% (medium) and over 45% (pronounced). The specific gravity is lowest (0.18-0.27) in weakly decomposed sphagnum peat. The specific gravity of well decomposed sedge peat is 1.5. An increase in the degree of decomposition and of the ash content entails the increase of the peat's specific gravity. The apparent density of lowland peat changes from 0.2 to 0.9 and higher. A decrease in moisture content entails a decrease of the peaty mass and an increase of its viscosity. The pore-space (in percentages of soil volume) fluctuates from 79-83 in sedge-meadow peat to 85-89 in birch-sedge peat and 88-92 in sedge-hypnum peat. When peaty soils are brought under continuous cultivation, their pore-space decreases considerably.

Peat exhibits various degrees of dispersiveness (g), this being governed by the ratio of the total surface of all the particles (s) of the hard phase of the soil to the normal volume of the whole mass (V). This ratio governs the so-called specific surface $g = \frac{s}{V}$. Peat possesses high hygroscopicity (20-30%), which is 5-7 times that of mineral soils. Agricultural practice has shown that peat litter, which possesses high hygroscopicity (up to 20-30%), is capable of absorbing 3-5 times more water than straw. As for the capacity to absorb gases (NH_3, H_2S), peat litter surpasses straw by 15-20 times. That is why peat constitutes a first-class bedding material in cattlesheds (at a rate of 3-6 kg per head per day), as well as a suitable charge for the air filters of gasproof shelters. Peat possesses special water properties. Maximum moisture corresponds to the water capacity of peat. This is distinguished from normal water capacity which, under normal conditions, corresponds to the amount of water which fills all the pores upon full swelling of the organic part of the peat, normal atmospheric pressure and full stability (consolidation) of the peaty mass under the effect of the hard particles' own weight. When more than normally saturated with water, the peaty mass is in a condition of unstable equilibrium.

The peats of lowland, transitional and high bogs are differentiated one from the other according to their capacity to give out water. The force needed for the removal of moisture goes up as the moisture content goes down. This force is governed by the yield of water coefficient, the ratio of the volume of water trickling down to the total volume of the peaty mass or to the total volume of the water saturating the peaty mass to the condition of normal water capacity. The average coefficient for the upper dusty horizon can be taken as having a value of 0.5.

Peat as found in nature contains 86-94% of water. The moisture content (gravimetric) of lowland peat amounts to 86-90%, that of transitional peat to 90-92 and that of high peat to 90-94%. Intense drainage can cause a lowering of the moisture content down to 82-88%. A moisture content of 88% corresponds to the average water retention of peats after removal from them of gravitational water. But the water given out by peats is generally low. As the degree of decomposition of peat and its dispersiveness increase, its water capacity goes down. When the dispersiveness of peat is fairly high and its moisture content does not exceed normal water capacity, almost all the water which fills the small capillary pores is subjected to the influence of surface forces. An increase in dispersiveness raises the apparent density and lowers the porosity of the peat. As the specific surface and surface forces increase, so does the association with water. Peat is capable of retaining water in the proportion of 25 g of water per 1 g of dry matter.

The water capacity of peat fluctuates considerably according to its origin. It goes up from 300-700% of the dry matter in lowland bogs to 1,000-1,500% and more of the dry matter in high bogs. The reserve of water in peat soils at full water capacity in lowland bogs is given in Table 31.

Table 31

Reserve of Water in Peat Soils of Lowland Bogs at Full Water Capacity

Type of peat	Full water capacity, %			Water unavailable to plants in a 1 m layer of peat, m³	Available water in a 1 m layer of peat, m³	Capillary water capacity	
	gravimetric	volumetric	m³ in a 1 m layer			m³	in % of full water capacity
Lowland sedge hypnum peat . .	444	86	8600	1950	6250	8200	97
Lowland birch-sedge peat . . .	395	83	8300	2170	5930	8100	98
Lowland sedge-meadow	195	80	8000	4230	3370	7400-7600	92-97

When the peat is moistened, its colloido-dispersive part swells, which entails a considerable increase in volume. When subjected to drying out, peat diminishes considerably in volume. The ratio of the volume of moist peat to its volume in the air-dried condition varies, depending on its composition and degree of decomposition, from 2.5 to 8. The surface of peat bogs settles considerably when dried out, with, as a result, a lowering of the absolute marks by tens of centimetres or 10-15% of the thickness of the peat layer, depending on the depth of the drainage ditches and on the properties of the peat. Settlement decreases as we move away from the ditches, i.e., as the effect of drainage becomes less pronounced. However, in the immediate vicinity of the ditches, settlement becomes somewhat less pronounced. Peat situated below the level of the ground water is not subjected to any settlement since its moisture content does not change.

The loss in volume exhibited by peat when its moisture content goes down does not correspond to the amount of the free water lost, due to the fact that part of the pores is occupied by gases (air). The air present in the pores of the peat prevents the removal of the whole of free water upon filtration. After it has been subjected to intense drying out, peat, especially sphagnum peat, loses the capacity to absorb water and swell, owing to the fact that its particles become coated with hydrophobic resinous substances. Capillary rise in peat soils does not exceed 120-150 cm. The water permeability of peat is insignificant, its coefficient of filtration being low, this changing in accordance with the direction of the filtration (Table 32).

Table 32

Direction of filtration	Coefficient of filtration of peat, cm/sec	
	of lowland bog	of high bog
Horizontal	0.00045	0.00075
Vertical	0.000475	0.000023

The water permeability of peat decreases as its decomposition becomes more pronounced and in the presence of streaks of strongly decomposed peat. The water given out by a layer of peat 1 m thick amounts to 200-600 m³/ha. The low coefficient of filtration (the average rate of filtration in a lowland peat is approximately 0.4 m/day), the high water capacity and weak water permeability of peat, condition the overmoistening of bogs even on elevated watersheds and on slopes (hanging bog). The water content of a reclaimed massif of peat bog goes down considerably at the end of a prolonged period, spread over years. The time required for the drying out of a given bog area may be approximately estimated from the water balance of the territory, the dynamics of the coefficient of filtration, and changes in the other physical properties of the peat.

The peat lying on the surface of bogs may lose 20-25% more water through evaporation than an open surface of water. The rate of evaporation is more than 1.5 times greater on peat than on sand. The dry upper layer of peat which forms on bogs subjected to drying out subsequently efficiently protects the whole of the lower lying stratum from drying out. The rate of evaporation from bogs is inversely dependent on the depth at which the ground water lies, in other words, the closer it is to the surface, the greater the evaporation. Evaporation is also much influenced by the transpiration of the vegetation growing on the bog.

The heat conductivity of peat is low, which accounts for the fact that peat bogs freeze to a considerably lesser depth in winter than mineral soils. In summer, they thaw slowly. The weak heat conductivity of peat is responsible for the overheating of the surface of bog soils in the day-time and the fairly frequent occurrence of night frosts. The heat conductivity of peat is all the lower as the peat is drier. The least favourable heat properties are exhibited by sphagnum peat bogs.

The various kinds of peat are recognised macroscopically from their aspect and under the microscope. Sphagnum peat constituting the final link in the development of peat bogs at the present time, lies on the surface of high bogs. It is a light yellow or rusty peat of low ash content. Young friable sphagnum peat is used as insulating material. Old sphagnum peat has a darker colour, due to the fact that its humification is more pronounced. Hypnum peat is light brown or cinnamon. As the decomposition becomes more advanced, it gets appreciably darker and turns black when exposed to the air. Hypnum retains its cellular structure for a fairly long time. Sedge peat consists of a dense fibrous rooty mass of a yellow or cinnamonish colour. When exposed to the air, it oxidises and turns black. It is easily recognised due to the presence of the long leaf veins of the sedges, which remain intact for a long time and appear as grey threads, and of seeds of characteristic shapes (bottle-shaped, elongated, more seldom rounded, etc.). Juncaceous peat has a coarse structure, is of a light yellow colour, with characteristic rhizomes and broad leaves. It contains an admixture of silt with, sometimes, a large amount of sulphurous iron. Horsetail compact peat is easily recognised under the microscope from the long cells of the horsetails. Alder peat is black with cinnamon-reddish woody remains. Other woody peats (spruce, larch, birch, pine and mixed) are recognised from the woody remains and especially from the bark, which is preserved intact for a long time. Under the microscope, the composition of peat is determined from the remains of the plants from which the peat was formed.

Peat is named according to the plants of which it is made up and their quantitative correlation. In the name figure the peat-forming plants which are present in a proportion of more than 20%, the chief peat-forming species with the highest percentage content being cited last.

The degree of decomposition of the peat can be determined under the microscope and with the naked eye, taking into account its plasticity, the quantity and state of preservation of the plant remains, the amount and colour of the water forced out of the peat, etc. Valuable data relative to the genesis and age of boggy areas are obtained by the pollinic analysis method, now widely applied. Pollinic analysis consists in the determination of the composition and percentage correlation of the pollen of the various species of plants present at different depths in the profile of boggy and other formations. The results of the analyses are expressed in the form of pollinic diagrams, which constitute a sort of chronicle of the physico-geographical conditions. The origin and age of peats is best determined from a study of the actual profiles of boggy areas.

More complex profiles are exhibited by the boggy areas of high turf-peats, which have passed through the stages of lowland and transitional bogs and which arose on swamped dry land or an overgrown water body. The profile of a turf-peat of this kind reveals the presence of several peat horizons, which reflect the evolution of the boggy area. Still more complex profiles are exhibited by boggy areas which were formed under sharply changing local physico-geographical conditions (change of the baseline of erosion, of swamping conditions and drainage, of climate and microclimate, of geological and biological conditions, etc.). The structure of peat deposits is quite varied. The upper layer of turf-peats of all types is composed of a live vegetative cover of varying thickness, which forms a rough matted layer. It is especially conspicuous in turf-peats of the high and transitional types. Under the rough matted layer lies a transitional horizon separating the cover of vegetation from the peat stratum. The rough matted layer and the so-called transitional horizon possess considerably higher water permeability than the underlying peat. The upper peat horizons are referred to as the active horizon. Its thickness depends on the peat-forming vegetation. The active horizon is characterised by a variable moisture content, alternation of aerobic and anaerobic conditions, fairly high water permeability, the presence of a live root system. The underlying, thicker, main layer of the stratum is referred to as the inert horizon. It is characterised by weak water permeability and permanent anaerobiosis. Changes in the level of the ground water are governed, in the main, by the water permeability of the active layer. The role of the inert horizon in the downflow of water from the boggy area is insignificant. The drainage of boggy soils takes place mostly as a result of the filtration of the downflowing water along the active layer.

Classification and Description of Bog Soils

The soils of the boggy type of soil formation are classified as follows:

I. Bogged up soils.
II. Meadow-boggy soils (soddy-gleisolic).
III. Mineral bogs:
  a) humus-gleisolic soils,
  b) uliginous-boggy soils,
  c) gley-boggy soils.
IV. Peaty-boggy soils (peaty-gleisolic).
V. Peat-boggy soils (peat-gleisolic).
VI. Turf-peats.

Bogged up soils are the constantly water-logged soils of the various native zones, characterised by the presence of a water-loving vegetation proper to these soils.

Meadow-boggy (soddy-gleisolic) soils do not possess a clearly marked peat horizon. In meadow-boggy soddy-gleisolic soils, immediately below the soddy horizon, lies a G gley horizon with numerous ochre spots and streaks. These soils are formed as a result of the swamping of meadows, pastures and other agricultural land. They occur at the periphery of lowland bogs. The soils of mineral bogs—humus-gleisolic, uliginous-boggy and gley-boggy—are characterised by a more pronounced gleisation and a high content of well decomposed organic matter. Uliginous-boggy soils are characterised by the possession of a thick uliginous, sometimes rooty-soddy, gleised viscous horizon, enriched with organic mater. In the soils of mineral bogs, just as in meadow-boggy ones, the peat has not yet had the time to form and in its place there is but an insignificant semipeaty layer composed of the above-ground parts of herbaceous plants, or a felt-like, rooty, humus-accumulative horizon.

Peaty-boggy soils are characterised by the possession of a permanent shallow peaty horizon, which does not exceed the average thickness of the arable layer (25-30 cm). The horizon of peat is often herbaceous, rooty, sometimes felt-like, semipeaty.

Peat-boggy (peat-gleisolic) soils are characterised by the possession of a well-marked peat horizon of medium thickness (up to 0.7-1 m). The root systems of the main boggy plants are still in contact with the mineral soil horizon underlying the peat. Differentiated according to the thickness of the peat horizon.

In turf-peats, the layer of peat reaches a thickness of more than 0.7-1 m. The root systems of the plants which gave rise to the peat (the indicators) are distributed in the upper layer of the peat; they do not reach down to its bottom and do not penetrate into the mineral horizon underlying the peat. Differentiated according to the thickness of the peat layer.

To the boggy soils also belong the soddy-podzolic-gleisolic and peaty-podzolic-gleisolic soils, arising upon the swamping of soddy-podzolic soils. As the latter become progressively swamped, they acquire, at first, a semipeaty and, later, a peaty horizon, the gley horizon getting gradually closer to the surface of the soil. When soddy-podzolic soils undergoing swamping contain hard ground water, they are converted into soddy dark-coloured gleisolic soils.

The above scheme of classification of boggy soils lends itself to further elaboration in accordance with the zonal and local particularities of soil formation.

Agricultural Significance and Utilisation of Bog Soils

Bog soils contain enormous agricultural reserves and are of great importance in the national economy.

Owing to their inherent capacities, bogs of the lowland type can make a most important contribution to agriculture. After they have been dried and subjected to the appropriate cultural and technical treatment, bog soils are converted into valuable agricultural land. The drying of bogs and the improvement of the water, air, heat and nutrient regimes which go with it can be achieved by means of various hydrotechnical installations (sluicing, mole draining, sprinkler irrigation) in conjunction with the proper agromeliorative and agrotechnical treatment. The treatment should ensure the maintenance of the ground water regime best suited to the nature of the soil and the crops grown on it. The level at which the ground water should lie is taken to be approximately 0.4-0.6 m for meadow grasses, 0.7-0.8 m for hay and pastures, 0.7-0.9 m for cereal crops, 0.8-1 m for industrial crops, root crops and vegetable growing, etc. The overdrying of bog soils leads to the pulverisation of the peaty mass and sometimes to salinisation and other undesirable results.

Bog soils should be given dressings of artificial fertilisers (NPK), organic manures (farmyard manure) and microelements (Cu, B, Co, Ni, etc.). Bog soils are usually deficient in potassium and phosphorus. These soils show quite a good response to liming and marling. The application of farmyard manure and compost to turf-peats enriches them with microorganisms. The yields obtained on reclaimed bogs, in metric tons, reach 3-4 for cereal crops, 40 for potatoes, 70-80 for roots, 8-10 and more for grasses (hay).

The peat of lowland bogs is a fertiliser of excellent quality; as for peat farmyard manure, obtained through the use of crumbled high-moor peat as litter, it can be favourably compared with straw farmyard manure. The same goes for the compost obtained through the aerobic decay of peat mixed with lime, ash, phosphorous fertilisers, farmyard manure or liquid manure, which contains a vast amount of ammonifying and nitrifying bacteria. Peat compost obtained in this way can also be favourably compared with straw farmyard manure and may even be better. The presence in peat of R_2O_3 and especially Ca accounts for the high absorbing capacity of lowland peat for phosphate ions, which is tens of times higher than that of soddy-podzolic soils. Hence the great importance of peat with respect to the mobilisation of phosphoric acid for plants, especially on the soils of the soddy-podzolic zone. Peat can be recommended for use on a large scale as a component of organo-mineral compound fertilisers. It improves the phosphate regime of the fertilised soils. The meliorative role of peat in raising the potential fertility of non-boggy soils is of paramount importance. The utilisation of peat as fuel is therefore to be deprecated and in those districts where it can be used to advantage as a fertiliser, it is inadmissible.

A radical improvement of boggy soils with a high ash content and high potential fertility can only be achieved through the activation of their biological processes, which are responsible for the accumulation of nutrients. The taming of boggy soils entails a more intense mineralisation of humic formations and an increase in the humification of plant remains.

To accelerate the reclamation of peaty and peat soils, one has recourse to the method of primary fallowing—repeated deep loosening of the soil in the year during which the new land is broken, with the application before the first operation of phosphoric fertilisers and bacterial inoculations or organic matter enriched with microflora. Deep tillage or roto-tilling promotes the accelerated development of ammonifying and other bacteria and fungi, which leads to an increase in fertility of the bog soils.

Free living nitrogen-fixing bacteria (nodule bacteria) make their appearance in the peaty soils of high bogs only after they have been tamed.

After they have been dried, bog soils lose their boggy character, acquire a good aeration, the protoxides are oxidised further, there is an increase in the mineralisation of the organic matter, the content of ash and nitrogenous substances goes up. The taming of bogs leads to a lowering of the ground water table, the disappearance of the boggy vegetation, the creation of a first-class plough layer. The colloids of a dried bog soil become coagulated, which leads to an improvement of the structure and sometimes even to the formation of cracks in the soil. The soil becomes more water permeable. Drainage of the soil promotes a better development and earlier ripening (by 7-10 days) of the plants. The plants of a properly dried soil are characterised by vigorous growth; the botanical composition of the meadow herbage is improved. The drying of bog soils brings about a rise of their temperature and an improvement of their heat regime as a whole. Drying is attended by a certain subsidence of the surface of the bogs and an increase of the apparent density of boggy soils, which, consequently, lose their excessive friability.

Old arable, drained, tame peat soils contain considerably more ash substances than untame ones (in percentages):

	Ash	CaO	R_2O_3
Tame peat soil	14	3.0-3.5	4.2
Untame peat soil . . .	10	2.4-2.5	3.4

Taming lowers the acidity, the pH rising from 4-5 to 6. But in the initial stages of draining, the acidity does not always fall and, in some cases, it even slightly rises. This is due to the removal of the bases, mainly CaO (sometimes up to 80 kg per hectare). That is why draining should be combined with liming. The application of CaO increases filtration. The sum of absorbed bases goes up and also the base-saturation, as well as the amount of available P_2O_5 and K_2O; the ferrous iron salts are eliminated. The taming of peat-boggy soils brings about an increase in the composition and quantity of microflora, butyric and nitrifying bacteria. The quantity of nitrates reaches 25-27 mg to 100 g of soil. This is due to an improvement of the water-air and heat regime in tamed boggy soils. Taming leads to a pronounced mobilisation of nutrient substances in available form. The content of calcium humates in the soil goes up.

With a view to improving the heat properties of dried peat soils, one sometimes proceeds to the mulching of turf-peats with a sandy-loamy or loamy shallow loose cover. Eventually, swamped and boggy lands, unproductive under native conditions, become converted into highly fertile soils, capable of bearing valuable crops of various kinds.

Deswamping of Soils

Deswamping constitutes a kind of soil-forming process, characterised by the conversion of bog and swamp soils into normal ones. This is brought about by changes in the native physico-geographical conditions or through the adoption of the appropriate meliorative and agrotechnical measures. It may be conditioned by the local topography, viz., the prominent forms of the land surface (ascending relief), as well as by a lowering of the base-line of erosion and of the ground water table.

Deswamping begins with a fall of the moisture content, which may, subsequently, become optimum and even deficient. But deswamping, as well as swamping, does not amount to a mere change of the moisture content and constitutes a far more complicated phenomenon. As a natural phenomenon, deswamping can be clearly seen at the southern boundary of the mass distribution of bogs in the belt of the secular shift of this boundary towards the north. Due to the increase of the average annual temperature, the lowering of the rainfall and the intensification of the natural drainage observed in this region, the swamp area is being appreciably reduced.

Deswamping, as it occurs under natural conditions, is accompanied by a progressive drainage of the soil and the replacement of the prevailing anaerobic processes by aerobic ones; it accelerates the decomposition of the organic remains, reduces gleisation, destroys the dense illuvium, brings the reaction of the soil solution closer to neutrality, etc. A soil undergoing deswamping retains the morphological features of bog soils for a long time. The boggy vegetation abandons the drained areas but slowly, maintaining for a long time the boggy soil-forming process. The artificial drainage of native bog soils eliminates, to begin with, the excess of moisture, but, here too, the essential characteristics of bog soils (gleisation, peat formation) persist for a long time.

The artificial deswamping of mineral soils which have undergone secondary swamping proceeds at a considerably faster pace.

When secondary swamping occurs in an alkaline medium, it does not, as a rule, lead to the formation of typical bogs, but is attended by the development of mineral bogs. Once the excess of moisture has been removed, these soils quickly return to their normal condition. The process of deswamping, as it occurs under natural and cultural conditions, does not, on the whole, repeat in the reverse order the stages followed by swamping but evolves more rapidly through its own stages in accordance with the particularities of the initial soils and depending on the drainage factors.

Deswamping should be regarded as a process of development of soils from lower to higher forms. But isolated stages of this process may, during certain periods of time, be diametrically opposed. In a number of cases, the processes of swamping and deswamping alternate, in accordance with changes in the local conditions. The same causes may lead to different results. Owing to its high rate of transpiration, a forest vegetation drains the soil, lowering the ground water table. But the same vegetation, by promoting the accumulation of water-absorbing organic matter, may promote surface swamping. Burning the water-retaining forest litter of bogged up forest soils and the semipeaty horizon of peat soils, brings about deswamping. Whereas, upon the existence of a high ground water table, the burning and felling of forests, by reducing transpiration, bring about swamping, the planting of trees on swamp land may, on the contrary, by increasing transpiration promote deswamping. An additional factor contributing to a lowering of the ground water table may lead to an acceleration of the process. Factors exerting an opposite effect may slow down the process of deswamping and lead to swamping. Even such a measure as the application of peat to clay loamy soils will, by raising the redox potential, counteract swamping, whereas the application of peat to swamp sandy soils will, on the contrary, promote swamping. Deswamping is promoted by ploughing and the intensive clearing of swamp land, in connection with a rise of the base-saturation of the soil, its neutralisation and structurisation.

In the permafrost zone, to begin with, taming leads to a certain increase of the moisture content and swamping of the soil. Thereafter, as a result of efficient agrotechnical and thermal meliorative measures, the permafrost is gradually lowered and the soil becomes drained.

Nowadays, enormous areas of swamp and bog soils are subjected to artificial deswamping as a result of drying. Just like any other soil-forming process, deswamping is amenable to planned regulation. It is indispensable to counteract swamping processes under native and especially under production conditions and to bring about the necessary conditions for the deswamping of soils. Whereas the process of swamping has been the object of relatively thorough investigations, the same can certainly not be said of the process of deswamping. In general, the processes of the progressive development of soils are invariably better known than the reverse ones. This applies to the solonetrisation, solothisation and regradation of these soils, their salinisation and desalinisation and especially swamping and deswamping.

Chapter XVIII

SALINED SOILS

Salined soils are widely distributed in the zone of dry steppes, semideserts and deserts; they are also found in the steppe zone and even the forest-steppe zone, i.e., where the annual amount of atmospheric precipitations is lower than the amount of water evaporating from the surface of the earth. They are formed in places where more salts reach the surface horizons of the soil than are leached out of them. But the process of formation of salined soils is not a mere quantitative accumulation of salts in the soil as a result of their being brought up to the upper horizons by ascending currents, it is much more complicated. Salined soils arise as a result of solonchakous and solonetzic processes of soil formation (Fig. 56). It is not necessary, thereupon, that the salts should be brought from elsewhere, nor that there should be a surplus of them in the soil-forming rocks or in the ground water. The prolonged process of the ascending movement of solutions even of low concentrations and the evaporation of water from the soil in amounts higher than rainfall lead to salinisation. The task consists in preventing the occurrence of this unfavourable process and, where it has already occurred, due to the incorrect utilisation of the land, in putting an end to it in its early stages and directing it towards a progressive improvement of the soils.

Origin of Salts and Salined Soils

The salts which cause the salinisation of soils arise as a result of processes of weathering of soil-forming rocks, as well as in consequence of the physico-chemical and biological processes occurring in the soil itself.

Forest and grassy vegetations, which penetrate deep into the soil-forming rock with their roots, extract a large amount of salts for building up their organic matter, but following upon their death and decomposition, these salts remain in the soil.

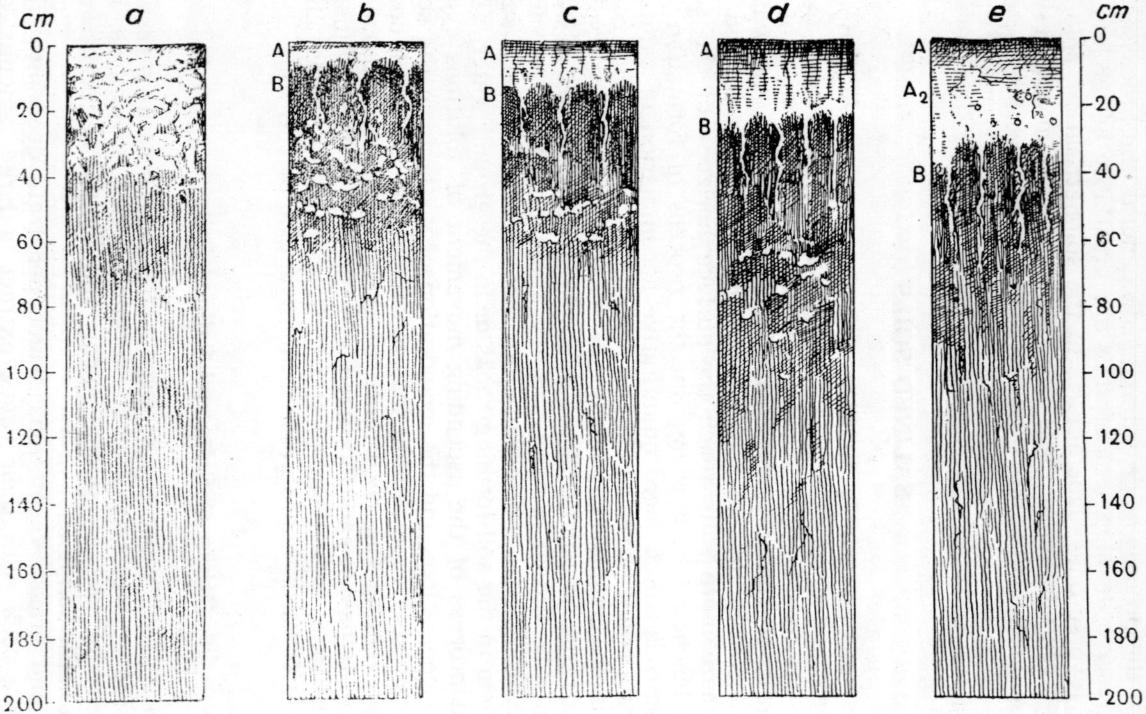

Fig. 56. Profile of salined soils:
a—sulphate-chloride solonchaks; *b*—crustal solonetz; *c*—columnar solonetz; *d*—deep-columnar steppe solonetz; *e*—solods of steppe depressions

Salts reach the soil from salined soil-forming rocks, are brought there with water from the elevated forms of the earth's surface, rise from the mineralised ground water, etc. Of some significance in maritime areas is the deposition of salts due to the wind blowing onto the land a fine spray of salt water when high waves batter the shore. Of no small importance is the direct pulverisation and translocation of salts from the surface of old saline soils (solonchaks). The amount of SO_4 and Cl present in atmospheric precipitations reaches 2-3 mg/l and sometimes more, especially in maritime regions. The burning of coal, combustible shale, oil, wood and other fuels in stoves and furnaces is responsible for the deposition of a certain amount of salts on the surface of the earth. Available data show that, in certain industrial areas, up to 10-80 kg of SO_4 per hectare settle annually on the soil. This amount plays a certain role with regard to the overall salt balance.

Salinisation of soils occurs as a result of the translocation of salts from the ground water or with soil solutions rising from the lower lying horizons. The prolonged accumulation of salts in the surface layers of the soil leads to salinisation even when the concentration of the ground water or soil solution is relatively insignificant. Salinisation may also occur as a result of the translocation and redistribution of salts within the soil. Under conditions of a bilateral circulation of solutions in the soil-ground, the soils become either salinised or desalinised. These processes may alternate in connection with changes in the physico-geographical and biological conditions.

The combination of readily, medium and weakly soluble salts and their differentiation through the soil section gives an approximate indication concerning the stage of salinisation of the soil at the given time. The presence of readily soluble salts close to the surface and their prevalence over other salts points to a beginning of salinisation, whereas their presence deeper down indicates, on the contrary, that the soil is being leached. The combinations of salts can be fairly varied.

The distribution of salts in the soil of the various native zones was spread over a fairly prolonged geological time, in accordance with the formerly existing physico-geographical conditions.

Apart from that, salts are distributed in accordance with the local geomorphological elements, from high watersheds to the thalwegs of valleys. Salts accumulate at the bottom of slopes and on the floor of valleys, in accordance with the surface runoff conditions and in connection with the decrease of the gradients. The same role is played by the ground water tapering out on slopes and on the bottom of valleys. In places where there occurs a discharge of, or underflooding of the surface with, mineralised ground water, not infrequently the soil undergoes salinisation. Soils of this kind may be found everywhere, even in the north, where the formation of saline soils of the zonal type does not take place.

Of great influence with regard to the distribution of salts in the soil are the forms of the local meso- and microrelief. The soils of meso- and microdepressions are usually leached, especially on water permeable soil-forming rocks, whereas on impermeable ones, not infrequently they undergo salinisation. Meso- and microelevations, on which the rate of evaporation from the surface is higher, favour the rise of the soil solutions and salts towards the upper soil horizons, causing their salinisation. There is an increase in the loss of water from fissured microhummocks and this intensifies salinisation. In the immediate vicinity, right under microsinks, hollows, patelloid depressions, are more often formed nonsaline dark-coloured limanic soils or soils of variable salinity—solonetzic-saline soils, whose salinity changes even from one season of the year to another. The presence of salined soils in southern regions, just like swamping in the north, is quite a normal natural phenomenon, reflecting the corresponding physico-geographical conditions. But salinisation, just like swamping, may be eliminated through agromeliorative measures. Both occurrences should be prevented or, where they are already in progress, actively counteracted. The fight against salinisation, just as against swamping, should be based on measures aimed at changing the soil formation of large areas, owing to the fact that the adoption of local measures aimed at removing particular causes proves inadequate or at the best constitutes a temporary palliative.

All salined soils are classified as follows:

1st series: a) saline soils (weakly-, medium- and strongly-saline), b) solonchaks (extremely saline); 2nd series: a) solonetzic soils, b) solonetzes; 3rd series: a) solodised soils, b) solods.

Solonchaks and Saline Soils

Solonchaks and saline soils differ from the typical soils of the native zone in that they contain a large amount of water-soluble salts. The excess of salts in the soils converts them into qualitatively different soils—viz., saline soils and solonchaks. These soils are easy to recognise by the salt efflorescences which appear in vertical section, without having recourse to analyses. The surface of these soils differs sharply from that of nonsaline or weakly saline ones. Saline soils contain water-soluble salts in such a proportion as depresses crops or prevents their normal development. The plants growing on saline soils are the so-called halophytic plants (Russian thistles—*Salsola*, *Salicornia*, etc.), which are not affected by the excess of salts. The protoplasm of the cells of Russian thistles can put up with high concentrations of salts, which accumulate in the plants.

The solonchaks occurring in nature are classified as follows: a) puffic, b) wet, c) black, d) meadow, e) "sors".

Puffic solonchaks are formed in the presence of an excessive accumulation of Glauber salts ($Na_2SO_4 \cdot 10H_2O$), which are capable of light dehydration and constant recrystallisation to $Na_2SO_4 \cdot 3H_2O$. In hot windy periods, when the air is very dry, the Glauber salt may be dessicated and converted to the anhydrous form, thenardite (Na_2SO_4). At the same time there occurs a salt splitting of clays, marls and other rocks to an extremely fine dust. These processes confer a certain puffiness to the surface layer of the soil. A soil of this kind collapses readily underfoot, the feet sinking into the puffy mass of the surface horizon. The amount of salts in the upper part of such soils reaches 5-15%; it is 1-2% in the subsoil horizons. At a depth of 1.5-2 m, the ground water is mineralised to a brine.

Wet solonchaks—typically hydromorphous—are salined to excess, mineral-bog soils. They are often covered with a glassy crust, which crunches underfoot; this is characteristic of chlorides, which cake from the heat of the sun. Owing to the high hygroscopicity of the salts, the soils are wet, especially when the ground water lies close to the surface.

Black, soda solonchaks are coloured in black due to the increased solubility of humic substances in the presence of soda. Soda increases the dispersiveness of the soil.

Meadow, carbonatic-calcic solonchaks develop under a meadow vegetation in conditions of underflooding with mineralised carbonatic ground water. They are enriched with calcium carbonates, to a lesser extent with magnesium carbonates and differ sharply from the other solonchaks in that they possess a relatively well developed meadow vegetation, a fairly high humus content and high fertility.

"Sors" ("shors") are saline muds forming on the seat of (temporary) salt lakes undergoing drying out.

For agricultural purposes, it has been found necessary to differentiate the solonchaks according to the cations and anions of the salts which they contain (Fig. 57). The salts contained in solonchaks and saline soils comprise mostly $Na^{\cdot}$, $Mg^{\cdot\cdot}$, $Ca^{\cdot\cdot}$ cations and Cl, SO_4, CO_3, HCO_3 anions. Hence the differentiation into sodium, magnesium-sodium, calcium-sodium, magnesium and calcium, as well as chloride, sulphate and carbonatic solonchaks. The listed cations and anions correspond to the following salts in the soils:

$$NaCl,\ Na_2SO_4,\ Na_2CO_3,\ NaHCO_3$$

$$MgCl_2,\ MgSO_4,\ |MgCO_3,\ Mg(HCO_3)_2$$

$$CaCl_2,\ |CaSO_4,\ CaCO_3,\ Ca(HCO_3)_2.$$

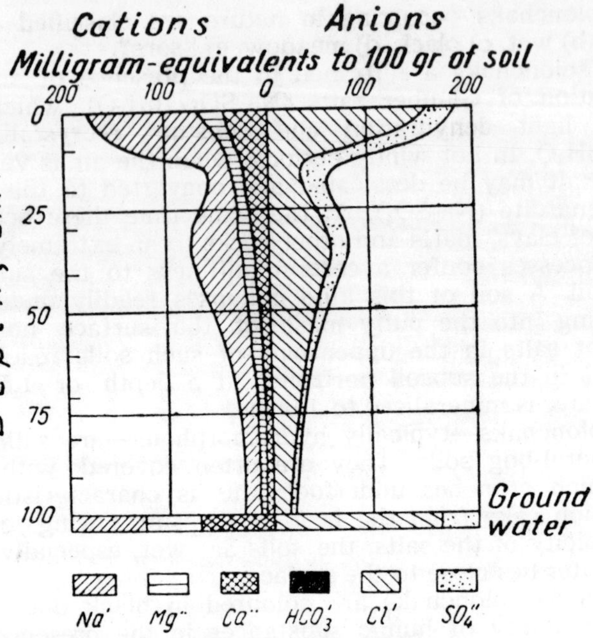

Fig. 57. Distribution and composition of salts in the profile of solonchaks

The above-named salts depress the normal development of plants—some more, some less. Plants may tolerate higher concentrations of some salts and yet not be able to tolerate relatively low concentrations of other salts. All the above-named salts can be divided into harmful (those situated above the broken line on the scheme) and relatively harmless ones (those situated below the broken line). The most harmful are all the sodium salts and all the chlorous compounds. The salts begin to exert a harmful effect when their amount reaches 0.7 g/l, but soda is already harmful upon a content of 0.3 g/l, the same content of gypsum being harmless. Solonchaks containing water-soluble chlorides and sulphates in various correlations, are classified according to the percentage content of chlorine:

sulphate	solonchaks . . .	less than 10% of chlorine
chloride-sulphate	" . . .	10-25% chlorine
sulphate-chloride	" . . .	25-40% chlorine
chloride	" . . .	over 40% chlorine

According to their soda content, the solonchaks are referred to as soda, sulphate-soda and soda-sulphate solonchaks. The type of salinisation is determined from the ratio (in milligram equivalents) of chlorine to SO_4:

$\frac{Cl}{SO_4}$=0.5—salinisation of the sulphate type,

$\frac{Cl}{SO_4}$=0.5-1.0—chloride-sulphate type,

$\frac{Cl}{SO_4}$=1.0-5.0—sulphate-chloride type,

$\frac{Cl}{SO_4}$=over 5—chloride type.

The harmful effect of water-soluble chloride-sulphates begins to show upon a content of approximately 0.1% of the soil's weight. Such soils are weakly saline. They can grow wheat, oats, millet, lucerne and other crops. The presence of water-soluble chlorides and sulphates in the proportion of 0.3-0.5% hampers the normal development of crops and for many of them, this amount proves fatal (especially for leguminous crops). On soils which contain up to 0.6% of water-soluble salts, cotton and barley can still thrive; meadow grasses and, among the cereals, such a salt-tolerant one as sorghum grow in the presence of 0.6-0.8% of salts. When the content of water-soluble salts reaches 0.8-1%, other meadow grasses (brome) develop and also beetroot, but they are much stunted. When estimating the harmful effect of water-soluble salts, one should take into account not only their absolute amount in soil but also the concentration of the soil solution, which changes with changes in the soil's moisture content. According to their content of salts and their composition, soils are classified as follows (Table 33).

Table 33

Classification of Soils According to Their Salinity

Degree of salinisation	Content of salts in a 1 m layer, %					
	from chemical analyses data				amplitude of the data (according to published results)	
	with predominance of					
	chlorides		sulphates			
	of all salts	chlorine	of all salts	chlorine	dense residue	chlorine
Nonsaline	$<$0.25	$<$0.01	$<$0.3	$<$0.01	$<$0.25-0.3	$<$0.01
Weakly saline . .	0.25-0.50	0.01-0.04	0.3-1.0	0.01-0.04	0.25-1.0	0.01-0.05
Medium saline . .	0.50-1.00	0.04-0.2	1.0-2.0	0.04-0.2	0.2-2.0	0.01-0.1
Strongly saline .	$>$1.00	$>$0.2	$>$2	$>$0.2	0.7-3.0	$>$0.1
Solonchaks . . .	$>$3.0	$>$0.1 any amount	$>$3.7	any amount	$>$2.0-3.0	any amount

Note. According to the depth at which lie the salt horizons (content of water-soluble salts $>$0.25-0.3%) and to the qualitative reaction for Cl and SO_4 are distinguished: nonsaline—deeper than 150 cm; deeply saline—100-150 cm; less deeply saline—70-100 cm; saline—30-70 cm; solonchakous soils—5-30 cm.

The harmful effect of salts is determined from a number of factors: the nature and salt-tolerance of the plants themselves, as well as their stage of development; the composition of the salts in a mixture (in a complex); the nature of the soils; the system of farming (agrotechnique) and the existing climatic (weather) conditions.

The degree and nature of salinisation exert an influence on the quality and quantity of the crop (low yields, shorter cotton fibres, etc.). When mixed together, water-soluble salts manifest an antagonistic interaction, so that in association, the toxic effect of salts is lower than that of one harmful salt in isolation. Of great importance is the combination of salts in pairs. If, for example, $CaCO_3$ and $Na_2SO_4 \cdot 10H_2O$ happen to reach the soil simultaneously, they react, with the inevitable formation of Na_2CO_3 and $CaSO_4 \cdot 2H_2O$, and, owing to the fact that the soil is an active body, the salts arising in it call forth the formation of other, usually less toxic, or harmless salts.

The interaction of salts in a mixture of them and their effect on plants is governed by the nature of the soil. Clayey soils may exhibit a more pronounced salinity, because they are capable of retaining more water than sandy ones. However, in practice it is the contrary that occurs, due to the fact that sandy soils are more easily leached by infiltrating water. It is easier to rid sandy soils of an excessive content of salts. By influencing the rate of solothisation, in connection with the amount and nature of the rainfall, climatic (weather) conditions may weaken or aggravate the toxic effect of salts. The rate of solothisation tolerated in a wet climate is higher than in the case of a dry one, owing to a lowering in a wet climate of the relative concentration of the soil's salt solutions.

Solonetzes and Solonetzic Soils

As a rule, solonetzes and solonetzic soils constitute a series of soils undergoing desalinisation, sharply differing from the series of soils undergoing salinisation—viz., solonchaks and saline soils. These soils are characterised by the content in the soil absorbing complex, in the absorbed condition, of a high amount of sodium ions. This accounts for their main properties, viz., their high degree of dispersiveness and high alkalinity.

Solonetzic soils, and after them solonetzes, often arise on the basis of saline soils and solonchaks. However, solonetzic soils do not necessarily pass through the saline stage. Solonetzic soils may arise even on weakly saline soils, if the soil absorbing colloidal complex contains sodium ions, which, owing to the relatively higher concentration set up, displace the calcium ions. Following upon the absorption of sodium, the soil becomes less stable with regard to the action of water and carbon dioxide. Due to the presence of OH ions, and later of Na_2CO_3, the soil acquires an alkaline reaction.

The process of the introduction into the soil colloidal complex of sodium ions and the acquisition by the soil of the above-mentioned specific properties and characteristics constitutes the solonetzic process, and soils possessing a high content of exchangeable sodium are called solonetzic.

When soils are relatively saturated with sodium, their structural aggregates are broken up, in contradistinction to saturation with calcium whereupon they become more stable. When saturated with sodium, soil loses structure and becomes water impermeable. In soils saturated with sodium, the amount of the fraction comprising microaggregates ranging in size from 0.25 to 0.001 mm decreases. The amount of the colloidal part ($d<0.22\mu$) increases several tens of times. The rise of dispersiveness affects the mineral as well as the organic part of the soils. A certain amount of the humus present in the soil becomes soluble. The uliginous part of a soil of this kind acquires the capacity to swell considerably and retain water, but when it dries up, it becomes packed and acquires the consistency of cement. The uliginous finely dispersive part of the soil acquires mobility, penetrating to a certain depth into the soil, where, becoming coagulated by electrolytes and partly retained mechanically, it forms an illuvial horizon. The physical properties of the soil are thereby sharply worsened.

The absorbing colloidal complex of solonetzic soils saturated with sodium acquires a high capacity for breaking down readily under the effect of water, especially if it contains carbon dioxide, this going on until the salts of sesquioxides and silicic acid are dissolved.

It has been established that the formation of solonetzic soils from solonchakous ones is tied with the entry of sodium ions into the soil absorbing complex together with the appearance of anaerobic conditions, the stagnation of water on the soil's surface and some leaching—displacement of salts and colloids. Leaching of the soil may be conditioned by a lowering of the base-line of erosion, a fall of the ground water table or, finally, by a certain increase in rainfall and a decrease of evaporation from the soil's surface. The gradual enrichment of the absorbing complex with sodium constitutes a stimulus for the development of alkalinity—leaching of electrolytes from the upper part of the soil with formation of soda. Further aggravation of the solonetzic process occurs as a consequence of its initial stages. The soil gels are converted into water-soluble sols. The high degree of dispersiveness of the mineral and organic matter promotes the inwash of suspensions and the formation of a clearly marked illuvial horizon above the salt horizon, in connection with the coagulation of the colloids, as well as mechanical absorption. When a soil-absorbing complex is saturated with sodium, it dissociates into R_2O_3 and SiO_2. The more stable R_2O_3 colloids are translocated into the B_1 horizon, whereas SiO_2 falls out on the seat of its formation; as a consequence, the A_2 horizon becomes relatively enriched with SiO_2, with the removal of mechanical elements and R_2O_3, which go towards the formation and enrichment of the B_1 and B_2 horizons.

In contradistinction to the A_2 horizon of podzolic soils which contains SiO_2 in the crystalline condition, the A_2 horizon of solonetzic soils contains much free amorphous silicic acid.

Solonetzes are characterised by the fact that their vertical soil section possesses a clearly marked binary character, being differentiated into the upper A_1+A_2 portion and the lower B_1+B_2 portion. These genetical horizons are quite clearly seen in the soil profile. Particularly clearly marked is the loose solodised A_2 horizon, which prevents the capillary rise of water to the surface of the soil. No less clearly marked is the B_1 horizon which, in contradistinction to A_2, is quite coherent and dense, being fissured and columnar in the dry condition. The presence of these two clearly marked genetical horizons makes it possible, as a rule, to recognise solonetzes and solonetzic soils in the field; and the detection in the laboratory of the presence in the soil of a high amount of absorbed sodium in B_1 points with certainty to the existence of a solonetzic process of soil formation.

The typical profile of native solonetzes presents the following average morphological features:

A_1—0-8 cm—grey, sometimes dark grey, soddy-humus horizon, loose. The transition to the underlying horizon is abrupt;

A_2—8-15 cm—whitish, solodised horizon, consisting entirely of a silicious mass in the form of a floury meal, of a lamellar structure, easily delaminated, the transition to the underlying horizon is very abrupt;

B_1—15-30 (50) cm—dark brown, brightly coloured, fairly dense, fissured-columnar, the upper extremities of the prisms or columns being rounded. The transition to the underlying horizon is gradual;

B_2—30-80 cm—brown, dense, column-like or nutty; carbonates are present at the bottom;

C—80 cm and lower—pale yellow, loess-like, carbonates are present.

All solonetzes are divided into: a) crustal (shallow) with an A horizon less than 6-7 cm thick, b) columnar (medium) with an A horizon 10-15 cm thick, c) deep-columnar (deep), whose $A_1 + A_2$ horizon reaches 20-30 cm.

In the crustal solonetzes of the forest-steppe zone, the absorbed sodium constitutes 50-70% of the sum of absorbed bases, whereas its proportion in the deep solonetzes is only 25-35%. In the dry steppe zone, the content of absorbed sodium is correspondingly lowered down to 25-35% in crustal and 15-20% in deep solonetzes. Crustal solonetzes are usually relatively younger soils, which later provide the basis for the development of medium and deep solonetzes. Crustal solonetzes may, however, arise as a result of the mechanical destruction, erosion and deflation of the A horizon of medium and deep solonetzes. Deep solonetzes are subdivided according to the general character and structure of the B_1 horizon into lumpy, prismatic, columnar and nutty solonetzes. There are also lowland-meadow and steppe solonetzes.

The natural typical vegetation of solonetzes is dwarfed and sparse and comprises black and white wormwood (*Artemisia pauciflora* and *maritima Bess.*), *Campharosma monspeliacum Richt.*, *Kochia prostrata Schred.*, quack grass (*Agropyrum ramosum Richt.*), thrifts (*Statyce Gmelini, S. tatarica*), etc. Crops grown on solonetzic soils are markedly stunted and sparse. When ploughed, solonetzes are easily recognised from the peculiar lumpy structure of the ploughed layer and its grey colour. Fairly clearly seen in the furrow bottom where they assume the aspect of a parqueted floor, are the cut polygonal irregularly shaped extremities of columns and prisms.

The chemical properties of the solonetzes are quite typical. Solonetzic soils are easily determined from data supplied by water extracts. The minimum amount of salts is found in the A_2 horizon and the maximum amount is in the B, especially B_2, horizon. The maximum amount of absorbed sodium is in the B_1 horizon which also contains up to 0.2% of water-soluble humus, this constituting up to 1/40 of its total amount. This is evidence of the high solubility of the soil humus in an alkaline medium, when the soil's absorbing colloidal complex contains sodium ions in the absorbed condition. A water extract from the B_1 horizon of a solonetzic soil has the colour of tea brew. The B_1 horizon is rich in colloids and contains plant ash nutrient elements. As it swells, this horizon considerably hampers filtration and the capillary rise of water. The clearly marked inwash horizon (B_1), which possesses high water capacity, low water permeability and water rising capacity, is responsible for the unsatisfactory physical properties of the solonetzic soils as a whole. The high coherence and density of the B_1 horizon hinders the deep penetration of the root system and prevents the plants from obtaining nutrients from the deep layers of the soil.

Solonetzic soils serve as a good example of the fact that morphological features cannot alone constitute a reliable criterion for the recognition of soils. Soils may possess the morphological features of solonetzes without passing through that stage; such soils are referred to as pseudosolonetzes. On the other hand, solonetzes forming as a result of the leaching of sodium solonchaks, acquire the specific morphological features of solonetzes with some delay. At the same time, a marked illuvial horizon may be formed without the participation of absorbed sodium, provided the soil is sufficiently dispersed. The solonetzes are sometimes regraded towards nonsolonetzic soils. The existing morphology of the solonetzes may be retained, but the absorbed sodium is displaced by calcium, i.e., the chemical nature of the solonetzes is annihilated before the morphological features. Such solonetzes constitute relict or so-called dead solonetzes. It can thus be seen that one cannot determine whether a soil belongs to the solonetzes and find out the degree of its alkalinity merely from its morphological features, the possession of the appropriate analytical data being indispensable. The most reliable indication that the soil is a solonetz is the presence of absorbed sodium in the solonetzic horizons (B_1).

According to their amount of absorbed sodium, soils are graded as follows:

a) weakly solonetzic, the content of sodium amounting to 10% of the absorbing capacity. These soils are characterised by a somewhat higher dispersiveness and alkalinity. They do not require any special ameliorations and are suitable for agricultural purposes;

b) medium and strongly solonetzic, the content of sodium amounting to 10-25 (30)% of the absorbing capacity. Need to be subjected to chemical treatment and removal of exchange products before they can be utilised in agriculture;

c) solonetzes, in which the amount of absorbed sodium is over 25-30% of the absorbing capacity. Require special chemical treatment and leaching operations. The presence in the B_1 horizon of absorbed sodium in the proportion of 15-20% of the absorbing capacity is harmful to plants, resulting sometimes in crop failures. The infertility of such soils is due to their extremely poor physical properties and the high alkalinity of the B horizon.

Should the sodium ions be displaced by calcium ions, solonetzes and solonetzic soils can be regraded and approach the normal typical soils of the native zone. Being sometimes subjected to secondary salinisation, they are converted into a particular kind of solonetzes: the solonchaks. Not infrequently, due to particular conditions in the water regime of the soils (in the flood plains of south-eastern rivers) or in connection with periodic changes in weather conditions, the solonetzic and solonchakous processes alternate.

Solods

Past a certain limit, the solonetzic process leads to the solothisation of soils and the formation of solods. Solods are distributed in the zone of forest-steppes and steppes, where they are found in depressions referred to as aspen bushes. They are also found in Western Siberia and further east.

Solods are formed as a result of a pronounced but not necessarily a through leaching of solonetzic soils. This leaching occurs in connection with a certain widening and deepening of microdepressions, a more vigorous stand of grass under steppe conditions, followed by the establishment in them of a woody vegetation. The depressions in the microrelief are often directly responsible for the formation of solonetzes.

By accumulating atmospheric precipitations (retention of snow), the vigorous grassy (sometimes weedy) vegetation growing in depressions in areas of deep solonetzes contributes to a more pronounced moistening in spring, which is attended by suffossion, leaching and desolonetrisation of the soil. This, in turn, favours the growth of bushes and a woody vegetation, which retain still more snow and cause the leaching of the soil and a deepening of the depressions. This change in vegetation leads to a change in soil formation, viz., the solothisation of soils of the solonetzic type.

The solothisation of solonetzic soils is tied with the destruction of the absorbing colloidal complex, occurring as a result of hydrolysis. This destruction is accompanied by the loss by the solonetzic soils of absorbed sodium, which is removed as bicarbonate and replaced by hydrogen ions. When the solonetzic process is still in a stage of full development, the sodium is almost entirely removed from the upper compartment of the soil (A_1+A_2), being retained only in the illuvial horizon, whence it is also gradually being removed. Due to the existing climatic conditions, not infrequently, leaching of the soil alternates with periods of drought. This conditions the upward migration of sodium salts, which promotes still more the subsequent destruction of the colloids in the upper compartment of the soil and their removal into the illuvium. The colloids suffer intense destruction at the contact of the A_2 with the B_1 horizon, this slowly spreading deeper inwards in a frontal band. Owing to the destruction of the upper part of B_1, there is an increase in the thickness of A_2. The maximum destruction of the humic and alumosilicate part of the soil occurs on the frontal surface. Water, which penetrates through the cracks of the B_1 horizon, removes calcium and colloids, causing the accumulation of the silicic acid on the boundary surface between A_2 and B_1. Overmoistening causes the small columns of the B_1 horizon to swell and brings about anaerobic conditions; as a consequence, ferric iron is converted to more soluble and mobile ferrous iron and is transferred further and further down from the boundary between A_2 and B_1, this boundary being also progressively transferred downwards. The products of destruction, which are being transferred below this surface, condition the uninterrupted formation of illuvium.

At the same time, the depression is deepened still further and conditions are set up for an increased wetting of the soil. As the moisture content goes up, so does the water table, until the soil water finally merges with the surface water. This aggravates the anaerobic processes in the soil and is accompanied by its gleisation and swamping. But periods of overmoistening of the soil alternate with periods of drying out. Sometimes the mobile sesquioxides present in the soil (especially the iron ones) migrate upwards with the solutions, and, getting oxidised, become concentrated in isolated spots and parts of the A_2 horizon, in the shape of concretions. This is further enhanced by the lateral inflow of R_2O_3. Ferruginous concretions, from tiny specks and small loose spots to large dense inclusions shaped like buckshots, sometimes form special subhorizons up to 20-30 cm in depth.

The destruction of the mineral part of the soil also affects the SiO_2, part of which passes into solution as sodium silicate, also periodically migrating upwards into A_2. As a result of drying and dehydration, the SiO_2 is precipitated here in the shape of an amorphous mass. Part of the amorphous silica present in soils arises as a result of the accumulation of silicic bodies which form in the tissues of plants (cereals, sedges) and due to the accumulation of the silicic covers of diatomae (kieselguhr). Solods can arise without passing through an intermediate solonetzic stage.

The solothisation of soils is accompanied by suffossion, which is engendered by the solonetzic process. Outwardly, solods resemble podzols. But their weakly acid reaction is limited to the upper part of the profile. In the B and C horizons, the reaction of the solods is alkaline, this constituting one of the characteristic features of the solods. In depressions where the soil is subjected to solothisation, the dense grassy vegetation may hamper the establishment of trees, reinforcing the soddy soil-forming process. The latter raises the amount of humus; as for the calcium, which accumulates with the decomposition of plant remains and migrates from below, it weakens the solothisation process. In that case, on the depressions subjected to solothisation, humus-meadow or soddy-meadow soils of a special kind are formed. Where the solothisation process has reached an advanced stage, in relatively northern regions, the solods may be converted into soils of the podzolic type of soil formation. Here, the solods sometimes evolve to peaty-gley soils, becoming markedly enriched with organic matter and ash nutrient elements, including available phosphorus.

The morphological features of solods, like those of solonetzes, are dependent upon the zonal and local conditions of soil formation. They can be represented by the following average scheme:

A_1—0-6 (10) cm—humic, dark grey, soddy;

A_2—6-20 (30) cm—light grey (whitish), leached from salts and silt, structureless or layer-imbricate, solodised horizon. Consists of amorphous silica. At the bottom remain traces of the previous columnar solonetzic horizon. There are many rusty spots and ferro-manganese nodules;

B—20-60 (80) cm—brown, sometimes reddish-brown, not infrequently mottled, lumpy (amalgamated), enriched with R_2O_3. At the bottom, here and there, where the B horizon passes into C, are found Ca and Mg carbonates. In the absence of the latter, gleisation is clearly marked.

Depending on the degree of intensity of the solothisation process and soddy soil formation, soils are differentiated into dark grey, grey and light grey ("beliaks" occurring under birch-aspen bushes) solodised soils.

The bond between solodised soils and solods and saline soils is unquestionable, since the BC horizon often contains a significant amount of water-soluble salts. As a rule, solods continue resembling solonetzes for a long time, being differentiated from them by their thicker profile. Present in the absorbing complex of solodised soils is absorbed sodium, whose amount goes up from A_2 to B_2. But the main role is played by the alkaline earths bases and partly by the hydrogen ions.

Solods and solodised soils are relatively poor. Their organic matter content is low and they are poor in mineral bases, especially in the A_2 horizon. These soils should best be left under forest, if it is already growing there. When ploughed, solods and solodised soils soon become puddled, with the formation of large clods; not infrequently, a soil crust and a plough sole are formed. These soils have an unsatisfactory water-air regime. They can be converted into useful agricultural soils provided they are enriched with organic matter (farmyard manure, compost) and dressed with slaked lime and mineral fertilisers (NPK). Good results are obtained from the spreading on the surface of the soil of humic earth hauled in from neighbouring areas. Ploughing of solods and solodised soils should be carried out without turning the furrows over, the alternative being to stir up the soil with a subsoiler.

Also included in soils of the solonetzic type are the takyrs, which were described in Chapter XIV.

Distribution of Saline Soils

Like bog soils in the north, saline soils are found as isolated areas or covering vast expanses of land. Changes in the climatic and biological conditions and in the dynamics of salinisation of the soils entail changes in their area of distribution. Particularly unstable in this connection are small isolated patches, which may become reduced in size as they are desalinised or, on the contrary, increased, should the degree of salinisation go up. Here and there, there even occurs a migration of isolated patches of salinisation.

Faulty irrigation in droughty regions usually leads to salinisation. To begin with, salinisation affects a band of soil 300-600 m wide lying along the sides of irrigation channels, due to a rise of the ground water table.

Salinisation can be seasonal, patchy or continuous. Continuous salinisation occurs in poorly drained areas, when the mineralisation of the ground water reaches 10-12 g/l. Continuous salinisation occurs also upon a high mineralisation of the ground water (15-30 g/l and more). This form of salinisation is more serious and difficult to get rid of. It is met with in the lowermost and least drained areas of irrigated lands.

Secondary Salinisation of Soils

What is referred to as secondary salinisation, as distinct from natural salinisation, occurs as a result of a sharp deterioration in the usual hydrological regime of soils of southern regions under conditions of production (rise of the ground water table). Secondary saline soils are a clear illustration of the evils attending the faulty management of land; most of them are the unwelcome heritage of a recent past. Not only does secondary salinisation sharply depress yields, it also lowers the productivity of labour and the effectiveness of the unit of volume of the water expended in irrigation. In properties and morphological characteristics, secondary saline soils bear some resemblance to the saline soils occurring in nature.

Secondary salinisation consists in the main in that the root zones of soils become enriched with readily soluble salts in amounts exceeding a critical limit, i.e., which depress and ruin crops. Secondary salinisation can occur with or without a change in the level of the ground water, mainly under conditions of irrigation, and sometimes also where the land has been dried.

In soils subjected to irrigation, secondary salinisation occurs as a consequence of changes in their water and salt regimes—disaggregation and packing of the soil, rise of the ground water table or formation within the soil of a water-bearing horizon, capillary rise of water solutions to the surface, evaporation of the water and accumulation of salts in the soil's root zone.

The optimum water regime of the soil deteriorates owing to an excess of irrigating water resulting from faulty watering. Salinisation of soils directly caused by the irrigating water is not a frequent occurrence but deserves great attention. As a rule, water containing more than 1 g/l of salts should not be used for irrigation purposes.

Secondary salinisation may arise without the participation of the ground water due to the redistribution of the salts already present in the soil and to the salts present in the soil water, without any rise of the ground water table. As the irrigation water infiltrates into the soil, it dissolves the salts, but during the long intervals between the waterings, the salt solutions may rise to the surface where, losing their water through evaporation, they will cause salinisation of the soil. Faulty irrigation and poor agricultural practices may result in the concentration of salts in the active root layer with a consequent disastrous effect on the development of crops, even though, to begin with (e.g., before sowing), the amount of salts in the soil was below the critical level.

Secondary salinisation may also occur in swamps with mineralised water subjected to drying. (The degree of mineralisation of the ground water fluctuates within a wide range. According to the dry residue in grams per litre, the ground water can be: sweet—1, weakly saltish—1-2, saltish—2-5, weakly salty—5-10, salty—10-30, strongly saline—30-80, briny—over 80). In that case, the process of secondary salinisation of soils consists in that, with the removal of the excess of water, upward currents of ground water are set up, which bring up dissolved salts. Upon drying out, soils become saline under special physico-geographical conditions—in the river valleys of the steppe belt and on the solonchakous lands of the forest-steppe zone. In such areas, drying out may cause an increase in the mineralisation of the water remaining in the soil, which results in the formation of solonchakous-boggy soils and, here and there, even in the disappearance of forests.

Where there is a general shortage of water, the removal of the surface water from saline marshes is considered as fundamentally wrong, due to the fact that all the available water should be utilised for watering and leaching operations. The drying of mineralised swamped areas requires therefore more complex measures and a most careful approach. Simply drying southern steppe flood plains may also result in their secondary salinisation. Damming up of flood plains may cause secondary salinisation. The soils of dammed up areas —isolated from flooding, give rise to steppes, they become enriched with salts, packed, lose their structure, etc. Under conditions of damming up, soils lose their flood-plain characteristics and get closer to the soils of the zonal type.

Of great importance with regard to their salinisation in general and secondary salinisation in particular is the mechanical composition of soils. The accumulation of salts is more intense in soils of a heavy mechanical composition than in lighter soils and the latter are more intensely leached of their salts than the former. The same applies to separate layers of soil. Clayey alluvial layers are not infrequently saline, whereas sandy ones are leached.

When the ground water table lies close to the surface, as a rule, flood plain soils lying on relatively water permeable loamy alluvium overlying drained sandy flood plain deposits, are not saline. The critical depth at which lies the ground water table is, in this case, determined not by the overlying loamy flood plain alluvium, but by the sandy flood plain deposits. The same result is given by sandy draining layers of adequate thickness. There is no progression in salinity even in the case of salined soils with mineralised ground water if a sandy draining layer underlies them at a certain depth. Most effective in this respect is a layer of sand of a thickness exceeding the height of the column of capillary rise of water in it. Such draining layers hamper the salinisation of overlying soil horizons or slow it down considerably. But sandy layers may bring about salinisation if they conduct water, especially under pressure. Lens-shaped sandy and clayey strata may condition a mottled (complex) distribution of saline soils in separate soil horizons, in vertical as well as horizontal section.

The dried flood plains of steppe rivers, isolated from the annual flood water and natural leaching by damming up, are, here and there, subjected to salinisation in a broad band, immediately beyond the bank. Such a secondary salinisation in the vicinity of the banks is due to evaporation and the removal of salts with the water rising under the effect of hydrostatic pressure along a depression curve (underwetting) (Fig. 58).

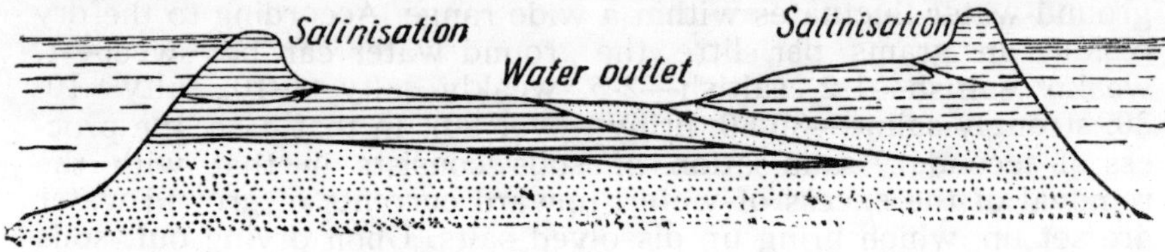

Fig. 58. Salinisation and discharge of water on a dammed up area

Faulty drying or irrigation may lead to the formation of solonchakous-solonetzic soils and solonetzes. On irrigated land, solonetzes are formed as a result of an increase in alkalinity. Irrigation and leaching operations with water containing sodium salts, as well as the irrigation of saline soils containing sodium salts, are not infrequently attended by a solonetzic process.

The solonetzic process can be used to advantage as a meliorative measure to fight the filtration of water from canals and reservoirs. Artificial alkalisation in conjunction with other measures, such as mechanical dispersion and packing of the soil at the bottom of canals and reservoirs, covering with water impermeable screens, etc., gives positive results. In essence, alkalisation consists in the saturation of the colloidal part of the absorbing complex with sodium and the intensive dispersion of the soil mass. This leads to colmatage and a sharp decrease of the filtration (by tens and hundreds of times). Artificial alkalisation requires the application of approximately 3 kg of NaCl per square metre. Repeating the alkalisation, even with small doses of dispersing agent, may give better results than the initial operation. Alkalisation is accompanied by the dispersion of the soil mass and increased mobility of the colloids as a result of the peptisation. Repeated alkalisations lead to the progressive increase close to the surface of the artificial soil illuvium. Not only does artificial alkalisation considerably reduce filtration in canals, it also prevents the bottom from getting overgrown and hampers the development of weeds.

Secondary saline soils are associated with recent and old irrigated areas (oases). Saline soils, especially solonchakous ones, are distributed from the chernozemic native zone to the sierozems and brown soils. In the chernozemic zone, solonetzes and solonetzic soils are found on flat drainless plateaus, in flat depressions. In the forest-steppe, solonetzic soils are associated with swamped depressions overgrown with aspen shrub, being disposed along their edges. In the centre of the depressions are found solodised soils. In the southern part of the steppe zone, solonetzes are formed in flat, wide suffosive depressions. Here, they are also found on river terraces and in the widened areas in front of gullies and ravines. Solonetzic soils are mostly formed and distributed in the chestnut soils zone. Light chestnut soils are almost all solonetzic. The soils here exhibit a very marked complexness, so that on relatively small areas are found soils of various degrees of salinisation disposed in a sort of overlapping fashion in association with solonetzes and nonsaline soils. This complexness is unstable, both in the form and dimensions of the contours and in composition. Here and there predominate nonsaline and very slightly saline soils, or, on the contrary, solonetzic and solonchakous soils occupy more than 50-60% of the total area. The percentage composition of the components of the complex is unstable and the areas of the contours are so small and their configuration so complicated that they are impossible to map using angular, even large scale, surveying. Their percentage correlation can be computed by laying traverse lines in various directions and calculating linear values—sections of lines at the intersection of the corresponding contours.

The best results are obtained by large scale aerial photographic surveys.

In the zone of brown soils, the areas under solonetzes are reduced and in the sierozemic zone they are usually almost nonexistent, and only modified solonetzic soils of a peculiar kind are found there.

PART THREE

IMPROVEMENT OF SOILS

Chapter XIX

IMPROVEMENT AND TAMING OF SOILS

Soil is the main means of agricultural production and, like any other means of production, it needs to be systematically, continually ameliorated and improved, or, to use a convenient term, tamed. Apart from raising its fertility, taming changes all the soil's properties which influence the growth and ripening of plants. It is obvious that the best soils yield the best crops with the minimum expenditure and labour. It is therefore indispensable to raise the quality of all soils without exception. Soils are improved (tamed) in the process of correct agricultural production and growing of higher and higher yields of agricultural crops.

Nowadays, the productive activity of man has become the most important factor of modern cultural soil formation and development of potential and effective fertility. This is achieved through the medium of plants, microorganisms, a system of manuring, high mechanisation, a system of tillage, amelioration, a system of special agricultural measures, etc. Whereas it required many thousands of years, in the past, to build up soil fertility under natural conditions, it has now become possible to obtain the same result with the minimum of delay, not exceeding man's lifetime.

In spite of the fact that the changes to which the soil has been subjected through man's productive activity began many centuries ago, it is only in our days that a firm course has been set towards its transformation and the controlled change of soil formation. Soil used to be regarded as the plants' substratum, given to man by nature from time immemorial and not liable to radical improvement. Man was chiefly concerned with the obtention of the current crop and not with the transformation of the soil with a view to ensuring subsequent high, progressively rising stable yields. Controlled soil formation cannot fail to lead to the radical improvement and transformation of soils.

Traces of spontaneous taming can be seen in the soils of all native zones, but incomparably better results are obtained through the improvement of soils and their controlled taming.

The ways and means of taming soil are quite varied, since all the most effective agrotechnical methods known may be systematically directed towards the creation of tame soils and a tame soil formation, without limiting oneself to merely raising their effective fertility.

Soil melioration in the broad sense of the term comprises all the aspects of melioration, including special agronomical soil meliorations and engineering hydromeliorations. An enormous role is played in particular by hydromelioration which, being one of the most important factors in the creation of a stable and progressively rising fertility of the soil, forms an integral part of the indivisible complex of factors of tame soil formation.

The regulation of the biological, salt, water-air and thermal regimes of soils opens up vast possibilities for the creation of highly valuable humous soils. The water regime of a soil is to a great extent determined by its nutritive regime, as well as by the conditions of the accumulation and decomposition of the organic matter. In turn, plants which are provided with an abundant supply of available food, utilise moisture and solar energy more efficiently.

On optimum irrigated land, under the conditions of a progressive system of agriculture, all the biological and physico-chemical processes may be directed towards the predominance of a soddy soil formation or the progressive accumulation of humus and plant nutrient elements. The additional amount of water (up to 500 mm and more during the vegetative period) which the soils systematically receive annually through irrigation in the zone of insufficient moistening increases their productivity and thereby contributes to their accelerated taming. This is tantamount to bringing them nearer to the soils of more humid native zones. As is well known, irrigation, by improving soils, exerts a favourable influence upon the microclimate, reducing the high and low temperatures of soil and air (by 2-4°) and raising the relative humidity of the latter. The humidity of the air over irrigated land is 2-3 times higher than over nonirrigated land.

Correctly irrigated soils are not subjected to salinisation. The systematic irrigation of saline soils may eventually lead to their desalinisation. Desalinisation is also achieved through special leaching operations to which strongly-saline soils are subjected. As the soil is being irrigated and desalinised, its humus, nitrogen and phosphorus content goes up, its structure is improved and its productivity gradually rises. With the irrigation of arid soils and the leaching of saline ones, new land areas can be reclaimed and brought under cultivation, which, here and there, have, until now, been lying in their pristine condition.

Water, which exerts a favourable influence on the course of soil formation—on the physico-chemical and biological processes going on in the soil, may, with faulty management, lead also to negative results. Faulty irrigation has, in places, resulted in the formation of tracts of land with an unfavourable irrigational-alluvial topography, with water-logged (swamped) and saline soils, which have in some or other degree lost their high fertility. Vast labour and financial resources as well as much time are required for their reclamation. Surface irrigation at high rates leads to a worsening of the structure of the upper horizons of the soil, with the formation of a uliginous crust, due to disaggregation and the dispersive effect of water. Complete flooding of the surface of the soil leads to a worsening of its biological processes, poor aeration, etc., whereas furrow and sprinkler irrigation preserves the structure of the soil and does not disturb the favourable course of the biological processes.

Positive results in the transformation of the soils of plains, gentle slopes and terraces may be obtained on a large scale through catchwork irrigation. This measure makes it possible to appreciably soften saline soils, structurise them and enrich them with humus. Catchwork irrigation consists in the formation of artificially flooded land, with a correctly regulated flood regime.

Of great importance with regard to the taming of soils are meliorative measures for drying soils, provided they are, together with the appropriate hydrotechnical measures, directed towards the elimination of the chief causes of swamping and towards the regulation of soil formation and provided that, once they have been dried, the swamped soils are immediately brought under cultivation and rapidly tamed. After having been dried, enormous tracts of swamped and bog land are converted into highly valuable, aerated, well moistened, warm, structural, extremely rich humous soils. Soviet agricultural practice has shown that in order to counteract swamping, it is not always necessary to install a dense drainage network and to resort to intensive drying. On mineral swamped soils, overmoistened from the surface, a loose drainage network (main, thalweg channels) is quite sufficient to drain the surface water from vast tracts of land. The same results are obtained in some areas by digging open ditches which draw off the surface and ground water responsible for the swamping conditions. Deep ploughing alone, followed by taming (including manuring) of soddy-podzolic semibog soils and swamped land is often sufficient to dry them, this being accompanied by the lowering and retreat of permafrost. Faulty drying operations may, however, have adverse consequences. The thorough drying of mineral bogs, for example, may give rise to saline soils, that of turf-peats, to pulverisation of their surface, etc.

Open drains interfere with the mechanisation of agricultural operations. This is obviated by resorting to covered pipe lines and also to what is known as mole drainage. The latter method shows great promise, which can hardly be overestimated if we take into account the fact that mole drainage may be used to advantage for irrigation and desalinisation purposes.

Modern achievements in the field of meliorative science make it possible to set up in soil the necessary optimum moistening, ensure its structurisation, a favourable temperature regime, as well as regulate the aeration of the soil, its reaction, the concentration and composition of the soil solution (salt regime). Hence, melioration is not only an engineering and technical undertaking but also, to a large extent, agrobiological. The role of melioration is to change and transform the nature of soils and not only to serve immediate agricultural ends.

The transformation of soils through meliorative methods can be achieved in quite a number of different ways. All of them need, however, to be investigated more thoroughly. The point is that the process of soil formation under concrete conditions of drying (drainage) and irrigation, and all the more so where both are undertaken simultaneously, proceeds in quite different ways. If, on the one hand, irrigation should be so conducted as to create favourable conditions for the accumulation of humus, counteracting the intensive mineralisation of organic matter, drying, on the other hand, requires the setting up of the necessary conditions for its moderate decomposition. Both require the existence of an optimum water regime and a combination of synthesis and destruction of organic matter, through the regulation of opposed biological processes, viz., aerobiosis and anaerobiosis. This, as we know, is best achieved in soils with favourable physical properties.

The prospects are also most promising with regard to the regulation of soil formation through chemical meliorations in conjunction with biological and water meliorations. Solonetzic soils and solonetzes, for example, can be improved through gypsuming, sowing down to lucerne, and through irrigation.

New possibilities are opening up for fixing and reclaiming vast tracts of sands at present undergoing deflation, as well as of sands advancing onto fields, meadows, gardens, canals and communities. New soils can be created on sands, clays, silty deposits, mudflow formations, sewage sludge, on newest alluvial deposits, proluvium, deluvium, etc. Finally, soils can be systematically improved in an indirect fashion, through the proper management of the land. We are all familiar with the variously tamed soils of fields, brush, pastures, forests, orchards, gardens and plantations now occupying the formerly uniform soils of some or other area of land. The boundaries of the contours of the variously tamed soils, arising on the basis of one and the same soil under otherwise equal conditions, coincide with the boundaries of the contours of the various lands. Careful management of the land at their disposal will serve as the basis for the radical improvement of all the soils of state and collective farms.

The ways and means of improving soils are fairly varied. This is conditioned by the zonality of soil formation. All the measures relative to the accelerated taming and radical improvement of soils should therefore necessarily take into consideration their zonal and local particularities.

What the soils of the podzolic zone need in the first place is a supply of organic matter, through the repeated application of farmyard manure, compost and peat, as well as through green manuring.

The enormous reserves of dredge peat found in this zone, especially in flood plain marshes, constitute an inexhaustible source of enrichment of podzolic soils with organic matter. That is why to burn peat of low-calorific value where many other sources of fuel are available, is wasteful and inadmissible. Apart from enriching acid podzolic soils with organic matter and plant nutrient elements, it is also indispensable to neutralise them, either through liming or, on sandy soils, through marling. In the podzolic zone, the main meliorative measure consists in fighting podzolisation and swamping.

When well-aerated through liming, subsoiling and deep ploughing, the soils of the soddy-podzolic zone—soddy-podzolic, semiboggy, humus-peaty and soddy-peaty-gley soils—can be radically improved.

Particularly promising with regard to transformation and radical improvement, are the modern soils of the chernozemic zone. The productivity of chernozems can be raised to unprecedented levels. Even high-in-humus chernozems, which already possess what might seem unsurpassable qualities, properties and productivity, can still be considerably improved. Vast possibilities are opening up for the further improvement of ordinary and southern chernozems, through the setting up of optimum conditions for the conservation and accumulation in them of humus, nitrogen and plant nutrients. This can be achieved in the first place through correct irrigation. The main meliorative measure in the forest-steppe and steppe zones consists in fighting the degradation and erosion of soils.

Quite promising is the systematic improvement of the soils of dry steppes and semideserts. These soils should be tamed through the provision of an optimum moisture content, the neutralisation of the alkaline reaction, the adoption of a progressive system of agriculture and irrigation, resorting to sound agrotechnical measures specially aimed at improving the soils, including their structurisation, enrichment with humus and plant nutrients, the creation of a deep plough layer, gypsuming, biomelioration, the regulation of the microbiological activity. There is also a possibility here of partially slowing down the mineralisation of the humus through increasing the moisture content and lowering the temperature of the soil, etc. Irrigated brown soils and sierozems can be transformed into soils appreciably enriched with humus. The main measure for the amelioration of the soils of the zone of steppes and semideserts consists in improving their water-salt regime and fighting deflation.

The main task of agriculture is to bring about the cultural conditions of soil formation best capable of raising the effective and potential fertility of the soil. By channeling soil formation in the suitable direction, it is possible to create highly productive soils. This opens up possibilities for the realisation of vast new agricultural reserves. Under the influence of the appropriate complex of meliorative measures, any soil can be converted into a highly tame, fertile one.

Agricultural Amelioration, Forest Improvement and Sand Fixation

The agricultural amelioration of soils is the process of their accelerated taming under conditions of production.

One of the important measures with regard to the improvement of the soil is the creation of a deep plough layer. Thus, by increasing the depth of the plough layer in the chestnut zone, the reserve of moisture in the soil is raised by 30-40 mm and even more. Snow retention following upon deep ploughing increases the reserve of moisture by another 40-50 mm. Provided it is properly utilised, such an additional amount of moisture in the soil may ensure high yields of spring sown crops even in arid conditions.

The increase of the depth of the plough layer is accompanied by an increase of the noncapillary porosity, which ensures deeper penetration into the soil of moisture and plant roots. At the same time, the processes of nitrification spread deeper down into the soil. Deep ploughing also enhances the effectiveness of fertilisers.

Soil tillage may improve the salt regime of the soil. Soil tillage exerts a favourable influence on the heat properties, the nutrient regime and the general biological conditions.

The soil is subjected to still deeper changes if it is fertilised and transformed through the use of chemical meliorations. Chemical meliorations take the form of a direct and indirect fertilisation of the soils. No soil can be said to be entirely free from deficiencies and all soils benefit therefore from the application of some or other fertiliser, in some or other amount.

Data supplied by agricultural experimental stations show that the systematic application of farmyard manure raises the soil's content of humus and with it of nitrogen and ash plant nutrients. At the same time, there is a rise of the absorption capacity and of the degree of saturation of the soil with bases, whereas the acidity goes down.

Quite important from the meliorative point of view is the practice known as green manuring, which consists in ploughing in plants grown specially for the purpose of fertilising the soil (lupins, sweetclover, seradella, field pea).

Apart from muck and green manuring, another valuable source of organic matter is peat, either in its natural condition or as compost. The best composts are those obtained from mixtures of peat, lime and minerals, peat-fecal composts, as well as composts of peat mixed with leguminous plants—vetches, lupins—and inoculated with bacterial cultures, etc. The application of peat to soils constitutes a meliorative measure, owing to the fact that the effect is long lasting. Furthermore, peat leads to the rapid taming of the soil. The presence of peat deposits in the immediate vicinity of podzolic and soddy-podzolic soils makes the application of peat a very commendable and inexpensive operation. The application of peat to soils promises to play an important part in the near future with regard to the improvement of agricultural land.

Other meliorative operations are liming, gypsuming, etc. Liming eliminates the acidity of the soil. Furthermore, the calcium of the lime constitutes an indispensable plant nutrient. Swamped, podzolic, soddy-podzolic, leached and other acid soils are in need of liming. The more acid the soil, the higher its relative lime requirements. But the actual requirements should be determined through the appropriate field and laboratory analyses. Guesswork with regard to lime requirements is inadmissible. Lime requirements depend upon the acidity (pH) of the soil and its base-status: soils with a degree of saturation higher than 70% are not in need of liming, those with a degree of saturation lower than 40% are, on the contrary, highly responsive to liming.

Occasionally, due to some deficiencies in trace elements (boron, manganese, copper, molybdenum, zinc, chromium, cobalt, iodine, etc.), liming does not give the desired results. On the other hand, the beneficial role of certain trace elements is actually lowered by liming. Some trace elements are present in organic manures, especially farmyard manure. That is why the application of farmyard manure raises the efficiency of liming and, conversely, the action of farmyard manure is enhanced through liming, owing to the fact that the Ca content of muck is not always sufficient to cope with the acidity of the soil.

Of great agricultural importance is the improvement of meadows, bogs and brush. Vast tracts of swamped tussocky shrubby meadows need to be dried and cleared, with subsequent ploughing, discing or preferably roto-tilling. Roto-tilling leads to a marked improvement in the aeration of the soil and raises the content of available nitrogen and phosphorus. After having been dried, such meadows should be subjected to accelerated grassing, using mixtures of highly productive grasses.

Much more difficult to reclaim are plavnis—vast swamped tracts situated in the lower reaches of rivers. Here, one resorts to colmatage; good results are also obtained by what is known as biological drainage (planting of cypresses, poplars). The reclamation of plavnis is tantamount to "making" new highly productive soils on the site of useless swamped areas. Related to the soils of plavnis are the uliginous-swamp soils overgrown with juncaceous thickets of the Volga delta and of the southern regions of the Volga-Akhtuba flood plain. Once they have been drained, ploughed to a depth of 30 cm and cleared of the rhizomes of rushes, with subsequent subsoiling, they are converted into useful arable land.

Stubbing affords a means of converting large tracts of shrubby land situated in nonchernozemic belts into productive meadows.

Of great importance is the artificial heating of soils, using industrial sources of waste heat (hot sewage, hot gases, steam, slags). The soil can also be heated through the application of heavy dressings of organic matter (muck, compost).

Of enormous importance from the meliorative point of view are forest shelter belts. The afforestation of sands and the planting of forest shelter belts in steppe and forest-steppe regions improve the microclimate and the water regime of soils.

Apart from afforestation, the reclamation of sands consists in their fixation through the planting of shrubs and grasses. The afforestation of sands is a good means to the rapid creation of sandy soils. Under forests and grasses, the upper horizon of sandy soils becomes considerably enriched with humus and fine earth. Fine-grained sands possess a higher limit field water capacity (up to 7%) and a more favourable water regime than medium-grained sands. Fine-grained sands are richer in mineral composition. Apart from quartz, these sands are composed of feldspar and sometimes calcite.

However, an excessive development of the vegetation on sands depresses the ground water table, which sometimes leads to the ruin, from want of water, of the tree stand.

Sands can also be fixed through the use of bituminous emulsions and other substances capable of conferring cohesion to sand. Emulsions obtained from the waste-products of the oil industry increase the wetting capacity of water and confer to sands certain colloidal properties.

When coated with bituminous pellicles, sand grains acquire the capacity to absorb bases and to coalesce into aggregates. This leads to an improvement of the conditions for the application to sandy soils of fertilisers.

The systematic fixation of sandy areas through the use of bituminous emulsions and the sowing of grasses affords a means not only of putting a halt to the deflation of sands, but also of solving the problem of their conversion into fertile soils.

Once they have been fixed by forests, shrubs, grasses, bituminous emulsions and in other ways, sands can be converted into fertile soils through subsequent operations aimed at bringing about their colmatage or silting. The colmatage of sands can be undertaken during spring floods, or at some other time, by the conveyance onto the surface of the sands of water enriched with suspensions of fine earth which become deposited there. The same results can be achieved in winter, when an uliginous crust can be created on the surface of sands, the expenditure of water containing fine matter in suspension being reduced to a minimum. The main part of the silty particles retained in the sand accumulate in the uppermost layer. But part of the clayey and colloidal particles penetrate deèper into sand, reaching a depth of 80 cm and more.

Apart from undertaking the colmatage of the soils, one should counteract their takyrisation.

As a rule, the water of irrigation contains silt, sometimes in fairly substantial amount. Viewed from the technical standpoint of irrigation, the presence of silt is a negative factor, inasmuch as it may lead to the silting up of the irrigation network and of the irrigated area. But, on the other hand, silt also plays an important meliorative role, both from the fertilisation and colmatage viewpoints. Silt may bring about favourable changes in the water-physical and biological soil conditions. By enriching light soils with colloids, silt stimulates their microbiological activity. The water of rivers flowing through sandy areas contains overabundant amounts of silt, which should be utilised for the colmatage of sands and sandy soils of vast adjoining tracts of land. The importance which silts can play in the rapid taming through colmatage of sandy soils can hardly be overestimated.

It should be noted that artificial colmatage shows exceptionally high promise: it consists in the deposition of silty sediments on the surface of poor lands (swamps, shingle, stony areas) with a view to raising their fertility or creating new highly fertile soils. Of enormous importance with regard to soil reclamation are hydromeliorative operations, which permit the realisaton of new agricultural reserves. Soil itself constitutes an inexhaustible reserve for raising the level of agriculture, provided it is subjected to a progressive improvement.

The Role of Hydromelioration in the Taming of Soils

Modern hydromeliorative operations are, in conjunction with agricultural amelioration, aimed at radically improving soils, at raising their effective fertility and the yields of all crops.

The object of hydrotechnical amelioration (irrigation, drainage, leaching operations) is to convert natural soils into cultivated ones, rich in organic matter and ash plant nutrients, with new favourable characteristics and high productive capacity. Hydrotechnical melioration may lead to the creation of highly fertile soils, in accordance with a carefully elaborated plan instead of relying on a spontaneous natural soil-forming process.

The drying of swamped land and bogs is destined to remove the main cause of swamping and to direct the process of the accumulation of raw organic matter towards humus formation. The drying of swamped land implies not only the lowering of the ground water table, but chiefly the creation of a favourable optimum nutrient and water regime in the soil. The latter is not necessarily tied with the level of the ground water table and the drying operations can therefore only approximately be governed by the level of the ground water. The level of the ground water at which the soil becomes waterlogged and plants are unable to develop normally, constitutes a sort of critical depth. It also changes in accordance with the characteristics of the soil and the requirements of the plants.

On land which was subjected to drying operations, one should create a type of soil formation which can be directed towards the development of soddy-humous soils with favourable water-air, heat, chemical and biological characteristics. Drying operations bring about radical changes in the water regime of the soil, whereupon bogs and swamped land are converted into cultivated soils.

The best results with regard to the melioration of mineral swamped land are obtained by combining a loose drainage network with the following agromeliorative operations: levelling of the surface, selective or ordinary ridging, mole draining, hoeing and subsoiling, ploughing with narrow furrow-slices along the incline, etc. For the drainage of soils of this kind, one creates a temporary regulating part of the irrigation system, which is periodically renewed at the same time as the soil is being worked.

Dried swamp soils should be ploughed with a marsh brushbreaker plough or be subjected to roto-tilling. The dried soils prepared in this fashion should be fertilised in the first place with potassium and phosphorus and, in a number of cases, also with nitrogen.

The inoculation of bacterial cultures stimulates the microbiological activity of bog soils. Bog soils show good response to the application of trace elements, especially of copper (at the rate of 25 kg of copper sulphate per hectare). With the object of providing the necessary cultures and nutrients to peat soils, they should receive dressings of 5-6 (25) tons of farmyard manure per hectare, and acid peat soils should be dressed with 2 to 6 tons of lime per hectare.

To begin with, dried soils should be sown down to preliminary crops (oats, vetch-oats mixture, rye grass, flax, etc.) owing to the fact that in the first year following upon the reclamation of bogs, the sod has not the time to decay thoroughly and still contains a large number of live roots and seeds of weeds. In the following years, the dried soils can be cropped with roots, vegetables, etc. Following upon the decomposition of the sod, bog soils are often sown down to grasses, this necessitating the reploughing of the soil to a depth of 25-35 cm in autumn, to be followed in spring by discing and the application of fertilisers. The seeds mixtures used consist of clovers with grasses (timothy, meadow fescue, meadow foxtail, alsike clover and red clover). Suitable seeds mixtures for flood plain marshes include reed canary grass, fowl meadow grass, slough grass, etc. Following upon the drying, peat digging and additional drying of turf-peats—deepening and cleaning of the drainage channels plus levelling of the surface—they are subjected to deep ploughing, after which one can undertake their taming. This is tantamount, here, to making new soils. In order to create new productive soils on the seat of peataries, one should leave a layer of peat approximately 0.5 cm thick. With the same object in view, the soil should receive heavy dressings of lime or marl, as well as of farmyard manure and mineral fertilisers, with, in addition, the inoculation of bacterial cultures.

Drying brings about radical changes in the soils. It leads to the intensive decomposition and humification of the peat. The soil becomes enriched with active forms of organic matter and ash plant nutrients. The colloidal complex of the soil is increased, which leads to an increase of the absorbing power and base exchange capacity of the soil. The base-saturation status of the soil is improved. The acidity of the soil goes down. During the first decade which after the drying of peat bogs, the thickness of the peat horizon is reduced by tens of centimetres. The taming of dried soils brings about changes in the natural vegetation: the typical bog plants disappear and new species make their appearance on the dried and adjoining lands. The drying of boggy soils leads to a distinct improvement of their water-air and heat regimes, which stimulates the biological processes. Being subjected to humification, peat forms humic substances (humic and fulvic acids), which, as the aeration improves, continue to undergo mineralisation. When subjected to taming, peat bogs are gradually converted from phytogenic geological formations into peaty-humous soils. The taming of dried peat soils deeply alters their physical and physico-chemical properties. The humification of the peat leads to the mineralisation of the organic matter. The mineralisation of the peat and humus entails a rise of the specific and apparent densities, whereupon the pore-space and the water capacity go down; the ash and mineral matter contents go up, as well as the overall amount of N, P_2O_5, K_2O, CaO, SO_3, SiO_2. The absorption capacity and the base-saturation status go up. The hydrolytic and exchange acidity are gradually lowered.

The content of $CaCO_3$ is almost doubled in tamed peat soils, especially in the plough layer. Conditions are set up in these soils, which favour nitrification, i.e., a widening of the nitrate balance. This consists in the accumulation of nitrates which are made available to plants and microorganisms, as well as in the chemical interaction of the nitrates with the absorbed bases, with the formation of readily soluble salts of nitric acid and their partial loss upon infiltration. The removal of the nitrates may result from their reduction, in connection with a lowering of aeration and a periodic rise of the moisture content. The process of nitrification is intensified by an intensification of the degree of dryness. The maximum amount of nitrates in the dried soil coincides with summer time. The process of nitrification is likewise intensfied by the application of farmyard manure and by deeper ploughing. With taming, the amount of nitrogen and phosphorus available to plants and microorganisms goes up in peat soils. The biochemical processes which occur in tamed bog soils lead to an increase of the amount of ash plant nutrients also in the horizons underlying the plough layer. Taming entails a gradual deepening of the plough layer and a rise of fertility. The drying of bog soils alters the course of soil formation: the aeration and heat regime are improved, anaerobic processes are replaced by aerobic ones, etc. In other words, there occurs an accelerated deswamping of the bog soils.

Soils undergo also quite important changes under the influence of the various systems of irrigation. Correct regular irrigation leads to the accelerated taming of the soil, which is radically improved, whereas flaws in the system of irrigation lead to a deterioration of the soil.

The improvement of soils through meliorative measures boils down, in the main, to the creation of stable, most favourable physico-chemical and biological conditions for the development of plants, i.e., to the creation of the best possible conditions of water and nutrient regime of the soil, this corresponding to the conversion of the natural soil formation into a cultural one. With this object in view, one has recourse to gravity irrigation, sprinkling, underground watering, i.e., one brings about a regime of moistening of the soil by far superior to the natural one. Particular emphasis should be laid on the enormous importance of regular irrigation in soil formation.

In order to plan such a system of irrigation, one should not limit oneself to the use of a soil-meliorative map and the quantitative expression of soil-meliorative indices; one should also take into account the dynamics of these indices. The packing to which the soil is subjected under the influence of irrigation and other factors decreases the pore-space and the limit water capacity, whereas structurisation and cultivation of the soil, on the contrary, increase the pore-space and water capacity. This entails changes in the rates of watering. The depth to which the soil should be wetted cannot remain at an arbitrary level either. Under natural conditions, the depth of wetting is interdependent with the depth down to which penetrates the main mass of the roots, so that in the existence of a deep root system, the soil is wetted to a greater depth and, at the same time, wetting of the soil to a greater depth leads to a deeper penetration of the roots into the soil. It has been shown by investigations that a shallow wetting of the soil may force the main mass of the root system to develop in the uppermost horizon. Wetting the soil to a great depth usually leads to the development of an abundant and much ramified root system and, consequently, to increased yields. This is due to the increase of the volume of soil supplying nutrients to plants. It is therefore advisable, whenever possible, to increase the depth of wetting (down to 1-1.5 m) and that of the root zone.

The depth of wetting down to 1-1.5 m occurring under natural conditions in the southern chernozems and the chestnut soils of the steppe regions appears quite sufficient to create and maintain a considerable amount of humus. As a result of artificial irrigation, here and there, the soil also becomes enriched with humus. In the region of desert steppes, we now find deep dark irrigational sierozems, which contain considerably more humus than light untamed sierozems.

Under natural conditions, in the presence of optimum moistening, there occurs an intensive accumulation of humus even in the semidesertic and desertic zones. A good example of the point in case can be seen in the flood plain soils rich in humus forming on the same alluvial soil-forming rocks as the steppe-like low in humus soils of high river terraces, where the existence of aerobic conditions leads to the destruction of the humus.

The amount of humus in the flood plain soils of the arid zone is 5-6 times higher than in the soils of the terraces above the flood plain. Flood plain soils are usually wetted once a year to the condition of limit field water capacity down to the depth of the whole root zone, remaining subjected, the rest of the time, to the influence of the factors of zonal soil formation. Nevertheless, they have a high content of humus even where the ground water table is relatively low. The same thing occurs in the soils of sinks and limans, in spite of the inadequate amount of rainfall, where, thanks to the additional moistening with water of the local runoff, the rich in humus dark structural soils of sinks are formed.

The high humus content and improvement of the soils are due, in these cases, to their additional moistening. The same results can be obtained through a systematic discharge of water into the soil. Such a discharge of water can be compared to the optimum moistening of the soil occurring under natural conditions. But the discharge of water should not be so abundant as to flood the soil and feed the ground water, because, in this case, apart from a substantial loss of water, the soil may rapidly become salinised.

No less promising is catchwork irrigation on watersheds, terraces, in ravines, depressions and flood plains. Catchwork irrigation permits the irrigation of gentle slopes, utilising the possible inflow of thawing and rain water. The regular optimum flooding of limans contributes to the development of a soddy soil formation. Intermittent and short duration flooding slows it down. Prolonged flooding (exceeding one month) leads to the swamping and salinisation of soils. When subjected to catchwork irrigation for tens of years, the chestnut soils of semidesertic steppes acquire the character of meadow-soddy soils. In certain limans, the solonetzic soils and solonetzes develop towards solothisation with a superimposed soddy process of soil formation. With optimum depth (25-50 cm) and duration (10-25 days) of flooding, the soils of limans undergo radical improvement and transformation. They become enriched with humus and acquire structure. Catchwork irrigation leads to a substantial increase in the soil of humic acids and available forms of humus. In the absence of a high water table, catchwork irrigation brings about the desalinisation of soils.

In the arid regions of the country, where the rainfall is exceedingly low, under natural conditions, soils degrade and undergo salinisation. The amount of humus goes down, the structure is destroyed, the soil becomes packed and enriched with salts. Processes of degradation may affect even meadow limanic and flood plain soils, should they be deprived of additional moistening. It is therefore most important, in order to prevent the possibility of a degradation and salinisation of the soils, to undertake in good time the irrigation of the flood plains of the regulated rivers of the arid zone no longer subjected to natural flooding.

By damming tributaries of rivers and creating artificial reservoirs in the mouths of ravines and gullies, supplying water to river terraces by means of gravity flow, their soils can be rendered nearly as productive as those of flood plains, as if returning to the terraces the favourable conditions of the regime which they possessed in the past, when they were flood plains, and thus radically halting the degradation of the terrace soils of the valleys of southern rivers. This compensates to an appreciable extent the losses of flood plain lands due to flooding by large reservoirs. At the same time, gravity feed irrigation on the terraces of rivers will lead to the formation of highly fertile humous soils.

Irrigation, which aims at improving soils, is combined with native (zonal) and cultural conditions and undertaken jointly with agricultural ameliorations. By increasing natural moistening, irrigation brings about the most favourable conditions for the development of the soil. In arid regions, it decreases the excessive aeration of soils and the intensity of the aerobic processes of soil formation. It compensates the shortage of water which affects the soil permanently or at certain periods during the development of plants in various regions. In the sierozemic zone, where the amount of rainfall is approximately 200 mm (2,000 m^3/ha), the normal development of crops can only be ensured through an additional provision of water supplied in the form of frequent waterings during the vegetative period. In the steppe zone, one single watering per day during 5-10 days, may prove perfectly adequate to ensure a high yield.

Thanks to irrigation, an additional amount of water and elements of ash and nitrogenous plant nutrients are brought from the major geological cycle of changes into the biological one. Irrigation increases the local internal moisture circulation, which entails an improvement of the climatic and hydrological conditions of soil formation.

The influence of irrigation on the course of soil formation is only known in general outline and requires further investigations. The strong influence of irrigation on the dynamics of the nitrates has been established. Immediately following upon waterings, owing to overmoistening, the heat conditions are somewhat worsened and the processes of nitrification are stifled. The nitrates of the soil are leached down to the lower horizons, where they are subjected to denitrification or become fixed within the cells of bacteria, being converted into proteid forms. Thereafter, when water and heat conditions approach optimum ones, the amount of nitrates in the soil rises again, this continuing until the soil dries out, whereupon the process of nitrification begins to fade once again. According to M. M. Kononova, the moisture content which best favours nitrification constitutes nearly 60% of the soil's full water capacity.

The nature of the moistening of the soil has a strong bearing on the amount and condition of the nitrogen present in the soil. It has been established that complete flooding of the soil leads to losses of nitrate nitrogen. No such losses occur with infiltration irrigation nor is there any interruption in the processes of nitrification. The same favourable results with regard to the melioration of soils are obtained through the fairly promising system of subterranean irrigation, which is accompanied by an appreciable condensation of the water vapour obtained from the atmosphere and from the deeper horizons of the soil.

The regulation of nitrification through cultivation and melioration is one of the most important means of raising the soil's fertility. Artificially set up fluctuations of the temperature, moisture and aeration of the soil create a high dynamism of the nitrification and allow its regulation. All cultivations, provided they are carried out when the soil is sufficiently wet, promote the accumulation of nitrates. The pulverisation and drying out of the soil, resulting from frequent tillage operations, depress the content of nitrates. The processes of nitrification are also regulated by the application of mineral and organic fertilisers. By applying mixed organo-mineral fertilisers to the soil, one can lower the high mobility of nitrate compounds, fixing them in the form of organic compounds of low solubility, which is of great importance under conditions of irrigation, reducing, as it does, the migration of nitrates.

Irrigation also appreciably changes the aggregatory composition and the salt regime of the soil. Owing to their being leached, the content of readily soluble salts goes down somewhat. When the soil is flooded and the microaggregates come into contact with dropping-liquid water, they tend to break up into minute aggregates and single particles. The collapse of the structure lowers the noncapillary pore-space of the soil and leads to the formation of a surface crust. Unfavourable phenomena are reduced to a minimum wherever one has recourse to the more perfect system of furrow irrigation. Apart from other causes, the structure is also subjected to destruction due to the effect of the explosive wave of adsorbed air which is ousted from the aggregates. This effect can be exerted by the air found in a compressed condition on the surface of the soil particles. As this air becomes displaced by water, it increases considerably in volume and destroys the microaggregates. When the soil's moisture content equals maximum hygroscopic moisture, the amount of adsorbed air is lowered to a minimum. It is therefore most important to see to it that the soil's moisture content does not fall below maximum hygroscopic moisture capacity. A moisture content of the soil not inferior to maximum molecular water capacity preserves the structure from destruction even better.

Sound irrigation improves the water, heat and salt regimes of the soil; moreover, soil formation becomes directed towards a rise in fertility. An important role in this connection is played by sound cultural practices combined with irrigation. With cultural practices at a high level, the fertility conditions of the soil and the processes of its taming follow a more favourable course in irrigated than in nonirrigated soils. A high level of agricultural practices and the presence of an adequate amount of colloids and absorbed alkaline earth cations (Ca, Mg) give rise to a water-stable structure ensuring an optimum water-air and nutrient regime of the soil. Irrigation raises the degree of dispersion of the soil's colloids, but due to the presence of absorbed bases, the edge of coagulation being low, they are hardly leached out at all.

The course followed by soil formation under conditions of irrigation depends also on the efficiency of the latter. The drier the soil, the more water is absorbed by its upper horizons and the less deep it is wetted. The higher the moisture content of the soil prior to watering, the deeper and the more uniformly it is wetted at the same rates of watering. The moisture content of the soil prior to watering may therefore prove a no less important condition of the wetting of the soil than its granulometric composition.

A sound system of irrigation does not exert any appreciable influence on the translocation within the soil of mechanical and microaggregatory soil elements. It stimulates the biological processes, brings about and maintains a soddy soil formation, thereby contributing to a progressive rise of the soil's fertility.

The successes achieved so far by meliorative science and pedology make it possible at the present time to direct soil formation towards the transformation of soils and the creation of better ones.

The improvement of soils by meliorative methods opens up new vast possibilities for raising the productivity of land. Of great significance in this connection is the planning of an agronomically well-founded hydromeliorative scheme, in strict accordance with the soils, the topography, the geological structure and the requirements of crops. Of no lesser importance is the sound application of hydromeliorative measures and the operation of hydromeliorative systems.

Taming of Soils Through Sewage Application

The sewage carried off by town drains can be used to advantage for enriching the soil with the fertilising substances it carries in solution and in suspension. 1,000 m^3 of sewage contain as much dissolved fertilising matter as 10-15 tons of farmyard manure. The suspended matter amounts to 2-3 tons per 1,000 tons of sewage and is precipitated in sumps as a deposit or silt. But sewage contains also harmful substances and organisms (chlorides, sulphates, carbonates, especially sodium salts, heavy metals, phenols and other chemical substances and among the organisms, eggs of helminths, bacteria, including pathogenic ones, etc.). Sewage has a neutral or slightly alkaline reaction (pH = 7-7.4) and a relatively high temperature.

Soil, acting as a filter and absorbent, rids sewage of suspended and dissolved substances as well as of its microflora: helminths and other organisms. In humid zones, sewage plays mainly a fertilising role and in dry regions, it serves, in addition, the purpose of irrigation. Sewage exerts an alkalifying and heating effect on soils. Should the appropriate preventive measures be delayed, faulty irrigation with sewage may, in northern regions, bring about the swamping and, in southern ones, the salinisation of soils, whereas correct irrigation and fertilisation with sewage enriches the soils with ash and nitrogenous plant nutrients, improving their structure and moisture-physical characteristics. Soil irrigated with sewage becomes the seat of an accelerated mineralisation of the organic matter, the rate of watering reaching 15 m^3/ha per 24 hours in the soddy-podzolic zone and up to 30 m^3/ha in zones where moistening is inadequate. The efficiency with which sewage is rendered harmless increases in a southward direction in connection with the longer duration of the vegetative period and the intensification of the biological activity of the soils. The suspended matter is absorbed in the upper horizons of the soil. It is also here that the colloids and a substantial part of the matter in solution are absorbed. A certain proportion of the sewage penetrates with its substances in solution and suspension deeper down into the soil, where these substances are retained and subjected to mineralisation. The influence of sewage irrigation makes itself felt throughout the whole of the soil's depth and even partly affects the soil-forming rock.

The natural biological purification of sewage is tied in the main with the oxidising processes which occur on the surface of the soil particles, microaggregates and structural units wetted by the sewage and harbouring a bacterial flora. The oxidation of the organic matter of the sewage occurring in the soil as a result of biochemical processes is attended by a lowering of the concentration and an appreciable change of the chemical composition of the substances reaching the soil. The concentration of ammonia in the soil goes down and there is a formation of nitrates and nitrites. The oxidation and nitrification of organic matter is most intensive in sandy and sandy-loamy soils, somewhat weaker in loamy soils and worse in clayey loamy and clayey, especially structureless soils. The application of sewage to the soils raises the humus content and the depth of the humous horizon. Systematic moistening of the soil with sewage, not exceeding limit field water capacity, leads to its radical improvement. The application of excessive dressings of sewage as if "exhausts" the soil; this can be seen on all communal irrigation fields, which receive 4 to 5 times more sewage than is necessary (12-20 thousand m^3/ha annually). On the surface of such fields appear bare patches devoid of vegetation, where the soil is fouled and overburdened with protoxides and harmful compounds. Excessive moistening with sewage is attended by anaerobic processes in the soil together with denitrification, due to the high expenditure of nitrogen on the decomposition of the cellulose contained in the sewage. In order to render sewage harmless and prevent losses of nitrogen, it is necessary to promote aerobic conditions of soil formation, stifling anaerobic ones. To preserve the soil from fouling and in accordance with sanitary regulations, sewage is subjected to compulsory purification in special sumps equipped with an arrangement for the destruction of harmful substances and microorganisms. The sludge resulting from the precipitation in sewage tanks of the substances thus rendered harmless is applied to the soil and ploughed in.

Sewage irrigation and fertilisation with sewage sludge allow the accelerated taming of sands, shingle, saline and other land, on which new soils can be created. Examples of this type of taming are afforded by the irrigation fields found around Odessa, Berlin and Paris, where cultural soils have been created on shelly marine deposits (Odessa), alluvial and fluvioglacial sands and light soils. Sewage played a decisive role in the taming of the sandy soils of the Ukhtomsk district of Moscow Region, and the year-round fertilising irrigation system used in a collective farm in the Noginsk district has led to the conversion of poor podzolised sands into highly productive soils with a deep humous plough layer.

The systematic taming of soils with sewage is of great economic significance and constitutes a very promising measure, which solves two most important problems at a time: one sanitary, the other agricultural.

A soil developing under conditions of directed soil formation, may also serve as the proper recipient of industrial sewage, thus avoiding the inadmissible pollution of water bodies, waterways and land areas.

Drainage and Its Significance

Drainage—the removal of soil and ground water—may be natural or artificial. The first type of drainage consists in the carrying off of water into the permeable rocks underlying the soil (coarse sands, gravel, pebbles). Artificial drainage consists in the carrying off of soil and ground water with the help of hydrotechnical installations (collector channels, underground drains—pipes or wells).

Swamped and bog soils are in need of drainage. Drainage is unavoidable upon the desalinisation of saline swamped (hydrohalogenous) soils, where the evacuation of ground water is hampered or prevented altogether.

Depending upon the existing natural conditions and agricultural tasks to be solved, drainage may be deep and rare, shallow and dense. Its effectiveness varies.

In order to lower the ground water table in the absence of a fall, the water is pumped up through wells (pump drainage). The water thus pumped up may serve as a most useful source for the leaching of saline soils and for irrigation, provided its mineralisation does not exceed a certain level.

Drainage ensures the so-called "salt ventilation" (leaching) of soil and ground water, as well as the lowering of the ground water table down to the wanted level. It makes it possible to bring about a permacidous leaching and to keep the soil-ground water flowing. But the role of drainage may vary in accordance with the salt and water regime of the land. It is sometimes erroneously thought that in order to achieve the desalinisation of soils, the ground water table should necessarily be lowered as much as possible and to no lesser a depth than a so-called critical level, so as to disengage the capillary fringe from the soil's root zone. In quite a number of cases, the main object of the drainage of saline soils consists in the provision of an uninterrupted downflow of saline water, this leading to the softening of the root zone and of the ground water.

When undertaking the melioration of swamped, bog and saline soils, one should avoid completely disengaging the upper boundary of the capillary fringe from the soil's root zone. If the upper horizons of the soil are in good (structural) condition, they will be adequately aerated. Should the capillary fringe be disengaged from the root zone, the plants may find themselves in a critical position, due to the fact that they will be entirely dependent upon the irrigation regime for their moisture supply. The existence of a contact between the capillary fringe and the root zone may bring about favourable conditions for ensuring an adequate supply of water to the crops and consequently the obtention of high yields.

The installation of so-called intercepting drains with the object of capturing sweet water from adjoining sources, such as submontane areas, rivers or canals, and thus bringing about the desalinisation of soils, is not justified either. Such waters may be used to better advantage. Erroneous and crude notions have led to the practice of installing drains in the vicinity of irrigation canals (irrigator drains). Like intercepting canals, these drains carry off sweet water, the mineralised water remaining untouched. Drainage is destined to carry off the mineralised water, so as to desalinise the soil. It should therefore be installed in the lowermost and most saline areas. It should constitute one single system with the irrigation network, the aim being to regulate the moisture and salt balance of the soil. It is indispensable to ensure the accelerated evacuation from the soil, via the drainage network, of the salt solutions, this involving a lowering of the level of the mineralised ground water and a certain increase of its head, due to the irrigation of the elevated elements of the relief. The drains are not meant to collect newly received surface water (irrigating water, rainfall) which has not yet passed through the soil.

When the ground water table lies considerably below the critical level, there is no need for a horizontal drainage network, especially if there is a lateral withdrawal of the ground water or if its level is lowered through other causes (evaporation, natural vertical drainage, etc.). Horizontal drainage is indispensable and comes into operation in the event of a rise of the mineralised ground water above the critical level. Prior to that, drainage may, should the local conditions warrant it, fulfil a prophylactic function.

The effect of drainage is more pronounced on soils of a light mechanical composition. In stratified grounds, it is advisable to lay the drains in the sandy strata. The life of a drainage system depends upon the soil conditions. Once it has fulfilled its special meliorative task, drainage may be utilised for the purpose of subterranean irrigation, similarly to the dual-purpose action of drying systems. A rational way of solving the problems relating to the moisture-salt balance of the soil is to cause deep irrigation canals to fulfil an additional, draining function. Wherever this appears necessary, horizontal drainage should be combined with vertical drainage, especially during leaching operations.

An important role with regard to the lowering of the level of ground water is assumed by biological drainage, founded on an increase of the transpiration capacity of the local vegetation. Plants with a powerful root system (trees, shrubs and grasses) penetrating deep into the soil may dry out the deep layers of the soil down to a condition corresponding to wilting point, whereupon the salinisation of the soil is slowed down. Such a degree of dryness cannot be achieved through ordinary drainage. One has recourse to biological drainage in all areas threatened by salinisation. To this end, trees are planted on the whole of the irrigation and drainage network, on roadsides, farmstead land, wasteland, in the tail ends of irrigated areas and other places.

The planting of trees along canals lowers the level of the ground water by 1 m and more, thus conditioning the existence of a fall from the surface of the water towards the canal. However, biological drainage exerts but little influence upon the salt balance of the soil, which remains practically almost unaffected. The soil and the ground water retain their salts.

When used in conjunction with artificial drainage, biological drainage may prove rather effective. But biological drainage cannot be put on a par with artificial drainage. It cannot replace hydrotechnical drainage, although its independent significance cannot be denied. In order to draw off saline ground water and bring about a steady desalinisation of the soil, one should have recourse to artificial drainage, whose rates can be lowered in the existence of biological drainage.

Drainage should be planned and its operation regulated in direct accordance with the type of salinisation of the land. Thus, seasonal patchy secondary salinisation does not require any special drainage installations. This can be eliminated through careful management, using the measures usually adopted for lowering the water table together with a high level of cultural practices, the most efficient possible use of irrigating water, selective leaching operations, etc. Under conditions of permanent patchy salinisation, its elimination can also be attained through the routine and agrotechnical measures usually adopted for lowering the level of the ground water. An increase in the accumulation of salt may necessitate the undertaking of an artificial removal of the salts through the installation of a drainage system and through selective leaching operations without drainage in autumn and winter, etc. Upon complete salinisation of the soils but low and medium salinisation of the ground water (up to 10-12 g/l), desalinisation is attained through leaching operations and the installation of a loose network of collectors and drains. Under conditions of complete salinisation and a high mineralisation of the ground water [12-30 (50) g/l], on soils of a heavy mechanical composition, it is necessary to install a permanent artificial dense and deep drainage network, to be combined with minor temporary drains aimed at accelerating the removal of the salts.

When the soil is fully saturated with salts, but the level of the mineralised ground water is low [7-12 (30) m], one should have recourse to leaching operations without drainage, relying on the internal water capacity of the soil-ground. The appearance of perched soil-ground water necessitates the installation of a temporary shallow drainage network. Where there is a real risk of secondary salinisation, the installation of a drainage system as a preventive measure appears necessary.

On weakly salined land, where the ground water tapers out (submontane plains, debris cones, continental deltas), a shallow dense drainage network is indicated. An excessive lowering of the ground water table here, may adversely affect the crops and necessitate additional waterings.

Highly saline soils require special leaching operations combined with drainage and waterings during the vegetative period. Following upon the special leaching operations, leaching should be systematically pursued by means of abundant waterings during the vegetative period, at rates exceeding limit field water capacity, so as to set up a permacidous leaching type of water regime. But under the usual irrigation conditions, a permacidous leaching type of water regime is not indispensable if it is found to lead to significant losses of water through filtration, with resulting undesirable consequences. But in certain cases, losses of water through filtration from the layer which it is intended to irrigate may even prove useful, provided they bring about a moistening of deeper layers of the root zone.

On land requiring drainage, losses through filtration are transformed into their opposite, i.e., they become useful. By feeding the ground water, these losses increase the head, bringing about the replacement of mineralised ground water by sweet water. The salt water is drawn off via the drains.

The theory and practical applications of drainage have only been inadequately studied so far and need to be investigated further.

Land Levelling

Levelling—giving a flat, horizontal surface to fields—plays an important role in agricultural and meliorative practice, particularly with regard to the improvement of the conditions desirable for the successful drying and irrigation of fields, and wherever it proves necessary to prevent or counteract salinisation and swamping. In the presence of an irregular surface, the distribution of irrigating water is not uniform: it accumulates in depressions, the elevated areas drying out and becoming focuses of accumulation of salts, which leads to the subsequent salinisation of the soils of the irrigated land.

As a rule, the object of land levelling is to eliminate irregularities of the microrelief (depressions and elevations), more seldom those of the mesorelief and even of the macrorelief. Levelling can be constructive and operational, light and capital, partial, selective or total. Levelling entails the cutting off of the elevated elements of the relief and the filling in of the depressions. The depth of the cutting off and filling in is greater or smaller, depending on the characteristics and structure of the soils. Deep cutting off and filling in can be undertaken on soils of a uniform mechanical composition. One should be careful when cutting, not to uncover underlying gravel, pebbles, sand, a salt horizon, or gleised and packed illuvium, since this can sharply lower the fertility of the soil.

Where the aim is to prepare the soil for leaching operations, levelling is usually undertaken in the autumn-winter time, prior to autumn ploughing, and ended in time for the leaching operations and reserve waterings. The levelling of new areas in preparation for irrigation can be undertaken at any time of the year, whenever the conditions permit it. During levelling operations, all depressions such as holes, furrows, unnecessary ditches, should be filled in, and mounds, banks, ridges, etc., should be eliminated. In places where elevations have been cut off, the soil should be ploughed and heavily dressed with organic and mineral fertilisers. Areas subjected to slight general levelling are cultivated with a chisel-plough or plough and when covered with a thick sod, they are disced. Prior to ploughing, one should eliminate all deep irregularities: fill in the gutters, split the ridges, fill the holes by hand or with a horse gutter shovel. Any irregularities remaining after ploughing and harrowing are levelled off with a smoothing harrow, a grader, a scraper, and special levelling implements.

Capital levelling is needed on areas with a pronounced microrelief, where the difference between elevated and depressed areas exceeds 5 m. Levelling of this kind may lead to variegated crops owing to the fact that the humus horizon of the soil is entirely removed from elevated places, uncovering the underlying horizon devoid of humus. In depressions, on the contrary, the humus horizon may become buried. In order to avoid this, levelling may be limited to certain areas or one may use the coulisse method (Fig. 59). With a view to preserve the humus horizon, the soil is ploughed deep, the furrow slice being turned over onto the places to be cut next (*b*). After that, the surface is levelled off and, as a result, in places where the soil has been cut off, the humus

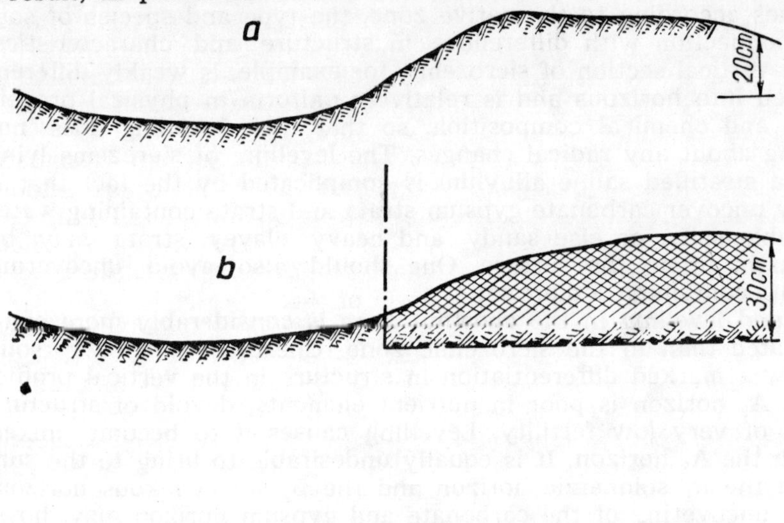

Fig. 59. Diagram illustrating the sequence of operations during land levelling:

a—original aspect of the surface; *b*—ploughing of microelevations with inversion of the furrow slice to a depth of 30 cm; *c*—cutting off microelevations and filling in of microdepressions; *d*—ploughing of microdepressions with inversion of the furrow slice to a depth of 30 cm horizon will come to lie on the surface and in depressions it will be buried (*c*). A second deep ploughing with inversion of the furrow slice onto the places of overlapping may partly or completely return the humus horizon from the buried position back to the surface (*d*).

Levelling off results in an inevitable disturbance of the soil profile. The influence of this disturbance on the soil's fertility varies according to the native zone, the type and species of soil, in connection with differences in structure and characteristics. The vertical section of sierozems, for example, is weakly differentiated into horizons and is relatively uniform in physical properties and chemical composition, so that their levelling does not bring about any radical changes. The levelling of sierozems lying on a stratified saline alluvium is complicated by the fact that it may uncover carbonate-gypsum strata and strata containing water soluble salts, or else sandy and heavy clayey strata may be brought up to the surface. One should also avoid uncovering strata containing sodium salts.

Land levelling in the chestnut zone is considerably more complicated than in the sierozemic zone. Chestnut solonetzic soils show a marked differentiation in structure in the vertical profile. The A_2 horizon is poor in nutrient elements, devoid of structure and of very low fertility. Levelling causes it to become mixed with the A_1 horizon. It is equally undesirable to bring to the surface the B_1 solonetzic horizon and the B_2 solonchakous horizon. The uncovering of the carbonate and gypsum horizon may, however, prove useful if it results in the self-melioration of the soil and improves the local supply of calcium salts. The depth of the humus horizon is small in chestnut soils, especially in light-chestnut ones (10-40 cm), so that levelling of the zone will uncover the underhumus horizon.

The solonetzic B_1 horizon increases in compactness from dark-chestnut to light-chestnut soils and the depth at which lie the readily soluble salts decreases. There is a risk that levelling might uncover the solonetzic and solonchakous horizons. Dark-chestnut soils are less adversely affected by levelling, provided they are not cut down to a depth exceeding 30-40 cm. In the chernozemic zone, the soil can be cut down to a relatively considerable depth without any appreciable changes in characteristics.

In the podzolic zone, the object of levelling is to improve the soil by mixing the A_2 horizon with the plough layer. The A_2 podzolic horizon is tamed through the application of farmyard manure, peat, mineral fertilisers (NPK) and through green manuring. Mixing the illuvial B horizon with the top soil or uncovering it does not lead to any special difficulties since it is better than A_2. The application of organic matter, mineral fertilisers and lime, under conditions of land levelling, is very effective.

In all soil zones, levelling should be accompanied by complementary measures aimed at raising the fertility of the uncovered lower horizons.

In mountain and eroded regions, thanks to the modern machinery at our disposal, levelling may lead to the creation of large new areas of useful land. In homogenous friable rocks (eluvium, deluvium), levelling boils down to the mere elimination of the irregularities of the mesorelief. Land levelling of this kind usually leads to the complete destruction of the natural soils. On strongly eroded and uneven tracts as well as on the seat of mountain excavations, this cannot be regarded as an obstacle to levelling, especially when it brings to the surface a homogenous friable rock endowed with potential fertility. A rock of this kind can give rise, within a relatively short time, to new highly productive soils. The levelling of stratified and compact rocks is considerably more difficult. It may prove more advantageous, in this case, to organise the mechanised carting of friable earth from adjoining elements of the relief.

Chapter XX

RECLAMATION OF SALINED SOILS

Salined and swamped lands have the tendency even to this day to enlarge their area of distribution. Salinisation and swamping result, in most cases, from the existence on the territory of a defective water regime duc to faulty irrigation or drying, a low meliorative technique, as well as the failure to take soil formation processes into consideration when drawing up meliorative plans and carrying out various meliorative measures.

The salinisation and swamping of soils should be energetically counteracted. The chief method of combating soil salinisation is to put in good order the moisture and salt regime of the whole of the territory subjected to reclamation, so that the fight against soil salinisation is, on the whole, a hydrological problem. But apart from a rise of the general standard of water utilisation, as well as good planning and management, preventive measures are also necessary, to forestall losses of water through filtration, coupled with levelling, and the installation of a network of collectors and drains, etc. Where the salinisation of the soil is due to the moistening of its salt horizons as they are drawn into the capillary fringe, or to a rise of salt solutions from the water table, the fight against salinisation should be directed towards reducing the capillary inflow of these solutions.

The fight against salinisation and its prevention should take the form of a system of agrotechnical, agromeliorative and hydromeliorative measures; the main ones being: a) proper utilisation of the territory and the setting up of a meliorative network; b) high agrotechnical standard and the choice of the correct system of farming; c) good management of the irrigation and drainage network; d) removal of the salts from the soil by means of leaching operations, etc.

The fight against salinisation can be conducted in two ways: without a lowering of the water table and with a lowering of the water table down to a depth precluding or slowing down the rise of salt solutions to the soil's surface. The most effective method for fighting salinisation is to lower the ground water table and prevent its rise. This is attained through a complex of measures (proper utilisation of the available water, accurate estimation of the watering and leaching rates, utilisation of the ground water for leaching operations and water supply, setting up a hydromeliorative and agrochemical control, putting the drainage system in good order, regulating the ground water, zoning the territory, etc.).

The fight against secondary salinisation and its consequences must be carefully planned as a comprehensive system of interdependent measures, which should include the technical exploitation of meliorative systems, special agricultural practices, agrobiology, chemisation, proper utilisation of the territory, etc. The measures for fighting secondary salinisation are subdivided in accordance with the stages of salinisation:

a) in areas suffering from secondary salinisation in its first stage (patches of salined soils against a background of nonsaline soils, the ground water table being low), the complex of measures should be founded on the prevention of salinisation and a high agrotechnical standard;

b) in areas suffering from secondary salinisation in its second stage (large patches of solonchakous soils against a background of weakly saline and nonsaline soils), apart from undertaking the above-listed measures, one should carry out a selective leaching of the soil at differentiated rates and autumn waterings aimed at eliminating the salts accumulated through the summer and autumn. One should bring about favourable moisture-salt conditions in the root zones of the soils. Collectors should be installed for drawing off the soil-ground water;

c) in areas suffering from secondary salinisation in the third stage (continuous areas of solonchaks, high ground water table, pronounced evaporation), the chief measures to adopt are leaching operations and the installation of a network of collectors with selective drainage.

Reclamation of Solonchakous Soils

The reclamation of solonchakous soils boils down, in the main, to their desalinisation and the creation of optimum conditions of the moisture-salt balance. Desalinisation or reduction of salinisation is achieved through various methods:

a) biological, through a lowering of the ground water table by means of intensive transpiration and withdrawal of the salts by plants, as well as through the improvement by plants of the soil's physical properties, etc.;

b) mechanical, i.e., raking and removing the salts. This method is not very effective and should be combined with other methods;

c) physico-chemical, desalinisation being achieved through changing the physical and chemical properties of the soil. The salts are dissolved and washed out from the soil with water. But the salts present in the soil do not constitute a mere mechanical admixture, so that it is difficult to achieve their complete elimination from the soil. They have to be dissolved first, after which they are forced out from the soil, in solution. A certain time is needed to dissolve the salts present in the soil and the degree of moistening should reach the condition of field limit water capacity. An additional amount of water is also needed to draw off the solution from the soil, this amount being all the larger as the soil contains more salts.

In agricultural practice, one begins by eliminating the salts from the upper soil horizons, after which the soil is sown down to salt-tolerant crops and subjected to further leaching. But in the absence of good underdrainage, the more the soil is leached, the more the ground water table rises and the greater the risk of secondary salinisation.

The success of the fight against salinisation depends upon the correct diagnosis of its causes and the adoption of the appropriate measures. Since salinisation under irrigation conditions is tied in the main with a disturbance of the normal hydrological regime of the soil and with a rise of the ground water table, it follows that the prevention of salinisation depends upon the maintenance of a normal water regime, i.e., measures capable of checking a rise of the ground water table.

Leaching of Salined Soils

Leaching is one of the ways of radically improving soils and consists in washing the upper salined layers through with sweet water in order to dissolve and draw off the harmful water-soluble salts. In most cases, the leaching of salined soils requires good underdrainage to ensure the evacuation of the salted water, softening of the ground water and a lowering of its level.

The leaching operation should be preceded by a thorough levelling off of the land, deep levelling ploughing to destroy fissures and burrows, and rolling of the level land so as to ensure uniformity of leaching. Such a preparation considerably reduces the expenditure of water. The leaching operation should be conducted according to a strict plan, at the rates and with the method best suited to the local concrete conditions. The plan should take into account the degree and nature of the salinisation (based on analytical data) and the condition of the crops.

The leaching rates are governed in practice by the degree and nature of the salinisation, the depth and dynamics of the ground water, the mechanical and microaggregatory composition of the soil, etc. Taking into account the level of the ground water table and the conditions of the underdrainage, the leaching rates should be so calculated as to bring the ground water to a level not higher than 1-1.5 m from the soil's surface towards the beginning of the vegetative period. The timing of the leaching operation is therefore of great importance. The most suitable time for leaching operations in cotton-growing areas is considered to be the period when the ground water table is at its lowest level, i.e., in the latter part of autumn and the beginning of winter (November-January) when the evaporation of water from the surface of the soil is at its minimum and the soil's moisture content increases owing to the condensation of water vapour. Highly salined soils require increased leaching rates, and the leaching operation should start in October. In the leaching of the soil (dissolution and removal of the salts) loosely bound (film), capillary and gravitational water takes part. The salt solutions of bound water decrease in concentration only by way of diffusion into mobile, less concentrated capillary water and all the more intensely as the velocity of the latter is smaller. When the soil is being wetted from above, the rate of diffusion is considerably smaller than the velocity of the water in the pores. In this connection, the water used for the leaching has not the time during its passage to get hold of all the salts dissolved in the loosely bound water, not to mention the salts coating the soil particles and contained within the aggregates, which have not the time to dissolve entirely. That is why the leaching operation does not, as a rule, rid the soil of all the salts, a considerable part of which remains behind. The water which freely circulates in the large pores, passages left by roots, wormtracks and fissures, takes an insufficient part in the leaching of the soil, so that all the passages apt to let the water pass through freely should be destroyed by deep preliminary ploughing and special cultivations. As a rule, the leaching of the salts from the soil is rather irregular. According to data supplied by some research stations, the upper 20 cm of a soil were washed following two leaching operations, whereas at a depth of 1 m, the necessary degree of desalinisation was not achieved even after seven such operations.

Salt solutions can be removed from the soil by way of repeated light applications of water, due to the fact that each time the first portion of the solution is drawn off, the soil retains salts both as solution left behind in the fine pores and on their walls as solids not yet dissolved. The object of repeated waterings is to reduce the concentration of the solution remaining in the soil and to eliminate the salts from the walls of the pores and the surface of the soil particles. As a rule, soil need not be completely desalinised to achieve a satisfactory degree of softening. Soils of a lighter mechanical composition are easier to leach.

Leaching operations sometimes lead to an increase of the Na_2CO_3 content, this being quite undesirable. In order to counteract the harmful effect of the sodium salts, especially soda, one is compelled to dress the soil with calcium salts, usually in the form of gypsum.

In winter, when there is a constant circulation of water in the form of water vapour from the lower warmer soil horizons towards the upper supercooled ones, the water resulting from the condensation of the vapour, having wetted part of the upper layer of the soil to the condition of limit field water capacity, percolates through the soil and leaches it. When the soil is mulched, the zone of condensation may be increased and brought up right next to the surface of the soil. This is confirmed by the accelerated desalinisation of solonchaks under various covers (buildings, hay and straw stacks, rubbish heaps, muck, etc.). It follows that, in order to fight salinisation, one should intensify and prolong the winter volatilisation of vapour, its condensation and the percolation of the water of condensation, by leaching the soil. The washing effect of the water of condensation may be enhanced by preliminarily wetting the soil to the condition of limit field water capacity, in which case, water supply irrigation acquires a particular importance.

Leaching operations are particularly effective when the soil is well drained and the ground water table lies considerably below the critical level. In practice, however, it is sometimes necessary to undertake leaching operations in the absence of any drainage, whereupon two cases may arise:

1) the ground water is mobile, so that there occurs a natural withdrawal of the salt solutions;

2) the ground water is stagnant and the overall salt balance remains either stationary or undergoes an insignificant change owing to the translocation of the reserve of salts throughout the soil horizons.

In the second case, the process of desalinisation is reversible, i.e., salinisation often recurs after the leaching operations, hence the necessity of their frequent repetition. The practicability of this repetition is dependent, however, on the level and dynamics of the ground water. Leaching operations of this kind can cause the translocation of the salts to a depth of only 0.5-0.75 m. In such cases, the leaching operation can only aim at delaying the recurrence of salinisation until after the current crop has ended its vegetative period and given a good yield. The amount of water used in the leaching operation should be in full correspondence with the water capacity of the soil-ground. This capacity is governed by the free pore-space of the layer (column) of soil extending from the initial ground water table up to the tolerable level. The amount of leaching water which the soil is capable of holding is determined experimentally. It is easy to calculate, knowing the pore-space and moisture content of the soil.

The theoretical aspect of the leaching of soils has not yet received the attention it deserves and needs further serious investigations. Experimental data (B. Konkov, E. G. Petrov and others) have shown that on land adjoining irrigated areas, the ground water table rises not as a result of diffluence, but as a result of hydrostatic pressure. From this point of view, the translocation of salt solutions is also regarded as an inwash and not as a lateral transference. But the existence of the latter cannot be denied. The soil is the seat of a continuous diffusion of salts.

The movement of strongly mineralised ground water occurs in accordance with the gradient of pressures. In the existence of sharp differences in the density of the water, the movement may take place against the hydraulic gradients, overcoming a drop equal to 0.8 m of a water column. As the flows of dense solutions present in the soil get into the ground water and raise its concentration, they cause less dense solutions to rise, here and there, and this makes for the variegation and dynamism of mineralisation of the soil-ground water. When the level of the ground water is high, its mineralisation depends on the salt solutions which it receives from the soil. Under solonchaks, the ground water becomes strongly mineralised, its composition corresponding to the chemical composition of the soil solutions and vice versa: the saline composition of salined soils reflects the mineral composition of the ground water.

When undertaking the leaching of salined soils, one should take into account the complexity of the movements of sweet and salt waters in the soil-ground layer. The Soviet scientist A. T. Morozov suggests to group the movements according to the nature and magnitude of the natural processes (zonal, regional, local, interaggregatory and intraaggregatory). To the class of known movements of soil water, comprising film, capillary water and flows of continuous media, A. T. Morozov proposes to add a second class—viz., the class of diffusive and convective* movements. Convective movements are responsible for the variegation and instability of mineralisation of the ground water, the instability of the soil's salinisation and the exchange of salts between soil and ground water. The presence of denser solutions over less dense ones leads to the formation of gravity currents. Convective and multilayer movements develop upon the interaction of salined fallows with leached desalinised fields. The effect of the translocation of Cl ions from fallow to field may reach 10-20 tons per hectare in one year. The highly mineralised soil-ground water exerts a saline head along the gradient of which the ground water may flow towards the fields and thus reinforce or cause secondary salinisation. The repetition, year after year, of the convective movements is responsible for the existing variegation of salinisation of the soils and the diversity of mineralisation of the ground water.

When undertaking the leaching of salined soils and their irrigation, one should take into account the critical level of the ground water. The critical level of the ground water is the level upon which the capillary rise of the water reaches the root zone and causes the salinisation of the soil. This level is governed by the water-physical properties of the soil, the mineralisation of the ground water and other natural and productive conditions.

All other conditions being equal, the critical level of the ground water is all the deeper as it is more mineralised. The critical level depends upon the agrotechnical standard and the meliorative treatment. It is equal to the height of the capillary rise of water in the soil, but is often lower than that. Irrigated soils which are usually softened by applications of water may become salinised only when the mineralised ground water lies relatively close to the surface (1.5-2 m). The critical depth of the ground water above which it cannot rise without danger, depends also upon the mechanical composition and structurisation of the soil. The salinisation of the soil is also influenced by the velocity with which the salts rise and the salt balance of the ground water. This balance may be in equilibrium when the increase of the salt content in the soil caused by the evaporation of the ground water is compensated by the removal of an equivalent amount of salts from the soil due to waterings during the vegetative period. It may also be positive (from the point of view of the accumulation of salts—salinisation) and negative (desalinisation). The maintenance in the soil of the appropriate salt regime may be regulated by the corresponding agrotechnical treatment.

To determine the critical level of the ground water responsible for salinisation, a distinction is made between the potential and

* Transference of soil water in connection with the uneven distribution of its density resulting from differences in temperature and concentration from one soil horizon to another and one area to another.

actual height of the capillary rise. The potential height of the capillary rise does not yet threaten the upper soil horizons with salinisation. This height does not yet characterise the critical depth of salinisation of the ground water. The actual depth (height) at which lies the ground water may get considerably closer to the surface of the soil than the potential height of the capillary rise. The decrease in height of the capillary rise is influenced by the depth of the evaporation taking place within the soil (2-3 m), and the critical depth is influenced by the softening effect of the irrigating water and rain. These influences depend upon the geomorphological and zonal conditions of soil formation. The critical level of the ground water can be calculated by deducting from the potential height of the capillary rise, the depth at which the evaporation within the soil ceases to occur and the softening influence of the irrigating and atmospheric water, under conditions of equal agrotechnical and geomorphological factors, is no longer felt.

The critical level of the ground water is a variable quantity, which fluctuates within a fairly wide range, so that it cannot be determined solely from the mechanical composition and one should also take into account all the water-physical, physico-chemical and biological conditions. During the vegetative period, the critical level of the ground water above which it cannot rise without risk, depends, all other conditions being equal, upon the degree of mineralisation of the ground water. The more pronounced the mineralisation of the ground water, the faster the soil becomes salinised and, consequently, the lower the height of the ground water table which can be tolerated. The critical depth of the ground water table varies with the degree of mineralisation (Table 34).

Table 34

Critical Depth of the Ground Water, in m
(during the vegetative period)

Soil-ground	Mineralisation of the ground water in g/l			
	1-3	3-5	5-8	8-10
Loess (after A. N. Kostyakov)	1.5-2.2	2.2-3.0	3.0-3.5	—
Irrigational loams of the Bukhara oasis (after D. M. Katz)	1.6-2.1	2.1-2.3	2.3-2.5	2.5-2.9

Conditions and methods of leaching. Leaching rates. The leaching of soils with a view to desalinise them is subordinated to a number of conditions. Leaching operations should be carried out in the autumn or winter, when the level of the ground water is at its lowest and the evaporation of water from the surface of the soil at its minimum, whereupon the thermal condensation of vapour occurring in the soil will intensify the effect of the leaching. Leaching operations carried in the early spring may delay the soil cultivations and sowing.

The leaching operation should begin with the central parts of the more elevated areas situated between the drains, using high leaching rates. When the salinisation is pronounced, the leaching operation usually begins with the most salinc and, as a rule, the lowermost areas. The sequence of the operations is governed, therefore, by the degree of salinisation.

Soils of a heavy mechanical composition are leached first (in late autumn or the early part of winter). Soils of a lighter mechanical composition are leached at a later date (in winter). Soils which become fit for working with the minimum of delay are leached last. The date of the leaching also depends on the crop to be sown.

Before undertaking the general leaching of the area, one should begin by leaching its solonchakous patches, so as to reduce their salinisation down to the overall degree of salinisation of the field and ensure the subsequent uniform leaching of the whole of the area. The effect of the leaching operation also depends on the leaching technique, the methods of application and passage of the water through the soil.

In a number of cases, the leaching operation is spread over several years. Repeated leaching operations soften the soil down to a great depth. Medium-salined soils of a light mechanical composition require 1 to 2 years for their leaching, strongly-salined soils, 2 to 3 years and heavy strongly-salined soils, 4 to 5 years and longer.

The low results not infrequently obtained from leaching operations under conditions of production are due to a delay in starting the operations, unsatisfactory levelling of the land, poor leaching technique, inadequate watering rates, underflooding of the network of drains and collectors, as well as adverse characteristics of the soil, such as low filtration and a water-logged condition. In the latter case, it is advisable to resort to deep plantage ploughing, subsoiling, mole draining, fissuring, etc. The effectiveness of leaching and drainage may be enhanced by a system of physico-chemical measures (coagulation of the colloids, thermal fallow—drying the soil and causing the efflorescence of salts along the fissures, salting out prior to the leaching operation, increase of the water permeability, etc.).

The level of the ground water and changes in its mineralisation should be subjected to a systematic differential supervision over the whole of the area undergoing leaching. The efficiency of the network of drains and collectors is estimated from systematic hydrometric observations consisting in the determination of the amount of water drawn off and in the analysis of the drainage water (halohydrometry). In other words, a check is kept on the course of the desalinisation, so as to avoid an excessive consumption of water, a rise of the ground water table, inadequate leaching and the occurrence of solonetzic processes.

In order to control the course of the desalinisation, samples of soil are collected from the same spots, before the leaching operation and after it. From the samples of soil, one determines the quantitative composition of the salts prior to and following upon leaching. In the course of the leaching operations, use is made of certain qualitative reactions, such as the reaction on chlorine of $AgNO_3$ (a white flaky precipitate), the estimation of the alkalinity from the normal carbonates from the pink coloration with phenolphthalein of soil extracts in the presence of Na_2CO_3, etc. The degree of leaching of the soil is determined from a series of indications of a morphological and physical nature. A well-leached soil gradually loses its dark colour and acquires a lighter colour with a reddish tinge. The fitness of such a soil for cultivation (tilth) comes earlier and when worked, the soil flocculates more easily. In contradistinction to a well-leached soil, a badly-leached one feels sticky underfoot. When a well-leached soil is fit to work, it becomes covered with a dense network of fine cracks, etc. The contents of chlorine in the upper one metre thick layer of soil should not exceed 0.01%.

In the first year following upon the leaching of strongly salined soils, it is advisable to sow a relatively salt-tolerant crop, such as barley. After the crop has been harvested, the soil should be levelled off for subsequent leaching operations to be carried out in the autumn-winter period. Well-leached land can be sown down to lucerne and subjected to further selective leaching and overdrilling. When the dense residue of the water extract exceeds 0.4-0.6%, the soil cannot be sown down to lucerne, even though this crop is highly salt-tolerant. In that case, the soil is sown down to pioneer crops such as Sudan grass, sorghum, Persian clover, green bristle grass, sugar beet, arhizomatous agropyrum, solonchakous foxtail, white sweetclover, ryegrass, etc. These pioneer crops are followed by lucerne and after that by cotton. After they have been leached and rid of their salts, formerly salined soils, low productive under natural conditions, are converted into improved fertile soils.

The best results with regard to the elimination of salinisation are obtained as a result of the systematic desalinisation of the whole of a salined area representing a natural physico-geographical complex (regions, basins, separate parts of these, etc.). Leaching operations involving only part of the territory of such salined areas can have but a palliative character, since they usually lead to a recurrence of salinisation. For the softening of a layer of salined soils 0.7-1.5 m thick, in which the ground water table lies at a depth of 3 m, the following rates are used (B. Fyodorov):

a) for soils with a salinisation strength 1-2 (from 0.01 to 0.1% of chlorine in a one-metre layer of soil): 2,000-4,000 m^3/ha;

b) for soils with a salinisation strength 3 (0.1-0.2% Cl): 6,000-8,000 m^3/ha;

c) for soils with a salinisation strength 4-5 (0.2% Cl): 8,000-10,000 m^3/ha and more.

Leaching rates include: 1) the amount of water necessary to dissolve the salts present in the given layer of soil ("saturation rates"). The saturation rate is equal to the soil's moisture deficit and is determined by difference, deducting from the limit field water capacity the amount of moisture present in the soil before leaching, and 2) the amount of water necessary to displace the salt solution obtained (rate of "displacement"). The rate of displacement is determined from the amount of water above the field limit water capacity necessary to displace the solution down to the needed depth.

It has been established that the amount of water necessary to dissolve the salts need not exceed the limit field water capacity of the soil.

The salt solution present in the soil can be displaced with a volume of water equal to that of the solution, i.e., to the field limit water capacity. It follows that the minimum leaching rate equals twice the field limit water capacity of the soil. This rate can be lowered, due to the fact that for practical purposes, one may limit oneself to leaching the soil down to a depth of 0.6-0.8 m instead of 1 m.

The leaching rate has to be applied in several portions, since it is impossible to displace the whole of the salt solution from the soil at one time. The salts are not leached in proportion to the volume of the water but require different lengths of time, depending upon their nature and distribution in the soil. The leaching rate is therefore applied in 3-4 (5) different waterings with intervals of up to 5-6 days, depending upon the degree of salinisation, the temperature of the water and air (the higher the temperature, the higher the dissolution of the salts, the shorter the intervals). The higher the degree of salinisation, the relatively more water is required for leaching the salts. The expenditure of water is not proportional to the salinisation. The leaching of the salts follows a slow curve, so that towards the end, the intervals should be somewhat increased.

The reserve of salts in the soil can be determined from the formula:

$$S = adh,$$

where S—the reserve of salts, in t/ha;

a—the dense residue of the water extract, in percentages of the weight of dry soil;

d—the specific weight of the soil;

h—the thickness of the layer, in cm.

The reserve of salts should be calculated separately for each soil horizon.

The leaching rates can be approximately calculated from L. P. Rozov's formula:

$$M = P - m + nP,$$

where M is the leaching rate, in m^3/ha;

P—the field limit water capacity of the given layer of soil, in m^3/ha;

m—the reserve of water in the given layer of soil before the leaching operation, in m^3/ha;

n—a numerical coefficient, which depends upon the degree and nature of the salinisation. It may be higher or lower than unity.

The n coefficient is only approximate (conditional), since it does not reflect the biological side of soil formation and all the changes undergone by the soil during the leaching operation.

$P—m$ (moisture deficit) is the difference between the value of the limit field water capacity of the soil and the available reserve of water in the given layer. Upon such a degree of moistening, readily soluble salts pass into solution. The withdrawal from the given layer of soil of the dissolved salts is achieved through the application of an additional volume of water equal to nP. After the additional moisture has percolated through the leached horizons, the amount of moisture which they retain always amounts to field limit water capacity with the residue of the salts. The leaching will take place for any value of nP, i.e., even when $n < 1$. If, after water, amounting to nP, with $n < 1$, has passed through the soil, the remaining concentration of salts proves harmless to crops, then such a leaching rate can be considered as sufficient. Such rates (approximately 2,000-3,000 m^3/ha) are sufficient in the case of weakly-salined soils. In the case of strongly-salined soils, the leaching rate or the value of the n coefficient should be increased. When $n=1$, the additional volume of the water supplied is equal to the volume of the salt solution in the soil. In theory, the whole of the salt solution should be displaced and the soil desalinised. In practice this does not happen, however, due to the fact that the soil is a complex porous natural body with pores of unequal and unstable sizes.

The movement of the water and salt solutions in such a body is fairly irregular. The salt solution is displaced first of all from the large pores. The fine pores and the micropores of the structural aggregates retain part of their solution. Furthermore, part of the solution—viz., that which is in the form of film water—does not participate in the gravitational movement. In narrow capillaries, in the presence of a film of absorbed water, desalinisation proceeds at a fairly slow pace, by way of diffusion. In order to remove the salt solution from the micropores and desalinise the loosely bound water, the soil should be repeatedly leached with applications of water amounting to nP, where n should be larger than 1.

This coefficient depends upon the changing physico-chemical and biological conditions. Its absolute value increases from soils of a light mechanical composition to heavy ones and from structural to structureless ones. In practice, the value of n lies between 0.25 and 3 (3), and the overall leaching rate fluctuates between 1,300 and 7,000 m^3/ha and more. The value of nP may be calculated according to the amplitude of fluctuation of the ground water table, the watering technique and the tilth. The ground water table is at its lowest in winter and at its highest in spring, with an amplitude of 1-2 m and more, depending on the local natural conditions. Investigations have shown that if the soil be watered in winter with a rate sufficient to raise the ground water table to its maximum level, no further rise occurs in spring. The maximum level goes down within a relatively short time. In the period during which the ground water table is at its lowest (late autumn, winter), the rate of application can be raised to its maximum, this leading to the best possible results. This rate does not worsen the usual hydrological regime of the leached areas with regard to the subsequent vegetative period, neither does it worsen the regime of the adjoining lands.

Interesting data have been obtained regarding the level of the water table in a yearly cycle in the delta of the river Murghab (Table 35).

Table 35

Averaged Depths of the Ground Water Table in a Yearly Cycle in the Delta of the River Murghab
(after L. P. Rozov)

Months	I	II	III	IV	V	VI	VII	VIII	IX	X	XI	XII
Level of the ground water table, m . .	2.7	2.3	2.0	1.75	1.6	1.7	1.9	2.3	2.6	2.8	2.9	2.9

The level of the ground water is at its lowest in November-December and at its highest in May, the amplitude being 1.3 m.

The periods of rise and fall of the ground water table are balanced, with a tendency towards a lengthening of the period of fall, which indicates that leaching and water supply irrigation are practicable. The graphical representation of these data illustrates the condition of the general water balance of the given territory (Fig. 60).

In the absence of underdrainage or when it is very slight, the leaching rates can be calculated from the following formula:

$$M = P - m + \frac{H - h}{r} 10{,}000,$$

where H is the level of the ground water before leaching, m;

h—the permissible level of the ground water immediately after leaching, m;

r—the ratio of the rise of the ground water to the height of a layer of water capable of raising the ground water table to that height. The value of r is determined from the formula:

$$r = \frac{100}{A-K},$$

where A is the overall pore-space above the ground water table, in volumetric percentages;

K—the capillary water content of that layer of soil, in volumetric percentages.

On strongly-salined soils of a heavy mechanical composition, where the usual amplitude in a yearly cycle or the permissible amplitude between the lowest and highest level of the ground water (h) is small, leaching may have to be continued for several years. In some cases, it may even be necessary to increase the successive yearly leaching rates on part of the land.

When the soil is subjected to leaching or irrigation, the water brings about the following changes in it (according to N. A. Kachinsky): as it wets the soil, the water activates the soil colloids; in the fine capillaries it plays the role of a bonding link between the soil particles; this role of the water gets weaker as the soil dries out and loses its plasticity, so that the formation of new aggregates in a soil of this kind when it is being worked is precluded, but the mechanical division of the existing aggregates is possible. An excess of moisture washes the soil particles apart. Under conditions of optimum moisture content (50-70% of the relative moisture content), the separate soil particles agglutinate, but stick only slightly to steel; hence, following upon leaching, it is when the soil is in this condition that it can be worked satisfactorily. In structural soils, the soil becomes sticky at a relatively higher moisture content and, in consequence, it can be worked during an appreciably longer period.

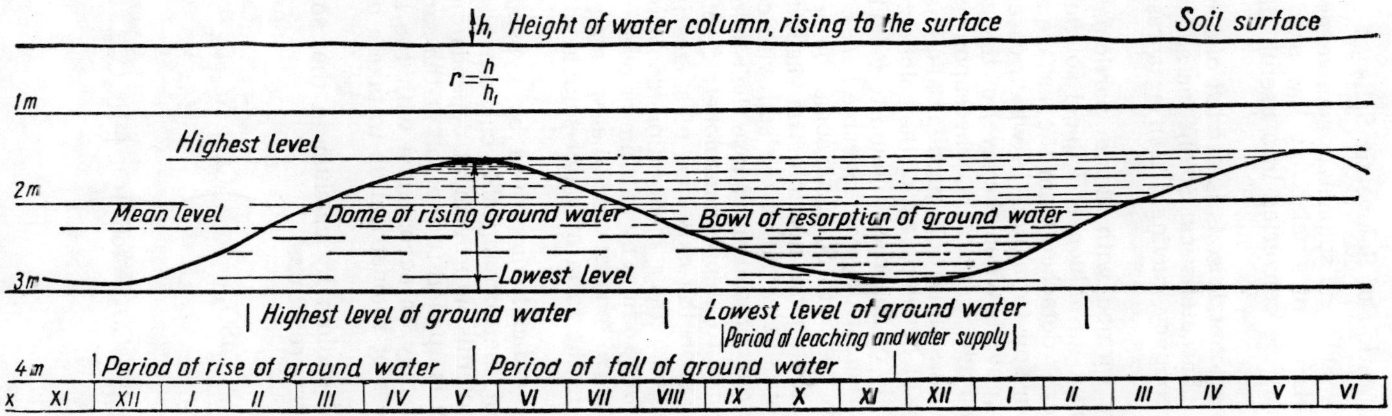

Fig. 60. Diagram illustrating the fluctuation of the ground water table in a yearly cycle in the delta of the river Murghab

Melioration of Solonetzic Soils

Solonetzes, alkaline soils and solonetzic complexes have a wide distribution. These soils are of a relatively low productivity. Yet all solonetzic lands may be converted into useful highly productive soils.

The melioration of solonetzic soils is carried out in accordance with their origin and characteristics. Solonetzic soils may be native (primary) and secondary, having arisen as a result of faulty meliorations.

Solonetzic soils are formed with some participation of the ground water or without its influence. Typical solonetzes develop when the ground water is detached from the surface and does not take any direct part in soil formation, in which case the process of desalinisation may prevail. When the level of the ground water lies close to the ground and is unstable, whereupon periodic desalinisation alternates with salinisation, peculiar alkaline-saline soils are formed. Each stage of development of the solonetzes is characterised by a certain ratio of absorbed sodium to calcium and magnesium in the composition of the exchangeable bases. The amount of sodium varies from 20-25 to 50-70% (soda solonetzes) of the absorption capacity in the illuvial horizon. The vegetation growing on alkaline soils may gradually enrich the soil with calcium, as a result of which the solonetzic soils may lose their solonetzic characteristics and get nearer to steppe nonsolonetzic soils.

Taking into consideration the existence of stages in the development of solonetzic soils, their melioration consists in the replacement of the absorbed sodium by calcium or other bivalent and trivalent cations (chemical phase) and the washing out of the products of exchange in the form of water-soluble salts (physical phase).

Solonetzic soils are radically improved by gypsuming, which leads to the increase of the exchangeable calcium at the expense of sodium. The result of this improvement depends on the meliorative substance which is applied to the soil at the same time. An adequate amount of it may, in accordance with the law of mass action, cause the process of solonetrisation to take a diametrically opposite course.

The displacement of sodium by calcium in the soil proceeds according to the following reactions:

$$\text{soil}\left)\begin{matrix}\text{Na}\\ \text{Na}\end{matrix}\right. + CaSO_4 = \text{soil}\left)\text{Ca} + Na_2SO_4\right.$$

$$\text{or soil}\left)\begin{matrix}\text{Na}\\ \text{Na}\end{matrix}\right. + CaCO_3 = \text{soil}\left)\text{Ca} + Na_2CO_3.\right.$$

The chemical melioration of solonetzic soils may be brought about by the application of any calcium salt or salt of iron and aluminium, as well as by the application of sulphur. Gypsum and sulphur are preferable to $CaCO_3$, due to the fact that, as a result of the exchange reaction, there is a formation not of soda but of Glauber salt, which is easier to wash out from the soil and is relatively less injurious.

The leaching of solonetzic soils leads to the displacement of sodium by hydrogen ions and the formation of base-unsaturated soils. In practice, this method is avoided, as it requires large amounts of water which, apart from other adverse conditions, it is impossible to pass through the solonetzes on account of their rather low water permeability. When the soil absorbing complex is saturated with hydrogen, it is easily destroyed, with, as a result, a radical worsening of the soil. In the first stages of the leaching of solonetzic soils, the alkalinity rises and there is a certain increase of the absorption capacity. But subsequently, as the leaching is pursued and the rates are raised, the absorption capacity diminishes, as a result of the breakdown of the absorbing complex.

The prolonged leaching of solonetzic soils leads to their solothisation.

The melioration of solonetzic soils boils down to the creation of a deep plough layer, to the removal from the root zone of injurious salts and absorbed sodium, to the elimination of the injurious physical properties of the B_1 illuvial horizon and the prevention of secondary salinisation. The irrigation of soils together with a higher Na:Ca ratio and a rise of the overall concentration of salts in the soil solution sometimes causes the appearance of solonetzic phenomena which have to be counteracted through the appropriate preventive measures. The risk of the appearance of solonetzic processes may be brought about by the irrigation of salined soils and the utilisation for watering the soil of mineralised water. Irrigating with water containing sodium salts causes the displacement of absorbed calcium by sodium and its removal from the soil in the form of $CaCl_2$ and $CaSO_4$, which reach the ground water. The soil, thereupon, becomes alkaline. The risk is particularly pronounced in the case of soils in which there is no reserve of free calcium. The irrigation of salined soils usually leads to their desalinisation, the type of desalinisation corresponding to the type of salinisation:

a) sulphate desalinisation is characterised by maximum leaching with the first watering, after which it follows a descending curve. This type of desalinisation does not condition any subsequent intensification of the solonetzic processes;

b) sulphate-chloride (subsolonetzic) desalinisation is characterised by minimum leaching of salts with the first waterings and maximum leaching with the subsequent waterings. This type of desalinisation leads to a qualitative change of the process towards solonetrisation and requires the application of chemical meliorative measures;

c) solonetzic desalinisation, characterised by a double maximum of leaching of salts, which are washed off with the first waterings and leached with the subsequent ones. Leads to solothisation and the breakdown of the soil absorbing complex. Chemical meliorative measures are indispensable.

The irrigation of solonetzic soils smoothes out their morphological features and alters their chemism. The irrigation of solonetzic soils is sometimes attended by an increase of alkalinity. Even a low concentration of sodium ions, leached out from the upper soil horizons as a result of irrigation, proves sufficient to bring about a change in the composition of the exchangeable cations in the lower horizons. Sodium increases the alkalinity and this raises the dispersiveness of the soil and sharply depresses the filtration capacity. Intensive leaching lowers the absorption capacity and causes the breakdown of the soil absorbing complex. Leaching is attended by a decrease of the adsorbed sodium, especially in the gypsiferous horizons, but the duration of the filtration is not lengthened, probably due to the migration of the SiO_2 and R_2O_3 colloids and the closing of the pores. Under these conditions, the removal of the water-soluble salts proceeds at a slow pace and their leaching requires too much water.

Thus, the main tasks regarding the melioration of solonetzic soils are as follows:

1. Elimination or lowering of the content of absorbed sodium, which is achieved through the application of calcium salts (gypsum) or sulphur in calcareous soils, in accordance with the amount of active absorbed sodium (chemical melioration).

2. Lowering of the salt horizons which are located in the root zone and, under conditions of irrigation, threaten to spread up to the very surface. This can be achieved through leaching operations combined with chemical meliorations. After that, the B horizon is gradually drawn into the plough layer.

3. Improvement of the physical properties of the solonetzes (physical melioration combined with biological melioration). This is achieved through special measures, such as growing crops with a powerful root system, mechanical loosening and deepening of the plough layer, drawing into it part of the illuvial horizon, etc.

The basis of the chemical melioration of solonetzic soils is the introduction into the composition of the colloidal complex of the soils of bivalent cations (mainly calcium), which ensure a normal correlation of cations in the soil solution, this conditioning the coagulation of the soil colloids and lowering the pH.

According to their effect on the displacement of absorbed sodium from solonetzic soils, salts are arranged in the following decreasing order: $CaCl_2$, $CaSO_4$, $Al_2(SO_4)_3$, K_2SO_4, $FeSO_4$. The major role in the melioration of solonetzes belongs to gypsum, however, because it is a more widely distributed and more accessible material. The application of gypsum in a quantity equivalent to the exchangeable sodium present in the soil, under high agrotechnical conditions, makes it possible to achieve the rapid desalinisation of the upper horizons of solonetzic soils. Better results are obtained when the gypsum is applied in a finely ground condition and thoroughly mixed with the soil.

There should also be an adequate supply of irrigating water, so as to ensure the dissolution and removal of the products of exchange.

The best results with regard to the lowering of the alkalinity are obtained when gypsum and chalk are applied in combination with sulphur. On calcareous soils, sulphur gives better results than gypsum and acidification. Under the influence of the sulphur bacteria present in the soil, sulphur is oxidised to sulphur trioxide. With water, the latter gives sulphuric acid, which converts carbonates into gypsum.

Well-aerated soils stimulate the formation of H_2SO_4 and the removal of by-products, whereas heavy soils with poor water-physical properties hamper these processes, the melioration of such soils being consequently delayed.

In solonetzic soils, a hydrolytical dissociation of Al and Fe compounds occurs. Al ions contribute to the precipitation of the soil colloids. The acid part reacts with the alkaline salts of the soil and, as a result, $CaCO_3$ is dissolved and the alkalinity decreases. The decrease of alkalinity is accompanied by a lowering of peptisation. The dissociation of the iron salts is attended by the same process, but then the soil pores become filled with Fe hydroxides, so that no positive results can be obtained from the effect of iron salts.

The best results with regard to the melioration of solonetzic soils are obtained through the application of gypsum in combination with farmyard manure, green manuring, irrigation and sowing down to perennial grasses. This complex of measures makes it possible to appreciably reduce the amount of gypsum required to counteract the exchangeable sodium. Gypsum eliminates the alkaline reaction of the soil, rendering it weakly acid; as for the grasses, they bring about the necessary conditions for maintaining this reaction for a long time. The sowing of cereal and leguminous grasses combined with gypsuming, confers to soils a stable cloddy structure and improves solonetzes. A suitable mixture of perennial grasses for solonetzic soils consists of yellow or hybrid lucerne with narrow-leaved *Euagropyrum*. Later, once the solonetzes have been improved, narrow-leaved *Euagropyrum* is replaced by broad-leaved leafy *Euagropyrum*, ryegrass and other grasses. Under grasses, the possibility of the degradation of solonetzes to solonchaks is excluded.

An important part in the melioration of solonetzes is played by woody plants.

Solonetzes are subjected to the usual type of irrigation, as well as snow retention and spring water retention of the catchwork irrigation type. In the region situated on the left bank of the Volga, a successful method for the improvement of solonetzes with a content of exchangeable sodium amounting to 20% of the exchange capacity in the illuvial horizon, consists in deep ploughing with inversion of the illuvium and leaching of the salts contained in the top 0.8 m of soil with rates amounting to 5,000-6,000 m^3/ha. The latter operation is undertaken without prior chemical treatment or upon the application of gypsum in doses not exceeding 50% of the amount corresponding to the exchangeable sodium.

Gypsum should be applied in frequent but small dressings, the total quantity applied reaching from 3-4 to 10-25 tons to the hectare. When the solonetzes spread as patches, the application of gypsum should be limited to the patches. The rate of gypsum is usually divided into two parts, one of which is applied before levelling off and ploughing and the other one, after ploughing. Ploughing should be effected to the maximum possible depth with a plough fitted with a skim-coulter. The rate of application of gypsum on land subjected to irrigation can be lowered by 20-25%. I. N. Antipov-Karatayev has put forward the idea of the self-melioration of solonetzes resting on the utilisation of the illuvial horizon rich in colloids and gypsum, which is brought to the surface through deep ploughing with completely inverted furrow-slices and mixed with the overlying solonetzic horizon. When the amount of this gypsum proves insufficient, the soil should receive additional dressings of it. Owing to the exchangeable sodium being redistributed and diluted, the alkalinity goes down and the chemical melioration requirements are decreased by 1.5-2 times in comparison with the estimated rates. The utilisation of the soil's own gypsum-carbonate horizon for the melioration of solonetzes deserves careful consideration.

Apart from gypsum, heavy (clayey) solonetzes are known to benefit from the application of sand and earth.

Under natural conditions and conditions of irrigation, the solonetzic process often leads to the solothisation of solonetzic soils, which is the extreme stage of destruction of the soil's colloidal complex. The melioration of solods and solonetzes undergoing solothisation consists in the consolidation of the colloidal complex and the displacement of the hydrogen ions by calcium. This is attained through the application to the soil of $Ca(OH)_2$ together with farmyard manure and other organic fertilisers. Apart from the application of calcium, high results can be obtained by enriching solodised soils with colloids, through the application of uliginous sapropel-like deposits (lacustrine and river silt). In order to avoid an aggravation of the acid reaction, solods are not subjected to gypsuming.

Chemical melioration gives high results provided it is coupled with a progressive system of farming and a high agrotechnical standard.

Melioration of Takyrs

Under natural conditions, takyrs are poor soils, owing to their adverse physical properties, caused by their alkalinity. This is aggravated by a high content of harmful readily-soluble salts located under the takyr crust.

The melioration of takyrs consists, in the main, in the elimination of their alkalinity, the improvement of their physical properties and the removal of the salts through leaching. The alkalinity can be eliminated through gypsuming the takyrs at the rate of 8-10 tons to the hectare, with subsequent leaching. Depending on the content of absorbed Na and as it becomes necessary, the dressings of gypsum are gradually increased. Should levelling off be unnecessary, the application of gypsum should be combined with that of farmyard manure. When levelling off proves necessary, the farmyard manure is applied after the levelling operation, together with the second half of the gypsum.

Good results can be obtained through deep ploughing, aimed at lightening the heavy mechanical composition of the upper crust of the takyrs, provided the clayey crust is underlied by easily accessible sand or sufficiently thick sandy strata.

Gypsuming is supplemented by the application to the takyrs of sand in quantities sufficient to improve their mechanical composition, provided there is a supply of sand in the immediate vicinity. The sand is applied before and after ploughing. This is followed by leaching in the usual manner. To begin with, takyrs should be sown down to corn crops (1-2 years), to be followed by lucerne (3 years), sometimes mixed with cereal grasses, after which comes cotton as part of a normal rotation.

The technical aspect of chemical meliorations and the methods for calculating the rates of application of the various materials are given in the practical course of meliorative soil science.

Chapter XXI

SOIL EROSION AND HOW TO FIGHT IT

Soil erosion is the process of destruction of the soil and the translocation of the products of destruction by water and wind. The activity of running water causes sheet and linear erosion, which constitute erosion by water. The blowing away or dispersion of friable soil by winds is referred to as deflation or wind erosion. As a natural phenomenon, erosion by water occurs where surface runoff is pronounced. Erosion is inevitably followed by the accumulation of the products of destruction of the soil. The intensity of soil erosion and, consequently, that of accumulation are closely dependent upon the natural and production conditions. Of great importance are the climate, the characteristics of the soil and soil-forming rocks, the incline, the nature of the vegetation and the agricultural methods used. Whereas under natural conditions, under a cover of vegetation, the extent of soil erosion is limited, it may, should these conditions be disturbed, lead to disastrous results and reach the proportions of a national calamity. In the first case, erosion only renovates the upper layer of earth whereas in the second case, it destroys it.

Sheet erosion leads to the removal of the upper genetical soil horizons on extensive areas. Eventually, sheet erosion is replaced by linear erosion, as a result of which the surface of inclines becomes cut by ravines, hollows and gullies, and valuable agricultural land is converted into wasteland. Here and there, linear erosion precedes sheet erosion or occurs simultaneously and in interdependence with it.

In the zone of insecure moistening, wind erosion takes the form of dust storms. Dust storms can occur with winds of any direction and at different times of the year. The foci which give rise to dust storms are ploughed fields and late winter sowings. The air becomes black due to the enormous amount of dust raised by the wind. Dust storms cause extensive losses to agriculture: they damage crops and sometimes completely blow away the humous arable layer. Wind erosion ruins or damages crops on tens and hundreds of thousands of hectares. The harm caused by wind erosion is further aggravated by the deflation of sands which invade valuable agricultural land. The deflation of sands and sandy loamy soils affects an area of some 196 million hectares.

The ablation and scouring of soils and the attending accumulation of deposits destroy arable land, meadows and pastures. Every year, soils subjected to erosion lose humus and with it nitrogen and ash plant nutrients in amounts far superior to the quantities of fertilisers applied to the soils spared by erosion. As the illuvial horizon gets closer to the surface or crops out, the soil's physical properties deteriorate and it loses its valuable microflora, etc.

Erosion is a menace to road building, built up areas, and causes the silting up and deterioration through sand deposition of reservoirs and river channels. Irrigational erosion destroys irrigated fields. The occurrence of destructive mudflows in mountain areas is a consequence of erosion.

Ploughing and sowing in drills along slopes, as well as ploughing steep (over 10-15°) slopes, brows and ravines and clearing sandy virgin and wasteland for agricultural purposes without prior application of appropriate antierosive agrotechnical measures, favour soil erosion. In spite of the fact that soil erosion constitutes a most serious menace, there is still a tendency to underestimate this natural phenomenon and the importance of keeping it in check, so that it continues to spread.

As a menacing natural phenomenon, present-day erosion is associated with the old river drainage—the main agent of scouring. It is governed by the energy of the topography—the difference in height between the highest and lowest land forms and the grades of the gullies, ravines and rivers, or a lowering of the general and local base-lines of erosion and denudation bases. Erosion is associated with the elevated elements of the relief of forest-steppes and steppes, especially with the regions situated alongside the right banks of rivers. The old river drainage in the form of rills and ravines (covering an area of approximately 60%), blind creeks (approximately 30%) and river valleys (approximately 10%) constitutes an old glacial and postglacial formation bound with the intense scouring activity of thawing water and local snow accumulations. The truncated and scoured lands resulting from present-day erosion now move from the old drainage onto watersheds. At the same time, the drainage of the eroded area increases and the level of the ground water there goes down. The alluvial cones bury valuable land and silt up river channels, thereby causing swamping of the adjoining land. Erosion is attended by the swamping of flood plains and the deposition onto them of material. A consequence of erosion is the replacement of the rich vegetative cover by a less exacting sparse vegetation. Present-day scouring processes are bound with man's interference with surface runoff, through the cutting and burning of forests and the destruction of the grassy vegetation and of the soddy soil horizon.

Useless or marginal lands are often formed where, at one time, there were highly fertile soils. At first, erosion destroys the rich and relatively loose upper soil horizons, thus uncovering the denser underlying layers which become, in turn, subjected to intense scouring. With the uncovering of the denser underlying soil horizons, the soil's productive capacity goes down, until the soil is completely destroyed and the soil-forming material crops out. The upper structural soil horizons offer more resistance to erosion. Once the upper, looser horizons with a higher water capacity and permeability have been destroyed, the rate of ablation sharply increases, owing to the fact that the packed lower horizons of the soil accelerate the runoff and are practically more vulnerable to scouring. Surface ablation is followed by the formation of gullies.

To each type of soil corresponds an erosive process of a certain nature. The soils of steppes offer a relatively low resistance to erosion, those of humid regions being more resistant, owing to the possession of a denser ground cover. The soils of the other regions occupy an intermediate position. On soils with a uniform profile, surface ablation prevails. On soils with a more complex profile, where there is a more pronounced tendency towards deeper ablation, erosion more often takes the form of gully formation. But within the boundaries of the steppe zone, the southern areas are less intensely eroded than the more humid northern ones due to the fact that the rainfall becomes less abundant and the water flows less powerful as we move southwards.

Depending on the incline, all other conditions being equal, the following degrees of erosion are distinguished:

Weak	erosion—angle	of	gradient	3°
Medium	"	"	"	3-5°
Pronounced	"	"	"	5-7°
Very pronounced	"	"	"	7-10°

A lengthening of the incline leads to a progressive increase of the destructive action of water, owing to its increased mass, greater velocity and *vis inertiae*. A certain role is played by the aspect and form of the incline accordingly governing the thermal, water and nutrient conditions of the soil. Erosion is considerably more pronounced on southern than on northern slopes. The same goes for convex slopes compared with concave ones. On convex slopes, erosion becomes more pronounced as the steepness increases down the slope, whereas on concave slopes, on the contrary, it diminishes, since the steepness of the slope decreases and so does the velocity of the thawing and rain water. That is why truncated soils predominate on convex slopes and washed-in soils on concave slopes. The effect of erosion on straight slopes is a replica in weaker form of that affecting convex slopes.

Soil erosion is the heritage of primitive agricultural practices and a predatory exploitation of the soil in the past.

The fight against erosion should, in the main, be directed towards the removal of its causes and not only of its consequences. This fight should be conducted not on separate areas of the earth, but on extensive territories, not be limited to the river drainage or part of a catchment area, for example, but involve large basins and river systems. The fight against erosion should be based on a harmonious complex comprising forest reclamation, agrotechnical and hydrotechnical measures, within the framework of a scientific management of the land. This complex and system of measures should be based on the fight against water and wind erosion involving the immediate protection of the soils of whole catchment basins through the creation of forest belts and meadow strips, the recreation of the soil's structure and the control of the cultural microrelief when working the soil, as well as through the undertaking of engineering works, viz., ditches, embankments, terraces, dikes, dams, drainage and water-absorbing installations (troughs, hollows, weirs, mole drains, slits, etc.) so as to redistribute the surface runoff and eliminate the ablation of the soil.

All meliorative measures connected with the fight against soil erosion are divided into two groups: passive—preventive, and active. But no clear-cut distinction can be drawn between them and, in practice, they are carried out simultaneously. It is most important to prevent soil erosion and where it has already begun, to put a stop to it and remove its adverse effects.

Active measures against erosion consist mainly of engineering works aimed at putting a stop to existing erosion and include the consolidation of slopes and of the bottom of ravines with sod, stones, fascines and even concrete, the installation of water drops, dikes, dams, etc., as well as the erection of protective ditches and ridges, the consolidation of the surface with vegetation, the levelling off of the eroded surface, etc.

Preventive measures comprise the correct management of the territory, the delimitation of the land as a whole and of its component parts, the arrangement of the fields so as to exclude the possibility of water or wind erosion, the disposition of the arable fields on levelled land free from the risk of scouring, leaving woods on variously inclined slopes, the disposition of pastures on land spared by erosion, etc. The afforestation of slopes threatened by erosion constitutes an important preventive measure, owing to the fact that the roots of the plants bind the soil together, contributing to the creation of a good structure. Under the protection of the forest litter, undergrowth and grass stand, the soil becomes more permeable and the surface runoff is almost entirely intercepted, the snow is retained and accumulated, the wind loses its force. The crowns of the trees weaken the destructive force of the raindrops and retain part of the atmospheric precipitations facilitating its evaporation back into the atmosphere. Under a vegetative cover, the depth of the winter frost penetration diminishes, the seasonal freezing ends sooner and the infiltration of thawing water begins earlier, leading to a decrease of the surface runoff in summer time. As a consequence, erosion is weakened and eventually ablation and scouring are halted, the entrance of silt and sand into rivers and reservoirs is curtailed, flood water runoff decreases whereas low water runoff increases. Consequently, the ground water table rises, the rivers receive a more abundant supply of water, the navigation conditions being thereby improved, and the life of reservoirs is lengthened.

Together with afforestation, quite an effective antierosive measure is the grassing of slopes and the culture of grasses. The culture of woody and grassy vegetation constitutes quite an important antierosive measure and one of the major phytomeliorations in current use. Wherever the vegetation develops well, erosion is weakened or excluded altogether. A measure deserving careful attention in the fight against erosion is the structurisation of the soil through the application of viscous substances, by means of green manuring and dressings of farmyard manure.

Preventive measures comprise also the interdiction to plough along slopes and alongside ravines and brows, where a strip of 20-30 m and more, depending on the slope and nature of the soils, should be left unploughed. The same goes for the interdiction to graze livestock on land subjected to soil erosion, due to the fact that it leads to the packing of the soil, a decrease of its water permeability and the aggravation of erosion even on relatively weakly inclined land.

The active fight against soil erosion entails the reduction to the very minimum of the runoff on the surface of agricultural land. The regulation of the surface runoff is achieved through: a) the retention of water on the catchment area, b) reducing the velocity of water, c) parcelling the streams and weakening their force by ridging the surface, hollowing it out, erecting ridges and digging ditches across the surface immediately after ploughing, sowing buffer grass strips, terracing slopes, creating a cloddy-granular structure and even resorting to chemical melioration, such as gypsuming, which leads to the structurisation of the soil and, consequently, to an increase of its water permeability. Special tractors fitted with mounted ploughs of a special construction are used for ploughing narrow areas across the incline.

Antierosive measures include the sowing of afterharvest and catch crops, crossed and close drill sowing, ploughing without inversion of the furrow-slice, the stubble being preserved as mulch, etc.

An effective antierosive measure, here and there, consists in sowing tall plants in the form of what are referred to as windbreak rows disposed across the slopes. The retention of snow on fields decreases the depth of penetration of frost and improves the capacity of the soil to absorb moisture. A way to put an end to the ablation and scouring of soils is to proceed to the damming up of late-autumn ploughed fields and fallows on slopes and watersheds (before the onset of winter) with earthen ridges 15-25 cm high, across the incline. This raises the yields and, in addition, it helps to counteract wind erosion. When ploughing and sowing the soils of steep slopes, buffer strips should be created or left across the incline, this leading, after a certain time, to the formation of terraces. Similar strips are left or set up along the brows of gullies and ravines, this being combined with damming up across the slopes.

Antierosive measures should be correctly matched and adapted to the local land forms and soil-climatic conditions.

A constant watch should be kept on the condition of the surface microrelief of the ploughed fields, which means the systematic levelling of open furrows and of wet-weather rills on ploughland by ploughing with inversion of the furrow-slice inwards and with the subsequent sodding of the hollows formed, the sodding of slopes threatened by erosion and of sands subjected to deflation. It is indispensable to avoid ploughing the water-conducting ravines leading to the apices and brows of the acting ravines and gullies and to proceed to their grassing, to apply organic and organo-mineral fertilisers, to lime acid soils threatened by water and wind erosion. Particular attention should be devoted to the creation on slopes of small-fruit patches, orchards and vineyards, disposing the rows along the natural contour lines and sowing down to grass the necessary number of buffer strips. Attention should also be paid to the urgent solution of a most important national economic problem, viz., the restoration of their fertility to the 30 million hectares of erosion-damaged arable land lying in the old agricultural regions of our country.

Of the utmost importance with regard to the fight against soil erosion is the choice of the correct rotation, taking into account the time (number of days in the year) during which the soil is protected from erosion by agricultural crops and by the snow cover, resorting, where necessary, to an antierosive organisation of the territory as part of the current land management.

Protective forestation in steppe and forest-steppe regions constitutes one of the most important measures for guarding soils against the harmful effect of droughts, deflation, ablation and scouring, for improving their water regime and raising their fertility. One should therefore resort to afforestation wherever the soil is in need of such a protection.

In order to prevent the deflation of sands and to assist in their fixation, one should abstain from ploughing large tracts of sandy soils, especially in open areas devoid of shelter belts. The ploughing of hilly sands should be avoided altogether, with the exception of narrow strips. When undertaking the afforestation of sands, it is indispensable to make use of the protection from winds afforded by the natural stand of grasses. When ploughing virgin land of a light mechanical composition, one should make sure to leave a 5-10 m wide buffer strip of unploughed land every 100-150 m.

Under mountain conditions, embracing whole catchment areas, one should have recourse to a complex of measures involving the organisation of the territory's economy, forest improvement and technical meliorations, protecting the land from mudflows and erosion. On slopes with a gradient exceeding 20%, as well as on slopes of a southern and western aspect, which are vulnerable to erosion, clear felling is ruled out. The same goes for forests growing on skeletal (gravelly) soils, which play an important role with regard to the regulation of the water regime and the protection of the soil against erosion. In order to protect the soil from erosion in the mountains, it is advisable to limit ploughing, for the purpose of growing annual crops, to slopes of a gradient of 10-15°. Field-protective forest belts here should be disposed across the slopes and combined with grass and tree and soft-fruit strips and terraces, as well as other water-retaining measures.

In the mountains, irrigated land should be watered by means of blunt flooded furrows (without faults) disposed either across the slopes or at a slight angle to them, or by means of small streams and flooding along narrow and relatively short strips, or else by sprinkling. A substantial role in the fight against soil ablation on irrigated lands belongs to the biological method, viz., soil-protecting strips, buffer strips, using mixtures of cereal and leguminous grasses. The Maltsev system of tillage offers good prospects in mountainous regions.

Forest improvement and agrotechnical measures play an important antierosive role in plains. The main measures should aim at weakening the force of the wind through a system of field-protecting forest belts consisting of fast growing species and of windbreak strips consisting of tall crops (maize, sunflower, etc.). Wind erosion is attenuated by protective belts consisting of the stems of intertilled plants. Windbreak strips retain snow on the fields, thus warming the seedlings, accumulate moisture and protect winter crops from deflation. To diminish deflation, winter crops should be rolled in autumn and winter. Another antierosive measure consists in sodding nonarable lands threatened by deflation.

It is indispensable to work out suitable methods for investigating water and wind erosion and to undertake a serious study of the role of the permanent factors of erosion: meteorological conditions, catchment areas, topographical conditions (dependence of erosion upon the gradient, aspect and length of the inclines), hydrological, geomorphological and hydrogeological conditions, the role of the soil conditions (changes undergone by soils upon taming) and the role of the variable factors, such as the nature of the natural and cultivated vegetation, man's productive activity (economico-technical and social moments) in its historical aspect.

In order to carry the fight against soil erosion to a successful conclusion, apart from scientific research, it is indispensable to intensify the explanatory work among the rural population and propagate ways and means of fighting soil erosion. In addition, a survey should be conducted to investigate the genesis and susceptibility to water and wind erosion of all the lands at our disposal. Cases involving loss of fertility or soil erosion should be treated with more severity.